his advice from many

in the way heard wishes

John F Kennedy

American Values

ALSO BY ROBERT F. KENNEDY, JR.

Judge Frank Johnson

The Riverkeepers

Crimes Against Nature

Thimerosal: Let the Science Speak

CHILDREN'S BOOKS

Robert Smalls: The Boat Thief

Joshua Chamberlain and the American Civil War

Saint Francis of Assisi: A Life of Joy

American Values

Lessons I Learned from My Family

• • • • • • •

ROBERT F. KENNEDY, JR.

HARPER

An Imprint of HarperCollins*Publishers*

HarperCollins books may be purchased for educational, business, or sales
promotional use. For information, please email the Special Markets Department
at SPsales@harpercollins.com.

FIRST EDITION

Front endpaper: Kennedy family collection
Back endpaper: Nat Fein/The New York Times/Redux

Library of Congress Cataloging-in-Publication Data has been applied for.

ISBN 978-0-06-084834-7

18 19 20 21 22   LSC   10 9 8 7 6 5 4 3 2 1

*To my grandparents, Joseph and Rose Kennedy,
their children, and all my Kennedy cousins of every
generation who bring me such joy.*

*And to my mother,
with admiration, gratitude, and love.*

CONTENTS

· · · · ·

American Values

· · · · ·

Grandpa

FROM MY YOUNGEST DAYS I ALWAYS HAD THE FEELING THAT WE WERE all involved in some great crusade, that the world was a battleground for good and evil, and that our lives would be consumed in that conflict. It would be my good fortune if I could play an important or heroic role. And from the very beginning this conviction was rooted in my family's fervent Catholicism, and deeply intertwined with the politics of the country.

There had been Irish politicians on both sides of my family for generations, so it was no surprise that we children talked politics from the day we could speak. Such passion came naturally to a people for whom the distinction between political and religious martyrdom had blurred during eight hundred years of British occupation. From their arrival in America, the Irish took to politics as the starving take to food, having been stifled for centuries by rules that forbade them from participating in the political destiny of their nation. As early as 1691, Irish law prohibited Catholics from voting, serving on juries, attending university, practicing law, working for government, or marrying a Protestant. In Dunganstown, near the port of Wexford, my Kennedy ancestors learned to read and write in an illegal "hedge school"; the priest who taught them was found out and hanged for the offense.

My progenitors were still in Ireland during that black February of 1847, when the announcement in England's House of Commons that fifteen thousand Irish were starving to death every day so moved Queen Victoria that she donated five pounds to the Society for Irish Relief. Britain produced, stored, and exported thousands of tons of grain and livestock from Ireland during the five-year famine—more than enough to feed the population—but the Crown refused to divert these money crops, so Ireland lost a quarter of her people. Skeletal corpses littered the countryside, their mouths green from eating grass like cattle. A million sons and daughters of Eire, including my great-great-grandparents, boarded the coffin ships sailing west, and the Atlantic became, in James Joyce's words, "a bowl of bitter tears."

Even in America the Irish encountered the familiar barriers of prejudice. Anti-Catholic sentiment would eventually drive my father's parents, Joe and Rose Kennedy, from their beloved Boston. Grandma would sometimes show us clippings from old Boston newspapers, where the acronym NINA (No Irish Need Apply) followed all the best employment listings. Still, wherever they landed, the Irish flourished. Fecund Irish mothers with little opportunity for personal fulfillment beyond childbearing produced an invading force that triumphed at law, politics, sports, literature, and business. No people were ever prouder than the prosperous Irish-Americans who returned to the Old World a generation later with heads high. During his triumphant tour of the British Isles in 1887, boxing champion John L. Sullivan, the apogee of New World cockiness, cordially greeted the Prince of Wales, "If ever you come to Boston, be sure to look me up. I'll see that you're treated right." My grandfather must have felt every bit as jubilant when he recrossed the Atlantic to attend the Court of St. James's as FDR's newly appointed ambassador to England.

Grandpa was the dominant figure during my childhood summers on Cape Cod. He made his home "The Big House," purchased in 1920 in Hyannisport on the shores of Nantucket Sound, where our house, John Kennedy's, and Jean Kennedy Smith's formed a tight compound surrounding it, and the Shrivers' and Teddy Kennedy's houses lay in a slightly more scattered orbit around the tiny seaside village. Grandpa bought the home when the Brahmins of brown-shoed Cohasset rejected his application to their golf club on account of his religion. Although the Protestant swamp Yankees who dominated Hyannisport were similarly bigoted, Grandpa's fellow Irishman Larry Newman, who owned the golf course, welcomed him.

During the first eighteen years of their marriage, Grandpa and Grandma produced nine children. God gave their firstborn, my Uncle Joe Jr., every gift but gray hair. He was born in 1915 and died in a secret Navy mission while flying a "drone" Liberator bomber that exploded over the English Channel in 1944, at age twenty-nine. Jack was born in 1917, Rosemary in 1918, and Kick, who married into Britain's greatest house, lost her husband in the war—and would herself perish in an airplane crash at age twenty-six—was born in 1920. Eunice was born in '21, Pat in '24, my father, Bobby, in '25, and the babies, Jean and Teddy, in '28 and '32. Our generation called them, collectively, "the grown-ups." We all spent each summer on Cape Cod at the family compound, where the twenty-nine cousins were raised communally and subjected to a daily regimen of athletic training supervised by a stout former Olympic diver, Sandy Eiler.

Three times a week we took riding lessons at my grandfather's farm in Osterville, or made long trots through the scrub pine forests and sandy marshlands with Grandpa astride his tall chestnut hunter, Shaleighleigh, whom he had imported from Ireland. Grandpa spoke soothingly to Shaleighleigh as he rode, calming the high-spirited gelding with his latest thoughts on politics and the economy, prompting the grown-ups to joke that this must be the smartest horse in the world. A posse of grandchildren followed behind him on horseback down the rough sandy trails across Barnstable County woodlands. Sometimes we children stopped to swim with our ponies in an expansive kettle pond on the northern fringe of Grandpa's farm, holding their tails to drag us, while he watched atop Shaleighleigh from a towering knoll along the wooded banks.

Every day we spent time on the ocean. My mother and father took us on the *Victura*, a twenty-six-foot wooden day sailor, for a picnic lunch on one of the nearby islands, where we fished for sand sharks, scup, flounder, puffers, and sea robins, gathered hermit crabs, periwinkles, and scallops, or dug for tasty steamers that betrayed their location on the tidal flats by squirting. With Captain Frank at the helm, we also took lunch outings on Grandpa's wooden cabin cruiser, the *Marlin*, crossing the Sound to Monomoy or Cuttyhunk to explore the Elizabeth Islands and gorge from picnic baskets of Grandpa's favorite foods—lobsters with hot butter and lemon, corn on the cob, strawberry shortcake, Boston cream pie, baked beans, and clam chowder. We children talked and caroused on the bow while Grandpa sat astern with the grown-ups—Uncle Jack, my father, Teddy, my mother, Aunt Eunice and Sarge Shriver, Jean and Steve Smith, and Pat and Peter Lawford. One day while cruising in Nantucket Sound, Uncle Jack and his closest friend, LeMoyne "Lem" Billings, assembled us on the green canvas cushions in the stern of the *Marlin*, where they sang "Heart of My Heart." Dave Powers and Jack sang "The Wearing of the Green" and taught us to whistle "The Boys of Wexford." It was Jack's favorite song and he knew all the words.

We are the boys of Wexford
Who fought with heart and hand
To burst in twain
The galling chain
And free our native land.

After returning from sailing, we played baseball on the field behind Uncle Jack's house, or touch football on the broad expanse below Grandpa's veranda. In the afternoons Grandpa sat in his chair on the great white porch, often holding hands with Grandma. They had fallen in love as teenagers, married seven years later, and remained hitched for half a century until his death in 1969. They were always open and demonstrative in their admiration and affection for each other. Typical of their tender expressions for each other is a letter from September 1960 just before Jack's election. She wrote him from Paris:

Joe dearest,
How can I have all this and you—
And still have Heaven, too!
Home for Christmas.

Love to all,
Rosa

Together they watched their children and grandchildren play on the sprawling green lawn bordered by sawgrass stretching into the sparkling sea.

Hyannisport was a magical paradise for me. I loved the endless palette of colors—the vivid blue of sea and sky, separated by rich green landscapes, peppered with ubiquitous roses, hydrangeas, and daffodils, each in their season, the gleaming white houses, and offshore a panoply of brightly hued spinnakers running downwind. The ocean was always changing, from blue to every shade of green, to gray and almost black, to match the moods of the wind and sky. Here, surrounded by my family, I could indulge my obsession with the natural world.

My cousin Bobby Shriver and I rode our bikes to the tidal inlets at Kalmus Beach to crab, or to the salt marshes at Squaw Island to catch fiddlers, killifish, and mummichogs in a wire trap. We went dip-netting for painted turtles and baby catfish from a dinghy on Anderson's Pond, or beach seining for eels, shiners, skipjacks, and Atlantic needlefish that hid in the floating sargasso weed and in the meadows of eelgrass that bracketed the harbor. We snorkeled for scallops, confining them in an underwater cage anchored in the harbor until we had enough to feed the whole family, a formidable task despite those bivalves' abundance.

When I was eleven my father gave me a motorized aqualung, a two-horsepower compressor wedged in a Styrofoam ring that bobbed at the ocean's surface, pumping air down a fifteen-foot umbilical hose into a mask, the perfect contraption for exploring the shallow waters off Hyannisport. I filled its tank with gasoline from the private gas pump adjacent to Grandpa's garage, and, wearing this apparatus, I swam with my little spear gun into dark caverns in the wrinkled rocks below the mile-long Hyannisport jetty. Pushing my way gingerly past keen-edged barnacles, I stalked giant tautog, considered a delicacy by the Portuguese fishermen who flocked around the jetty lighthouse each weekend from New Bedford and Fall River to cast for scup and flounder. Their fishing rigs, baited with sea worms and squid and anchored with pyramid sinkers, could not tempt the tautog, who preferred barnacles and crabs. So I sold the Azore men fish for comic-book money.

During the years when the Kennedy family compound served as the summer White House, the faint sucking *thwuck, thwuck, thwuck* each Friday afternoon summoned everyone to assemble for the landing of my father's and Uncle Jack's green-and-white Marine Corps helicopters on the big lawn between Grandpa's house and the ocean. Grandpa's gardener, Wilbur, hoisted the presidential flag up the pole and we cheered and waved as my dad and my uncles Teddy, Steve Smith, and Sarge Shriver climbed off the choppers. Leaving his own helicopter last, Uncle Jack would go kiss Grandpa and Grandma on the front porch of the Big House, then all the cousins would pile onto the golf cart as Jack took the wheel for a spin. A waiting fleet of limousines would whisk away Jack's special assistants, Kenny O'Donnell and Dave Powers, the "Irish Mafia" pols who hitched rides on Marine One for weekend visits with their own families. Police boats and a Coast Guard cutter bobbed just offshore, and fire engines stood poised among the flood of a hundred reporters at the end of the driveway, adding to the excitement.

On Saturday evenings my cousins and the older generations, their weekend guests, and the household staff gathered to watch movies in Grandpa's basement theater. An ardent film buff, Grandpa—who owned RKO Pictures, founded Pathé Studios, and presented and produced around a hundred features—could get first-run films, and wanted everyone to enjoy them. Grandpa also loved sports, and excelled at baseball, swimming, riding, and golf. A six-handicap with a fiendish putt, he often played with Boston Red Sox pitcher Eddie Gallagher, boxing champ Gene Tunney, and with my mother, who also had an excellent game. Shortly after his appointment as ambassador to England, Grandpa delighted his golf-loving British hosts and made headlines across Europe when he scored a hole-in-one with a 120-yard drive at Stoke Poges Golf Club west of London. "Where was Houghton when the ball got into the cup?" his sons Jack and Joe inquired mischievously in a congratulatory cable— Arthur Houghton being Grandpa's roguish protégé and golfing companion who had accompanied him that day.

Grandpa was just over six feet tall, and I remember him always smiling. He had big, white, perfect teeth of which Grandma was immensely proud, and bright blue eyes. He was gentle and affectionate, and loved to tell stories and roar with laughter. He read Latin and Greek, and fancied classical music. He relished the company of children, and always took time to play or to take us to the kitchen to visit his beloved cook Nellie and sample her famous angel food cake.

Yet even with his grandchildren, Grandpa had no tolerance for self-pity. A whining child would get him to clapping his hands rhythmically—"I don't want any sourpusses around here." He wanted us to understand that we should be grateful for our good fortune; it was spoiled to mope or complain. Grandpa also led the

Kennedy adult chorus in its constant demand for industry: we children were to fill all the interstitial spaces of our lives with some useful activity. If he busted us watching television other than the news, he turned the set off and sent us outside to play, rain or shine; he considered the TV a catastrophic waste of time.

When I was very little I would rest on Grandpa's lap as he read stories to me in his library, or sit with him on the porch and look at the sea. On his birthdays we children dressed in costumes to perform little plays in his theater and sang ditties we composed for him. During the winter we visited him at Palm Beach, where he fielded phone calls while sunning himself in a poolside wooden bunker called the "bullpen." He would swim with us in the waves and he took me for long walks, helping me corral crabs. He showed me how to pick up the Portuguese man-of-war without getting stung. On Grandpa's boat, the *Marlin*, Captain Frank brought us out on the Atlantic, where, in southeast Florida waters, I caught my first glimpse of flying fish, as well as cormorants, which, along with egrets and ospreys, had been largely extirpated north of the Mason-Dixon line by DDT. During one visit we watched green turtles hatch on Grandpa's beach.

Against Imperialism

Grandpa has been much maligned, as powerful men often are, accused of everything from being a German appeaser during World War II to a bootlegging confederate of Al Capone and Frank Costello during Prohibition. Urban legend has it that he conspired with those capos to fix the 1960 election by manipulating the Illinois vote to give Uncle Jack his electoral victory. None of these slanders is remotely true, yet those myths are riveted even more solidly to the American consciousness than the (equally untrue) Muslim affiliation and fabricated birth certificate of Barack Obama.

The topic of bootlegging follows Grandpa like a pilot fish. The notion is largely rooted in Grandpa's purchase—with the help of the president's son, Jimmy Roosevelt—of the British company that owned White Horse scotch and Dimple Pinch just as Prohibition was ending. According to historian David Nasaw, who researched the issue exhaustively for his definitive biography, *The Patriarch*, "Kennedy neither imported nor sold any liquor during his years in Brookline or at any time during Prohibition."The slander debuted in the mid-1960s as part of a partisan campaign to tarnish John Kennedy's name, which had become a powerful steroid for progressive politics after his death. Incapacitated by a stroke, Grandpa was unable to defend himself. Of course, if there had been any truth to those rumors, Grandpa's many enemies would have raised them to wound him in his four Senate confirmations for high-level posts, and Jack's enemies would have wielded them against him during the 1960 campaign. They didn't. Nora Ephron, who spent several years of her life researching a book on the liquor industry, came to the same conclusion. "All sorts of biographers and journalists casually insist that Kennedy was a partner of Frank Costello and Al Capone during Prohibition," she said. "It's so not true. And I happen to be an expert on the subject. No one seems to care about [the truth] but me. Frankly, I'm not even sure his own family cares." Well, actually we do—at least I wince every time I hear it. But when I was a child, the grown-ups taught us to ignore malicious gossip, and never to dignify it with a response.

I recall a time when there were so many malevolent libels circulating about our family that several of the cousins considered challenging this family practice. But who wants to waste a lifetime, or even spoil a moment, replying to an endless stream of fictions? As Teddy wrote in his memoir, *True Compass*, "With exceedingly few exceptions, we have refused to complain against the speculation, gossip, and slander. Some have viewed our refusal as excessive reticence, even as tacit admission of the innuendo at hand. In my view, it is neither. At least for me, it's the continuing assent to Joseph Kennedy's dictum: 'There will be no crying in this house.'"That said, I hope Grandpa will forgive me, in this case, for defending him.

The even more venomous Nazi accusation is equally without

merit. Despite his Irish heritage, Grandpa was a shameless Anglophile who urged the strongest support for England against Hitler. He asked FDR to increase shipments of all aid to Britain short of war, breaking with Charles Lindbergh, William Randolph Hearst, and the America Firsters. He abhorred Nazism, which he called "the new paganism." He condemned the fascist persecution of the Jews as "the most terrible thing I have ever heard of." His outspoken support for a Jewish homeland led the Arab National League of Boston to brand him a "Zionist Charlie McCarthy," and he made tireless efforts to rescue Jewish refugees from the Nazis. After a speech I gave in Minneapolis, a young woman approached me, introducing herself as Lisa Brenner. "Your grandfather got my grandmother, Mary, out of Germany before World War II," she told me. "I wouldn't be alive if it weren't for him." I can't count the times I've heard similar stories. In his exhaustive biography of my grandfather, historian David Nasaw chronicles how Grandpa's frantic efforts to find safe havens for German and Austrian Jews after Kristallnacht ruined his relationship with the British government—which lodged an official complaint against him to Secretary of State Cordell Hull, and ultimately with Franklin Roosevelt. As the Roosevelt administration turned on Grandpa in May 1938, Rabbi Solomon Goldman, president of the Zionist Organization of America, cabled him on May 9: "Feel it duty [to] say American Zionists have always regarded you as devoted friend. . . . [Be] assured we feel indebted for earnestness with which you have furthered interests [of] American Jews and cause of people whose fate must be of deepest concern to you." Grandpa's lonely campaign helped earn him an invitation from his Jewish friends to be the sole gentile member of the Palm Beach Country Club, an honor he cherished.

DESPITE HIS COMPASSION FOR THE PLIGHT OF THE JEWS, GRANDPA, like the vast majority of Americans, opposed U.S. intervention in World War II prior to the Japanese attack on Pearl Harbor. Like most of the world, he underestimated Hitler's barbarism.

Americans wouldn't learn of the Nazi death camps until after the war. He had fiercely opposed World War I, believing it would benefit only banks and munitions dealers, which was precisely the outcome. As an Irishman, he viewed the First World War as Britain's struggle to preserve its colonial empire—of which Ireland was the oldest colony—and saw no reason for American boys to die in that cause. He refused to join the national orgy of jingoism against the Kaiser, nor was he seduced by promises of a "short war and glorious victory." He lost several treasured friendships on account of his convictions. He rebuilt those relationships when his friends returned from Europe traumatized and disillusioned.

The outcome of that bloody conflict—three-quarters of a million British servicemen dead and casualties across Europe exceeding 40 million—fortified Grandpa's convictions. World War I left the continent's population starved and grieving, while many European nations collapsed into totalitarianism: Russia to the communists, and Spain, Germany, and Italy to the fascists. The yellow press had sold World War I to Americans as "the War to End All Wars," yet the Versailles Treaty served only to plant seeds for the next world war; Europe waited just one generation—long enough to grow a new crop of human cannon fodder—before ramping up the continental abattoir.

Grandpa believed that America's prosperity was its most reliable source of stability and national security. He was suspicious of foreign adventures that would spread our troops along distant frontiers and exhaust our nation's wealth. Grandpa understood what the founders had also known—that we must make a fundamental choice between democracy and imperialism. "We cannot dominate the world," he said, "but we can make ourselves too expensive to conquer. This is not a plea for isolation, but it is a plea against imperialism."

Grandpa's antiwar sentiments prior to World War II put him squarely in the mainstream. FDR had won the presidency on a peace and noninterventionist platform. The postwar period has made Grandpa seem prescient about the dangers of foreign entanglements. He recognized the appeal "to the best . . . of men and the best of nations," of spreading our democratic system across the planet, but he cautioned against falling into a foreign policy of "minding other people's business on a global scale" and counseled against "world-wide meddling under the guise of benign interest." In speech after speech, he warned his country that democracy at home was incompatible with empire abroad. In this he agreed with America's founding fathers. Washington, Adams, and Jefferson had all urged against "entangling alliances" that might embroil us in Europe's regular-as-rain conflagrations. In words we might wisely heed today, sixth president John Quincy Adams said of America, "Wherever the standard of freedom and independence has been or shall be unfurled, there will her heart, her benedictions and her prayers be. But she goes not abroad, in search of monsters to destroy." Military adventures abroad, Grandpa presciently argued, would hemorrhage money needed for infrastructure, education, and economic development at home, and weaken America. It would hollow out our economy and turn America into a garrison state. "It is not too difficult to predict that our democratic institutions cannot long survive such a strain," he said.

In his concern that a bellicose foreign policy would bring our country financial ruin and international enmity instead of peace, Grandpa also anticipated modern scholarship on the subject. In his seminal 1988 book *The Rise and Fall of the Great Powers*, Paul Kennedy (no relation), the British historian and Yale professor, argues convincingly that great nations are built on strong economies, but then invariably destroy themselves pursuing the seduction of empire. Expanding on this idea, the late Chalmers Johnson, one of the most influential right-wing foreign policy gurus of the late twentieth century, observed that "there is no more unstable political configuration . . . than . . . domestic democracy combined with a foreign empire." *In other words, restraint in the use of power is the best way to preserve democratic governance and national greatness.*

Grandpa's relatively mild isolationism was rooted in his cynicism about war, that it was as a political strategy benefiting monarchs and industrialists, at a cost, in blood and treasure, borne by the common man. Most of all, he found the thought of losing a child unbearable. Decades after Uncle Joe's death, the mere mention of his name would reduce Grandpa to tears. Following Uncle Joe's death, Jack assembled a collection of remembrances in a book titled *As We Remember Joe*. In that book, my father wrote of his revered big brother, "It is the realization that the future held the promise of great accomplishments for Joe that made his death so particularly hard for those who knew him. His death, coming before the worldly success that was so assured, seems to have cut out the natural order of things." Jack created the book for Grandpa. But Grandpa never read it. "I have to put it down after the first couple of pages," he explained to a friend many years later. "I never got over that boy's death." In his private correspondence, Grandpa imagined the same unspeakable desolation endured by millions of mothers and fathers around the world. He found it nearly impossible to conceive of any political dispute that would make such a sacrifice worthwhile. Grandpa was devastated when Truman dropped the atomic bomb on the civilian population of Hiroshima. He rushed to see Henry Luce and New York's Cardinal Spellman to beg both the publisher and prelate to implore President Truman to give Japan time to surrender before dropping the second bomb. Both Luce and the conservative bishop ignored his appeal. When my Uncle Teddy was eight years old, Grandpa wrote him from London during the height of the Blitz, describing the terrible suffering, adding, "I hope when you grow up you will dedicate your life to trying to work out plans to make people happy instead of making them miserable, as war does today." His passionate conviction on this account destroyed Grandpa's political career, yet he continued to urge that our military policy should invest in the defense of "Fortress America," and to make our economy the strongest on earth.

In his 1946 "Fortress America" speech at the University of Virginia, Grandpa presciently predicted that our Cold War military buildup, rather than bolstering national security, would serve to unify the Communist Bloc. Grandpa saw communism as a hateful ideology but not a threat to United States security. He wanted us to build an inviolable citadel at home, but to shed no blood on foreign soil.

In Grandpa's last speech, in 1952, he declared that by engaging in an endless arms race we had ceded to the Russians the power to dictate "how much we can and dare extend for social purposes." In that sense, we had given the Russians a victory. He complained that President Truman was squandering our national treasure "trying to remove communism from the globe," a dream that, "worthy though it might be, was impossible of accomplishment." He warned against Cold War ideologues who would engage America in disputes over the governance of developing nations. The Korean War confirmed Grandpa's worst fears. General Douglas MacArthur's decision to invade North Korea in defiance of a very specific warning by the Chinese not to cross the Yalu River prompted a massive assault by Red Chinese forces, driving the American army back across the 38th parallel and escalating the conflict. Our meddling in Asia was artificially preserving communism as a unified force; but if we left it alone, Grandpa forecast, communism would breed "within itself internal dissensions" that would lead to its collapse.

He was, of course, correct in all his predictions. In 1969, Russia and China engaged in the first of several bloody wars with each other, and Russia refrained from nuking China in part because of pressure from the United States. Following our 1975 withdrawal from Vietnam, the Red Monolith we had been there to stop quickly crumbled. In 1979, four years after we left Vietnam, the Red Chinese and Vietnamese fought a bloody border dispute, with China losing 30,000 troops in a single month. "The great irony of Vietnam," Utah's former Republican governor and the former United States ambassador to China, Jon Huntsman, recently observed to me, "is that in leaving, we achieved virtually everything we failed to achieve through massive military intervention; Vietnam threw out the Russians and Chinese. It's a thriving capitalist state at peace with its neighbors and friendly toward the United States. As it turns out, we just needed to leave them alone!"

Integrity and Horse Sense

Joseph Patrick Kennedy, my patriarchal grandfather, was born in East Boston on September 6, 1888. His father, Patrick Joseph Kennedy ("PJ"), was a stevedore and saloonkeeper whose parents had immigrated from the Irish hamlet of New Ross during the height of the famine in 1848. PJ parlayed those professions into a long career as East Boston's political boss, representing his district in the Massachusetts State Legislature. Contemporaries knew PJ as a discreet man with sound judgment and a compassionate heart. Grandpa attended the Boston Latin School and captained its tennis team as well as its championship baseball team in 1907, winning the Mayor's Trophy for the best batting average among the city's high school players. Entrepreneurial from the start, Grandpa sold newspapers as a child and organized sandlot baseball games, hawking tickets to spectators. While attending Harvard—he was one of the rare Democrats in that institution—Grandpa purchased and operated a bus, charging sightseers for guided tours of Boston's historic sites.

After graduating, he worked as a government bank examiner, learning the business inside and out before entering the banking trade, where his sharp mind found him quick prosperity. In 1914, Grandpa thwarted a hostile takeover of a neighborhood bank, the Columbia Trust Company, and at age twenty-five he became its president. The Hearst newspapers profiled him in a series of articles titled "The Youngest Bank President in the World." By 1926 he was a millionaire, but still an outsider; Boston's State Street was a Protestant gentlemen's club for oligarchs, where social pedigree trumped financial ingenuity. He eventually ditched the stuffy Brahmins of Boston for New York's Wall Street, hoping to find more of a meritocracy. By mastering every aspect of business and finance, Grandpa was able to make money almost casually, but the subject bored him. "Dad never talked to us about money," Eunice Shriver told me the summer before her death in 2009. "He considered the topic vulgar and tedious, and wanted us to focus on subjects that were elevating and enlightening."

While Grandpa's acolyte, the liberal Supreme Court justice William O. Douglas, would later describe him as a "crusty old conservative," Grandpa's populist sympathies and mistrust of big business prompted him to campaign for Montana senator Burton K. Wheeler of the Progressive Party, who ran for vice president in 1924 alongside Wisconsin's "Fighting Bob" La Follette. Wheeler, a contemporary version of Bernie Sanders, had helped expose corruption in the Teapot Dome scandals. Wall Street branded Wheeler a radical, but Grandpa admired his views on limiting corporate power, and helped finance his efforts, furnishing Wheeler with a car and a chauffeur to campaign across New England and New York. After his loss, Wheeler remained Grandpa's friend and they talked often about the need for a progressive political agenda. In 1934 and '35, as chair of the new Securities and Exchange Commission (SEC) and adviser to FDR, Grandpa worked closely with Senator Wheeler on legislation to break up the powerful trusts that controlled 75 percent of America's electric utilities. Wall Street characterized the bill as communism and warned shrilly that it would bankrupt millions of American shareholders. Instead it brought fair competition to the utility industry and better prices for consumers.

Grandpa's associates knew him for his judgment, integrity, and horse sense. During World War I, Grandpa took charge of the Fore River Shipyard just outside Boston, a cesspool of corruption and mismanagement. He muscled United States shipbuilders into submitting reasonable bids, and supervising construction of a record fifty ships a year for a decade. Grandpa's boss, Vice Admiral Emory Scott Land, called him "the best administrator and the best executive I have ever known in my eighty years." In 1929, Grandpa preserved his fortune by getting out of the stock market just before the crash, an exit famously hastened by a youthful bootblack who offered him investment tips along with a shine. "When a shoeshine boy knows more about the market than I do," Grandpa reportedly remarked, "it's time to get out."

His performance at Fore River won him the respect of the secretary of the Navy, Franklin Delano Roosevelt. In 1932, Grandpa campaigned hard for FDR, and contributed generously to his presidential campaign. Grandpa's endorsement helped divide the Catholic vote during FDR's primary against Al Smith, and Grandpa played a key role in breaking a convention deadlock by leaning on his friend William Randolph Hearst, who controlled the largest delegations, Texas and California. After the election, Grandpa became part of FDR's inner circle of advisers, "entrusted with some of the most important parts of the New Deal program," according to Roosevelt's son James. Grandpa was the only major New Dealer with a business background at a time when our national banking and financial systems had cratered following the 1929 stock market crash, and faith in the system had evaporated. During the Great Depression, some nine thousand banks went under and Americans stashed what they had left under their mattresses. Suicide rates tripled, people stopped having children, and millions of families were homeless or living in shantytowns.

By the time Roosevelt took office, in March 1933, "Many Americans believed that capitalism and democracy were failed experiments and that we would move to some other economic system," historian Michael Beschloss told me. On the left lay communism, the control of business and property by government, and on the right was fascism, the control of government by business. Germany, Italy, and Spain veered right when confronting the same economic depression, passing austerity budgets and handing their ministries over to industrialists and militarists who suppressed labor unions and workers' rights and promoted monopolies. They clamped controls on the press, repressed civic freedoms, deregulated markets, and hiked police and military budgets. While most Americans have forgotten fascism's dangerous appeal in 1930s America, the era remains an important object lesson today, as private financial powers once again reassert dominion over American democracy. The current hostile takeover of American government by the same energy, chemical, pharmaceutical, and military-industrial cartels that flourished in fascist Europe in the 1930s has been hastened by the weakening of the labor movement and sold to the public by artfully packaging the fascist ideology of unfettered corporate dominion beneath a moral and patriotic gloss.

At that time in history, the daily broadcasts of the right-wing radio priest, Father Charles Coughlin, enchanted millions of Americans with tantalizing divinations of moral restoration through authoritarian rule. American business generally applauded the rise of the Third Reich and lavished investments on both Hitler and Mussolini. A 1934 *Fortune* magazine cover story lauded Mussolini's fascism for breaking unions and transferring wealth to corporations and the wealthy classes. As FDR pushed through his New Deal proposals, America's leading business lawyer, John Foster Dulles, the senior partner at Sullivan & Cromwell, encouraged his clients— many of them Wall Street's biggest players—to defy the president. "Do not comply. Resist the law with all your might, and soon everything will be all right," he said. Wall Street funded efforts to impeach Roosevelt, and even to depose him. In 1933, high-ranking officers of some of Wall Street's richest corporations, representing Goodyear, Bethlehem Steel, JP Morgan banks, DuPont, and others, attempted to overthrow the American government. Only the heroic action of General Smedley Butler, one of the most highly decorated Marines in United States history, averted the coup. The tycoons had invested an initial $3 million toward enlisting Butler to lead a paramilitary veterans march on Washington and overthrow FDR. Instead, Butler exposed the "Business Plot" conspiracy to the press and later revealed its details in testimony before Congress. In 1944, Vice President Henry Wallace predicted in the *New York Times* the dangerous rise of American fascism, which he warned would be coterminous with the ascendance of right-wing media. "With a fascist the problem is never how best to present the truth to the public but how best to use the news to deceive the public into giving the fascist and his group more money or more power," Wallace warned. "American fascism will not be really dangerous until there is a purposeful coalition among the cartelists, the deliberate poisoners of public information, and those who stand for the K.K.K. type of demagoguery." Wallace outlined a blueprint for escaping the kind of fascist takeover that was already devouring Europe. "Democracy, to crush fascism internally, must . . . put human beings first and dollars second. It must appeal to reason and decency and not to violence and deceit," he said.

FDR rose to democracy's existential challenge by confounding both fascists and communists. He managed, in a relatively short time, to inspire hope in the American people. Steering a middle course, he created reforms to restore functioning democratic capitalism, keeping excessive corporate power at bay with his left arm while warding off excessive government power with his right. He raised taxes on the rich and big corporations. He secured the nation's banks and invested in public works. He gave assistance to miners and small farmers. He found jobs in the Civilian Conservation Corps for young hobos riding the rails. He put two hundred fifty thousand people to work in a three-month period.

Even though the United States didn't get out of the Depression until Roosevelt launched his massive stimulus package in preparation for World War II, he gave America a sense of forward momentum that restored people's faith in the future. FDR's reforms expanded and fortified the middle class, launching America into an era that economists would call "the Great Prosperity," fifty years of steady expansion. By 1945, one of every three American workers belonged to a union. America had a healthy minimum wage and a social safety net that included Social Security, public education, and health care. FDR made large investments in public projects. Roosevelt rewrote the social contract to anchor American democracy in humanity and fairness, qualities that inspired admiration for America around the globe. The GI Bill of Rights subsidized higher education for veterans returning home from World War II, and the Glass-Steagall Act stopped commercial banks from speculating in stocks, a practice that had caused millions of Americans to lose

their life savings during the crash. And he created the Securities &
Exchange Commission to regulate Wall Street and the stock exchange. In 1934, he appointed Grandpa its first chairman.

Grandpa was a shrewd pick. He had prospered in the unregu-
lated stock market during the Roaring Twenties. Taking advantage
of the absence of rules, he earned a reputation as a pirate among
Wall Street's elite old guard—the Morgans, the Whitneys, and the
Rockefellers—who presided over a world they wished would remain closed to "cheeky Micks" like him. Senator Burton Wheeler,
who strongly supported Grandpa's appointment, told Roosevelt
that Grandpa "had played the market long and short, knew the
game inside and out, but that he had a wife and nine children and
would not do anything that would bring discredit upon himself
or upon them." When Wheeler later repeated the conversation to
my grandfather, Grandpa grew serious and said, "You never told a
greater truth in your whole life."

Grandpa knew that the stock market, rigged as it was to favor
people of a certain background, provided perilous waters for out-
siders; having learned all the tricks himself, he knew how to fashion
rules to protect the small investor. And it was obvious to Grandpa
that markets would not self-regulate. Left to their own devices, corporations tended to concentrate into monopolies with the power to
eliminate smaller competitors and to manipulate prices and wages.
This trajectory could be kept in check only by external forces—
labor unions and farm cooperatives that could demand fair competition, and, most importantly, healthy democratic governance.

At Grandpa's twenty-fifth Harvard reunion, he shocked his former classmates, including some of Wall Street's leading tycoons,
by chiding industrialists for trying to crush the labor movement,
warning: "We are reaping the whirlwind of a quarter century mis-
handling of labor relations." Grandpa felt that government could
play a critical role in protecting small investors by requiring trans-
parency, encouraging fair negotiations between industry and labor,
and stopping giant corporate syndicates from selling adulterated
food, polluting the air and water, or using unfair market practices

to crush small businesses. But he was by no means anticapitalist. He simply believed that integrity and fair dealing were needed to ensure the survival of the markets. He loved America and the free-wheeling marketplace that had given him the chance for a new life. When sentimental Irish patriotism infiltrated the conversation, he'd scoff, "I'm not Irish, I'm American. Ireland never did anything for the Kennedys!"

While Grandpa loved capitalism, he wanted to save it for the many, not the few. And his reforms helped save Wall Street from itself. At the time of his appointment, the mission of the SEC was only vaguely defined, and he would play a leading role in crafting the character of the agency that would revitalize the American marketplace. In 1935, for example, Grandpa ruled it illegal—for the first time—to trade on insider information. The SEC went on to become a model for stock exchanges and financial markets worldwide. Tony Neoh, a historian of the Great Depression and chief adviser to the China Securities Regulatory Commission, told me, "Joseph Kennedy designed the system that all of us operate by today. It was robust, reliable, and effective. Today, we all look to America for leadership in rational market regulations." He added, "When Phil Gramm and the Republicans dismantled the Glass-Steagall Act [in 1999], the reaction in Asia was to do the same, since America had always been our model. I had to tell them, 'Hold on, we don't want to follow the Americans this time.' This helped [the Chinese] avoid the collapse America experienced in 2008."

When Grandpa resigned from the SEC in 1935, his friends as well as his former critics showered him with praise. In its cover story on Grandpa, *Time* called the SEC "the most ably administered New Deal agency in Washington." A columnist in the *Wall Street Journal* wrote that "Mr. Kennedy happened to be in a class alone" among Roosevelt appointees. He earned the respect of Senate Republicans and Democrats for his hard work and integrity. However, despite his role in saving the markets, Wall Street's moguls would always consider him a "traitor to his class"—a "Judas," in the words of one Wall Street broker. Their unhinged hatred for Roosevelt attached itself to my grandfather, despite his role in restoring the fortunes of those who now reviled them. At the outset, Grandpa would argue with his Wall Street friends. When Grandpa's pal William Randolph Hearst dismissed the New Deal as a Communist plot, Grandpa wrote him, "There has been scarcely a liberal piece of legislation during the last sixty years that has not been opposed as Communistic." His reply did little to mollify Wall Street. Hearst continued to deride FDR as "Stalin Delano Roosevelt."

An intense loathing for Roosevelt among the wealthy classes was a dominant feature in the American political landscape during the decade of the 1930s. Historian William Manchester observed of the era that wealthy Roosevelt haters "abandoned themselves in orgies of presidential vilification." Searching for a shred of reason behind the craven revulsion for the president, a bewildered *Harper's* magazine editor published an inventory, in 1937, of the most vocal Roosevelt haters, including the CEOs of Phillips Petroleum, National Steel, DuPont, General Foods, Monsanto Chemical, and General Motors, with a side-by-side accounting showing the tremendous growth in their stocks that had resulted since the implementation of FDR's New Deal policies. "The majority of those who rail against the President," *Harper's* concluded, "have to a large extent had their incomes restored and their bank balances replenished since the low point of March 1933," before Roosevelt came to power. "That is what makes the phenomenon so incredible. It is difficult to find a rational cause for this hatred." *Harper's* went on to describe the reaction among the wealthy as a "fanatical hatred" so intense and unreasoning that it could only be explained as the product of "abnormal psychology." By 1936, the *Harper's* article continued, "The Roosevelt haters had developed into a well-defined cult among the nation's business elite, their lackeys in the press and on the editorial boards." Roosevelt compared their vitriol to a rich man who, after being rescued from drowning, complained that the lifeguard lost his silk hat.

Grandpa stopped going out in Palm Beach because of the poisonous disapprobation of his well-to-do neighbors. In his 1936 book *I'm for Roosevelt*, Grandpa summed up, with some bewilderment, the rage toward FDR among the wealthy titans whose fortunes he had saved. His analysis of their wrath against the president sheds light on waves of calumny toward Presidents Clinton and Obama in recent years. "Anyone who has moved in the circles of so-called fashion and actual wealth," Grandpa wrote, "can testify to the truth of my contention that 'hatred' is the only word that properly defines the attitude toward Roosevelt of thousands of men and women among the more fortunate American classes. Theirs is not a political disagreement or even a moral dispute. It is an unreasoning, fanatical, blind, irrational prejudice against President Roosevelt and his plans for a fairer national economy." Roosevelt, he argued, had saved capitalism from the corporations and titans whose conduct in recent years had completely shattered "the belief that those in control of the corporate life of America were motivated by honesty and the ideals of honorable conduct." But in doing so, FDR had triggered Wall Street's bile by debunking the presumption that wealth is a divine reward for good behavior, and accumulation an indication of civic virtue. Of the corporate titan, Grandpa said, "His moral prestige is gone. He is being judged by new standards, which are quite unfamiliar to him. He feels exposed. He has been shaken in his own faith in himself. He is made to doubt secretly that he represents the American system in its most perfect expression. All this is unconscious and for the most part frightful. Being unable and certainly unwilling to analyze with coolness the cause of his anxiety, he seeks relief in vituperation and hatred." Young people looking vainly for reasons why Clinton and Obama inspired such venom may take some comfort in knowing that it has pursued every liberal president beginning with FDR.

Despite his successes at the SEC, at the U.S. Maritime Commission, and as ambassador, Grandpa's outspoken opposition to World War II, and his maverick, near manic, campaign to galvanize the president to secure a safe haven for Germany's Jews after Kristallnacht ultimately poisoned his friendship with Roosevelt, ending any political aspirations Grandpa may have had for himself. For the remainder of his life, his children would be the centerpiece of his existence.

For the Good of All

Grandpa had a fierce sense of family, taking great interest in all his children's activities and encouraging their engagement with the world. The dinner table was a forum for discussion and sometimes heated debate. Prior to supper he gave his children research assignments—Algerian independence, for example, or socialism. Then he goaded his children into lively disputes to satisfy himself that they had both brains and backbone. Uncle Jack's best friend, Lem Billings, a frequent visitor at the house, told me how Grandpa often took extreme positions to incite his offspring—boys and girls—to argue and defend their points of view. Describing those dinner table dialectics, my dad wrote about Grandpa, "My father has believed we could think and decide things for ourselves. . . . There were disagreements, sometimes violent, on politics, economics, the future of the country, the world." Grandpa wanted his children's minds unshackled by ideology. He groomed them to question authority, and to be beholden to nothing but the enlightened open minds God gave them.

Even as youngsters he sent his children on trips to learn from teachers or political leaders whose views clashed with his own. To broaden their perspectives, he sent both Jack and Joe to study under the socialist philosopher Harold Laski at the London School of Economics, even though, according to Lem Billings, Grandpa considered Laski a "screwball." Joe returned enthused by Laski's left-wing worldview, which informed his dinner table advocacy for the merits of communism, until he crossed his father's pain threshold. "I don't want to hear any more on that subject," my fuming grandpa told him, "until you've sold your sailboat and car and donated the proceeds to the proletariat." Grandpa sent my father and Kick and Joe to the Soviet Union, and Uncle Joe visited Fascist Spain to chronicle its civil war. Grandpa sent Jack to Spain, Italy, and Germany, and my father went to Czechoslovakia following its fall to communists. Grandpa also encouraged my father to attend lectures by the popular fascist priest Father Feeney, and applauded my dad's willingness to debate the charismatic cleric.

Grandpa was as interested in the intellectual development of his girls as he was of his boys. "While many fathers at the time pushed their daughters to land husbands, Grandpa insisted his daughters find jobs," Eunice recalled, and this, no doubt, explains why all the Kennedy girls married relatively late. Eunice tied the knot at thirty-two. After college she got a job in the State Department, assisting returning POWs, then moved to the Justice Department. For a time she worked undercover, investigating corruption in a West Virginia prison, where her fellow inmates included Tokyo Rose and the gun moll of Machine Gun Kelly. My Aunt Pat, who had a gift for figures, went to Wall Street, to work for an investment firm, and then to Catholic Relief Services.

Concerned lest they join the idle rich, Grandpa encouraged his children to experience the wide world without regard for comfort. He hoped they would learn to appreciate the value of money without coveting it. He himself was generous to the Church, to people in need, and to his staff. When his employee Bill Sullivan at RKO died suddenly, Grandpa paid Bill's widow his full salary for the rest of her days. But when it came to his children, Grandpa considered austerity a virtue. He wanted his children to benefit from the best schools, teachers, and training, but without any special privileges. Uncle Teddy recalled, "He wanted us to have a good time, but never a frivolous one. Ski, fine, but why do you have to go to Europe to do it? Enjoy a restaurant, but stay out of the nightclubs. . . ." When Uncle Joe wrote Grandpa from England that he had developed an interest in racing yachts, Grandpa replied sternly that he would "much prefer you to study and travel until time to return home."

According to my mother, Grandpa rode Teddy the hardest. While his elder brothers were all involved in serious endeavors, fun-loving Teddy felt the tug of the playboy lifestyle. My mother recalls Grandpa yelling at Teddy, "No son of mine is going to play polo!" Grandpa cautioned in a brief missive: "Dear Teddy, If you're going to make the political columns, let's stay out of the gossip columns. Love, Dad." When Grandpa somehow learned that Teddy, then at Harvard, had a horn on his car that bellowed like a mournful cow, he addressed the matter in another letter. "I don't want to be complaining about things you do, but I want to point out to you that when you exercise any privilege that the ordinary fellow does not avail himself of, you immediately become the target for display and for newspaper criticism. It's all right to struggle to get ahead of the masses by good works, by good reputation, and by hard work, but it certainly isn't by doing things that people could say, 'Who the Hell does he think he is?'"

While he was toughest on Teddy, Grandpa had a special soft spot for Aunt Pat. Pat was Grandpa's favorite; she was beautiful, loved Hollywood as he did, and had the best head in the family for money. She sat beside him at meals, and she was always able to make him laugh.

Grandpa believed in hard work, despised shams, and loved the truth. He stood up for the underdog and believed in giving people a second chance. When West Point expelled some hundred cadets for violating the honor code, he arranged with Notre Dame's president, Father John Cavanaugh, that any expelled cadet who chose to attend Notre Dame could do so at Grandpa's expense. His sense of propriety was so strong that it even eclipsed his inclination toward personal loyalty. When a relative by marriage told Grandpa that he had been retained to appear before the SEC on behalf of a client, Grandpa told him firmly, "Get out of the Commission and never come back." When Danny Kaye, an actor whom Grandpa much admired, asked him to use his influence to obtain a draft deferment, Grandpa ejected him from his office.

Grandpa was a devout Catholic, but he didn't wear his religion on his sleeve. He was neither pious nor sanctimonious, but prayed quietly in the back pews of St. Francis Xavier Church in Hyannis, to which he had donated a beautiful Italian marble altar in memory of Uncle Joe. He attended mass weekly and never missed a Holy Day, and, like my father, Grandpa considered profanity and off-color stories taboo. I never heard him curse. Nor did Lem Billings, who became a fixture at Grandpa's home. Grandpa's friend Thomas Campbell recalls, "He had tremendous charm. He could charm a bird out of a tree, but on the other hand when aroused and angry he could blister two coats of paint off a brick wall without using a profane word." Grandpa's Catholicism was anticlerical and rooted in the ethical teachings of the Gospel, rather than the rituals, formalism, and orthodoxies promoted by the conservative Church hierarchy. Among his closest friends, Cardinal Richard Cushing shared Grandpa's skepticism of papal infallibility and fundamentalist dogma.

Grandpa came to rely on Cardinal Cushing, one of the Church's leading progressive voices, for spiritual inspiration. Cushing, the son of an Irish blacksmith, who grew up on the tough streets of Southie, was a genuinely holy man, beloved in Boston by people of all faiths for his tolerance, kindness, and good works. He believed the economic system needed a moral compass to serve the interests of all society, not just those at the top of the economic ladder; social justice, he believed, should be a central preoccupation of the religious mind. The centerpieces of Cushing's theology were the dignity of work, workers' rights, and the union movement. During the extraordinary confrontation, which ultimately forced the Vatican to change its own teachers, he excommunicated the popular Boston demagogue Father Feeney for preaching that Jesus could not enter heaven.

Grandpa funneled much of his personal charity through Cardinal Cushing in order to keep it anonymous. The two men called each other "Joe" and "Richard," and the Cardinal sometimes picnicked with Grandpa on the *Marlin*. President Kennedy trusted Cushing more than any other Church leader. A frequent visitor in our house, Cardinal Cushing married Sarge and Eunice Shriver, baptized my brother Chris, married and buried my Uncle Jack, and presided over my father's funeral in New York in 1968, and over

Grandpa's in November 1972, at St. Francis on South Street in Hyannisport.

In 1961, Grandpa suffered his stroke. For the rest of his life he was mainly confined to a wheelchair—his right side paralyzed and one hand twisted on his lap. He could not laugh or talk, but he remained interested and alert, loudly grunting to show his pleasure or disapproval. His handshake was firm and his eyes sharp. He brightened when my father and Teddy entertained him with political stories or funny anecdotes, and he loved seeing his grandchildren. In the morning, we read to him in his den. Our cousin Anne Gargan wheeled Grandpa onto the porch in the afternoons to watch us play touch football and to look at the sea. At the end of the day we went over in our pajamas to kiss him good night in his living room, surrounded by family photos. Amid the many sterling frames on a red walnut coffee table was my favorite memento—a small glass box displaying the shell casing of a German firebomb bearing my grandfather's initials: JPK. A Nazi bomber had dropped the engraved explosive onto the ambassador's house in London. Luckily, the incendiary failed to detonate. Grandma assured us that the attack did not frighten Grandpa.

I can't personally vouch for Grandpa's character with much authority, since I was only seven when he suffered his stroke and eighteen when he died, but my memories of him are warm and joyous. He was a powerful figure who commanded both respect and love from his children, and all of them felt his love in return. We grandchildren overheard the grownups caution each other not to get Grandpa "cross"; they jokingly called him "the Bear," a reference to both his imposing presence and his stock market prescience. Jack was forty-three years old when he skipped over the back fence one Sunday to avoid being rebuked by Grandpa for missing daily mass. On another occasion Grandpa bawled out Jack, Lem, and my dad when he caught the three of them standing on their heads in the "bird room" at the Palm Beach home of stuffy Texas oilman Jack Reisman, amid an overly precious display of hundreds of fragile porcelain eggs and birds. As late as 1961 he scolded Jack and his youngest daughter, Jean, for being late for lunch. "Get your tails up to the table right now, you're ten minutes late." As they jogged up the beach toward the big house, Jack turned to Jean and said, "Do you think he knows I'm president of the United States?"

Popular books, movies, and documentaries portray Grandpa as a kingmaker who drove his children relentlessly into public life in order to satisfy his own hunger for power. It's hard to square that characterization with the brood of children he and Grandma produced, idealistic and devoted to our country's least powerful citizens. He once told Cardinal Cushing, "My ambition in life is not to accumulate wealth but to train my children to love and to serve America for the welfare of all people."

In this central ambition, he largely succeeded. Grandpa sired nine children, including three U.S. senators, one of whom also served as attorney general and another as president of the United States. His oldest son completed more than twenty-five missions over Germany and France as a bomber pilot in World War II before dying heroically after volunteering for a near-suicidal mission: the nation honored him with the Navy Cross and the Air Medal. The presidency of Grandpa's second son, John F. Kennedy, was synonymous with idealism. Grandpa's daughter Kick served in the Red Cross for two years during World War II and, despite her Irish ancestry, married Billy Hartington, the (Protestant) heir to the title of Duke of Devonshire, the greatest house in England. Billy died a month after Uncle Joe, fighting Nazis on the Maginot Line, and Kick died in an airplane crash in France shortly thereafter. She is buried in Chatsworth, her husband's ancestral home. The headstone on her grave, written by her mother-in-law, the Duchess of Devonshire, reads: "Joy she gave—joy she has found."

Grandpa's third daughter, Eunice, founded the Special Olympics and many other philanthropic enterprises that have changed the lives of millions of intellectually disabled citizens around the globe. His third son, my father, Robert Kennedy, died as candidate during one of the most optimistic presidential campaigns in American history. Grandpa's youngest son, Teddy, served half a century in the U.S. Senate and left behind a larger portfolio of progressive legislation than any senator in history. Grandpa's youngest daughter, my Aunt Jean, served her country as a highly respected ambassador to Ireland and helped broker the Northern Ireland peace agreement, which finally ended eight hundred years of hostilities between England and Ireland.

There is no doubt Grandpa had a dominating personality. He must also have had a strong gift of leadership to inspire the devotion and ambition that enabled his family to achieve such historic success. As his friend Carroll Rosenbloom pointed out, "The Kennedys could not have accomplished so much in the service of their country unless the head of the family had a deep love for that country and communicated it to his children." My father wrote that Grandpa's overarching purpose was to engender in his children a social conscience. "There were wrongs which needed attention. There were people who were poor and who needed help. . . . And we have a responsibility to them and to this country. Through no virtues and accomplishments of our own, we had been fortunate enough to be born in the United States under the most comfortable conditions. We, therefore, have a responsibility to others who are less well off."

My father went on to say of Grandpa, "What it really all adds up to is love—not love as it is described with such facility in popular magazines, but the kind of love that is affection and respect, order, encouragement and support. He loved all of us. Our awareness of this was an incalculable source of strength, and because real love is something unselfish and involves sacrifice and giving, we could not help but profit from it. His feeling for us was not of the devouring kind, as is true in the case of many strong men. He did not visualize himself as a sun around which satellites would circle, or in the role of a puppet master. He wanted us, not himself, to be the focal points."

CHAPTER 2

Grandma

Being American implies the *obligation* to both know and understand history.

—ROSE FITZGERALD KENNEDY

AS A LITTLE BOY, THERE WAS NOTHING I LOOKED FORWARD TO MORE than our winter trips to Grandma and Grandpa's house in Palm Beach. I stayed with my brother Joe on twin beds in the room where Uncle Jack wrote *Profiles in Courage*. With Grandma's blessing, I would catch anole lizards and blue-tailed skinks under Grandpa's bullpen and keep them behind the window screens in her cabana until we packed to leave. We stored pints of frozen shrimp in Grandma's refrigerator and fished from the flying balustrade that reached out over the surf from her seawall, casting for pompano, jack crevalle, small bonito, and spots, while watching nurse sharks cruise through the glassine faces of cotton candy–blue waves.

Grandma never objected that we threw our live catch into her saltwater swimming pool, where they darted about in small schools as we swam among them, wearing goggles and flippers. For me the high point of our Palm Beach vacations was our annual visit to Trapper Nelson in the nearby Loxahatchee Swamp.

Patrolling in his airboat, we peeled soft-shelled turtles, alligators, and bullheads from his trotlines. One year he gave me a baby possum, which, with Grandma's help, I raised on infant formula from a tiny baby bottle.

Even now I can hear Grandma's clipped Boston accent and airy laugh, and picture her colorful sweaters and blouses, and the wide-brimmed hats that shaded her dark, thick hair, her regal nose, and her kind smile. Of medium height, she was a towering personality, and the wellspring of values that flowed through generations. My father, who was the most powerful figure in our lives, would worry about what she thought of his haircut, or his performance on *Meet the Press*. Laughing, he told us how she used to paddle him with a coat hanger.

My grandmother Rose Elizabeth Fitzgerald was born in Boston's North End in 1890, the eldest of six children, and lived to be 104. From the age of five she campaigned with her father, John Francis "Honey Fitz" Fitzgerald, who was New England's first Catholic congressman and Boston's first ghetto-Irish mayor. His nickname described both his famous oratory skills and the sweet tenor voice that in our family line was inherited only by my uncles Jack and Teddy, and disappeared altogether in my generation.

Honey Fitz published his own newspaper and reigned as Boston's North End boss from 1891 to 1914. He married his second cousin, Josie Hannan, and quit Harvard Medical School a year later to support his new family. Known also as the "Little General," "Fitzie," and "Young Napoleon," Honey Fitz became embroiled in a famous political rivalry with James Michael Curley, immortalized in Edwin O'Connor's classic American novel *The Last Hurrah*. Honey Fitz was also a political adversary of Grandpa Joseph Kennedy's father, East End boss P. J. Kennedy, whose quiet reserve and rigid propriety provided a stark contrast to the brash, impetuous, and noisy Honey Fitz. A brilliant stump speaker and a cocky, flamboyant retail politician, with his gift for blarney and his genuine love for people, Fitzgerald was a thorn in the side of the Brahmin swells and machine politicians who preferred a tame, business Irishman in City Hall. Following his retirement from politics, his favorite diversion was riding the train from Boston to Cape Cod and back to banter with the passengers.

As mayor, Honey Fitz was a populist who took pleasure in rattling the Yankee establishment and fighting for Boston's poor immigrants and working stiffs. A heedless administrator, his mayoralty was an imbroglio of good works combined with cronyism and "honest graft," all Boston staples. In Congress he was a defender of Catholics, Jews, and African-Americans. When Republican Henry Cabot Lodge filed a bill to end immigration, Honey Fitz opposed it, and when it passed, he persuaded President Grover Cleveland to veto it, leaving America's immigration door open for another quarter century. His own proudest achievement in Congress was passing legislation allowing Italian-American citizens to vote even if they did not speak English. As a child I was especially proud that he founded the Franklin Park Zoo and the Boston Aquarium, and today I'm happy to remind my children that Honey Fitz commissioned the Cape Cod Canal, cutting the New York–Boston ferry shuttle by sixty two miles.

In the Big House at Cape Cod, Grandma showed me old clippings of her father waving jauntily as he took off in an early biplane, his thick, dark hair slicked back from a handsome face framed by a square jaw and lit by sparkling eyes and a wide grin. He referred to himself as "the greatest mayor Boston ever had," and Lem Billings told me that in his dotage my great-grandfather would laugh so hard at his own jokes that he sometimes wet his pants. "Sweet Adeline" was his signature ballad. At a conference of Western Hemisphere mayors in Rio de Janeiro he swore to the Latin American delegates that the song was the new American national anthem, and had them memorize and sing it to a confused FDR at a presidential reception. From then on, FDR ever referred to Honey Fitz as "El Dulce Adelan."

In retirement, Honey Fitz rented a house in Hyannisport each summer. He irked Grandpa Joe Kennedy to distraction with his habit of telling filthy jokes to the Kennedy boys until they roared with laughter. He further nettled Grandpa by sunbathing nude on the beach for hours in front of Grandpa's house, after first wrapping his body in seaweed—a ritual he considered to bestow prodigious medical benefits. As mayor he had banned the turkey trot and tango as immoral, yet he would dance a jig through the Big House hallways with arms held high, index fingers pointed toward the ceiling, singing, "I love the ladies! I love the ladies! The fat ones, the slim ones, the tall ones, the short ones. I love them all!" His voracious affection for the ladies—in particular for a blond-haired cigarette girl named Toodles Ryan—had ultimately derailed his political career. A shoo-in for reelection during Boston's 1914 mayoral race, he intended next to challenge Senator Henry Cabot Lodge in 1916. But Honey Fitz's archenemy, James Michael Curley, discovered the alleged dalliance and sent Josie Fitzgerald an anonymous blackmail letter, while publicly announcing a series of museum lectures, titled, respectively, "Great Lovers of History—From Cleopatra to Toodles" and "Libertines in History from Henry VIII to the Present Day." Honey Fitz swore to Josie he had done nothing but kiss Toodles at a public event at which Josie was present, but Josie insisted he abandon his mayoral race rather than damage their daughter's marriage prospects.

Honey Fitz died in 1950, four years before my birth, so I didn't know him, despite my children's persistent efforts, when they were little, to summon him up on the Ouija board. However, I learned about the cocky bantam's titanic personality, and heard many entertaining tales from Lem and Teddy. Teddy told me that, walking with Honey Fitz, as a young teen, they crossed paths on Beacon Hill with Fitzie's lifelong nemesis James Michael Curley a few weeks before his sentencing to the Danbury Penitentiary on a fraud conviction. Honey Fitz was carrying Teddy's green flannel book bag. "Do you have yah burglar tools in that satchel?" Curley maliciously inquired. Honey Fitz raged at the insult to Jack and Teddy during the two-hour drive to the Cape that afternoon.

Lem told me of Honey Fitz's special love for Uncle Jack, his namesake. "You've got the goods," Honey Fitz told him when my uncle launched his 1946 run for Fitzie's former congressional seat. "You are going to play on a much larger stage than I ever did." And Uncle Jack idolized him in return. Jack ran his campaign out of Honey Fitz's retirement address at the Bellevue Hotel, where his vigorous eighty-three-year-old grandfather advised him on strategy and worked with as much enthusiasm as the former mayor had mustered for his own campaign. At Jack's victory celebration, Honey Fitz climbed on a table and sang "Sweet Adeline," danced the Irish jig, and predicted that one day Jack Kennedy would be America's first Irish Catholic president. Shortly after fulfilling that prediction, Uncle Jack renamed the presidential yacht the *Honey Fitz* in honor of his beloved grandfather.

Honey Fitz's wife, Josie, outlived her husband by fourteen years, dying in 1964 at age ninety-eight. She kept a tiny shingled summer cottage in a pine grove near Kalmus Beach, a mile from the Kennedy compound in Hyannisport. My mother took us to visit after Sunday mass. As she combed Josie's thick hair, Kathleen, Joe, and I took turns reading to her. Josie's parents had emigrated from County Limerick during the famine, and she was the fifth of their nine children, only four of whom survived childhood. Josie died a year after Uncle Jack. She had no TV, and by then her eyes had dimmed so she couldn't read the newspaper. Everyone was under strict orders not to mention Jack's assassination, because she worshipped her grandson, whom she always referred to as "my boy." The grown-ups feared the news of his murder might kill her.

Grandma Rose was the eldest of Josie and Honey Fitz's six children, and the apple of her father's eye. Outgoing and beautiful, she became a public surrogate for her pious mother, who steered clear of politics, tending to her brood instead. Grandma represented Honey Fitz at events ranging from Women's Improvement Society gatherings to steamship christenings, and she helped lead his legendary torchlight parades when he ran for mayor in 1906 and 1910, playing an upright piano atop a bunting-draped flatbed truck. Grandma accompanied Honey Fitz on the keys as he crooned outside campaign halls, Irish saloons, and at neighborhood picnics.

Even in her nineties she played beautifully. On summer nights and family birthdays our family crowded into the living room on the Cape, where her grand piano groaned beneath a forest of framed family photos. Grandma pounded the ivories as Teddy led our sing-alongs with patriotic ditties like "I'm a Yankee Doodle Dandy," the show tunes of Irving Berlin and Cole Porter, and Grandma's Irish favorites—"When Irish Eyes Are Smiling," "MacNamara's Band," "Molly Malone," and "My Wild Irish Rose." Grandma Rose beamed when Uncle Teddy serenaded her with "Sweet Rosie O'Grady." Afterward, her children and grandchildren knelt around her bed for the nightly rosary.

In the summer of 1908, when she was eighteen, Grandma and her sister Agnes took the Grand Tour of Europe with Honey Fitz, after which Grandma attended finishing school in Holland for a year, becoming fluent in both German and French. Grandma Rose continued to study and practice those languages past her eighty-fifth birthday. She was missionary about the importance of communicating with people from other cultures, and often forced the grandchildren to speak French with her. I was hopeless, but she gave me some leeway because my Spanish was tolerable. She genuinely loved scholarship, and regretted not going to college. A good women's Catholic college did not exist at the time, and despite her wishes, her father would not allow her to attend Wellesley.

Lack of a formal education did not dampen Grandma's love of learning, however. She regarded an inquisitive mind, well furnished with history, philosophy, and culture, to be the hallmark of character and respectability. "As a young girl," she would later write, "I had a complex that I should read everything, and that I should hear the good opera singers and the good violinists, and it was rather a disgrace if I did not." She graduated from Dorchester High School at the top of her class, and even after her mind began to decline eight decades later, she recalled with painful clarity the name of the only girl who surpassed her in French. She joined a local theater group, acting in plays and musicals. She taught catechism and Sunday school. She did settlement work and sewed for the poor. She was the youngest member of the Boston Public Library Investigating Committee, which developed reading lists for the city's children.

Grandma and Grandpa met as teenagers on the Maine seashore, where their families had rented summer cottages, and immediately fell in love. She often spoke to us grandchildren about Grandpa and their life together, the fun they had, and how dearly she cherished him. "He was the only man I have ever kissed," she told me. When they met, Grandpa was president of his class at Boston Latin School and on his way to Harvard, but Honey Fitz discouraged the relationship. He believed Grandma was too young to go steady, and that she should consider other suitors, a considerable list, she assured me, which included Sir Thomas Lipton, the Irish cabin boy who became a self-made tea and grocery magnate. Honey Fitz reassessed Joe Kennedy when he became the youngest bank president in the country. Grandpa and Grandma were married in October 1914 and loved each other deeply for fifty-seven years.

Curiosity Transcendent

Grandma mellowed with age, but all the Robert Kennedy children recall the rigid discipline beneath which we chafed when she came to supervise us during our parents' prolonged African tour in 1964, following Uncle Jack's death. She fine-tuned our table manners and put us to bed an hour early. We, who had driven dozens of robust German, French, and British governesses to the brink, were soon wearing shorty-shorts and knee socks to afternoon tea, attending two daily masses, saying the morning rosary, and arriving punctually for meals, hair slicked and nails clean. By force of will she cured my cowlick, and switched all of us from southpaw to right-handed before we could say *"Je ne veux pas."*

Grandma instilled in her children and grandchildren the habit of daily devotions. She drilled us in the rosary, our catechism, the Holy Days of Obligation, the Meditations of Cardinal Newman, and the Stations of the Cross, all of which continue to be part of the daily lives of many members of my generation. We ate fish on Friday and made our sacrifices during the forty days of Lent. Not even my uncle Jack when he was president would lightly consider missing Sunday mass or a Holy Day of Obligation. During her final decade, Grandma's frailty prevented her from attending church, so every Sunday the priests brought mass to her home in Hyannisport. All the cousins knelt and prayed together, shoulder to shoulder, on a carpet embroidered with roses, some of us still groggy from Saturday night.

At Cape Cod, Grandma alternated her afternoon lunches and evening meals with different sets of grandchildren. It took courage to be late for a Big House dinner. Like Grandpa, she was prompt, and rigid about decorum—elbows off the table, hair groomed, fork placed neatly next to the knife on our plate after dinner, etc., etc. Her love of learning along with her deep religious faith were principal remedies against materialism, which, she declared, eroded everything of value. She was an exemplar of the platonic love of knowledge and beauty without the need to possess. Grandma wanted us to be well rounded, interested in every aspect of life, including politics, music, art, science, religion, architecture, history, sports, and languages. Over dinner of chipped beef, asparagus, and angel food cake, she inquired about our summer reading, or instructed us on how to distinguish Doric from Ionic and Corinthian columns. She led us in discussions of topics ranging from local politics to African topography to the Gaelic Renaissance. She would accost us on the compound lawn and grill us about entomology, multiplication tables, or the Stations of the Cross, or give us spot quizzes on history, astronomy, or religion. "What happened in today's Gospel? What was the sermon about? Can you name the capital of Poland? What are you studying in French? *Comment allez-vous?*" She taught us about art; not surprisingly, her favorite painters were both American: Grandma Moses and Thomas Cole.

If you ran into Grandma she might hijack you for a trip to confession, or invite you into the Big House to watch the news, or drag you into the basement to explain the British partition of Pakistan, or make you locate Zanzibar, Burma, or the Rift Valley on the great map of the world that adorned the wall outside Grandpa's theater. At night she would escort us onto the Big House porch and make us name the constellations, and find the Big Dipper or Polaris. She would explain why the North Star alone remains stationary in the night sky and ask us about the vernal equinox, and from there launch into a lecture on Latin roots of the English language.

Grandma loved language, regarding it as the vehicle for clear thinking. She corrected our grammar, and made sure we always used the appropriate pronouns. She returned our notes and letters with red-penciled editing marks and suggestions for improvements. Part of her legacy to her grandchildren was an intense love of poetry. At her urging we memorized her favorites, Longfellow's "Paul Revere's Ride" and Emerson's "Concord Hymn," as well as excruciatingly long passages from the works of her great hero, Henry David Thoreau. We always came to dinner ready to recite a poem.

Grandma was curious about my generation's attitudes and ideas. She wanted to know all about our opinions and ambitions, our clothes and our hobbies. What were we reading? What were we studying at school? What subjects did we discuss with our friends? She listened to Bob Dylan and read the poems of Rod McKuen. Once she wandered into my cousin John Kennedy's riotous high school graduation party and quizzed his suddenly subdued guests about their life plans. Then she got them all to solemnly study their cuticles by inviting them to church.

One night we told Grandma we were headed for a Rolling Stones concert at the Cape Cod Coliseum. She must have mulled it over before deciding to witness the spectacle herself. She came alone, but word of her presence rippled through the Coliseum as she searched for her seat, and the young audience rose spontaneously to greet her with a long ovation. That gesture touched her deeply. There were, however, limits to her embrace of the cultural revolution. She drew that line at the musical *Hair*, telling an interviewer, "I can see plenty of naked bodies in my own basement when people are changing into their swimsuits."

Personal dignity was high on Grandma's list of priorities. She once scolded Tom Brokaw for drinking in her theater, and I still recall her ferocious scowl when my sister Courtney's boyfriend, Dan Dibble, stripped off his shirt to expose a giant rose tattoo while performing a rendition of Dion's "The Wanderer" for Grandma's ninety-fifth birthday party. I suspected that, had she been able to leave her wheelchair, there wouldn't have been a coat hanger left in the closet, nor an unviolated strip of flesh on Dan's body.

As a youthful prank, Uncle Jack routinely took his friend Lem's sneakers from his bedroom and placed them on the living room piano, in order to prompt a lecture by Grandma about neatness and propriety. When she finally learned that it was Jack's doing, she apologized to Lem and scolded Jack. When Lem was visiting the Summer White House in 1961, the family filed in to dinner to find his sneakers on the piano. Grandma could hardly contain herself as she chided her son. "You are the president of the United States, and you still put Lem's sneakers on the piano?"

Hygiene and physical fitness, along with manners and personal appearance, were her reliable preoccupations. Were we brushing our teeth? To promote daily dental care, she showed us a World War II painting of Uncle Jack towing his injured shipmate toward a moonlit island, the end of a tether clenched in his strong, white teeth. With surprising superficiality, she told my brother Joe that Grandpa's gleaming ivories were his greatest asset.

Sporting excellence was likewise a priority. She was intensely interested in our sailing, swimming, and tennis contests, and our successes at waterskiing, fishing, and scuba. In her nineties she was still spry from her own daily calisthenics, long walks, and daily swims. Although she maintained a pool at the Cape for Grandpa to exercise in after his stroke, Grandma discouraged dips by other family members. The ocean, she insisted, provided a more vigorous swim. She herself swam even in the coldest weather. From the shore, we could spot her bright bathing cap bobbing like a rogue lobster buoy in the waves far out at sea.

In her eighties, when I stayed with her at Palm Beach, Grandma employed a hirsute motorcycle enthusiast to carry her past the breakers for her morning splash. They were an unlikely pair. He stood at six foot one and weighed three hundred pounds, his tangled beard and thick mats of brown body hair nearly obscuring the savage tattoos on his chest and massive forearms. She was tiny and pale in her one-piece swimsuit and her blue latex bathing cap. One day a photographer walking the shore asked to snap their picture. I know Grandma enjoyed it when the photo ran in a national magazine, showing the two of them standing together in the sand, dripping wet, arms laced across each other's shoulders, the elderly matriarch and the heavily foliaged biker. It looked like a Kodiak bear landing a silver salmon.

With her straight-as-an-arrow golf shot, Grandma was a legend at the Hyannisport Golf Club. Parking her car daily on the dirt road off the back nine, near the old caddie shack, she marched the fairway, briskly gripping three clubs. She played solo with two balls simultaneously, consistent with her philosophy of maximizing opportunity. Time being a precious commodity, she would often do two things at once, listening to French records while having a massage, studying vocabulary while being driven on an errand, or saying the rosary if it was too dark to read.

An incessant walker, she urged us to never go to bed without a twenty-minute after-dinner stroll. Her own evening rambles covered a good two miles, even in her late eighties, and she didn't reserve them just for that time of day. She walked regardless of pouring rain or blazing sun. My cousin-in-law Arnold Schwarzenegger once wrote that he "was quite shocked at how much walking she did and how the family always rotated who would go on walks with her." To preserve ourselves from exhaustion, we each took hourlong shifts as she meandered around Hyannisport's seaside golf course, or on the towpath along Lake

Worth in Palm Beach, or on the sandy shores of Nantucket Sound. In preparation for these promenades she pinned to her sweater snatches of poetry, clippings from the morning papers, and noteworthy quotations or passages from books she was reading. As we walked she might depart from this syllabus to ask about our own reading, or to talk about our family and its history.

Grandma consistently reinforced our sense of identity. We were Americans first, yet she wanted us to remember the lessons of our Irish heritage. We learned about the bitter history of British conquest and oppression, the holocaust of starvation and exile that depopulated the Emerald Isle, and our tribal struggle for liberation. She told us about the gallant figures of the Irish Rebellion, especially Michael Collins and Éamon de Valera. She recited Yeats's poem celebrating the famous love affair between the revolutionary heroes Charles Parnell and Kitty O'Shea. Once, at dinner, she recited a poem penned by the grief-stricken mother of Irish revolutionary Pádraig Pearse who, like Grandma, lost three sons in the service of her country. In Grandma's clear Boston accent there was heartfelt sadness, but her eyes also glimmered with mischief. She closed with a wink, as her listeners fought back their tears.

But Grandma loved American history best. She wanted to be sure our parents were taking us to sites of historical significance, as Honey Fitz had done with her, and as she had done with her children. My parents faithfully continued that tradition; and until his death in 2009, Uncle Teddy would take the twenty-six living grandchildren and fifty-plus great-grandchildren on regular camping expeditions to our nation's historical battlegrounds: to Bunker Hill, to Boston's Old North Church to learn about Paul Revere, to Plymouth Rock, Walden Pond, Gettysburg, and Manassas, sometimes accompanied by Civil War historian Shelby Foote or documentary filmmaker Ken Burns.

Two years before his death, Teddy took three generations of Kennedys to see "Old Ironsides" where she lay at the wharf. Here in Boston Harbor, he reminded us, American patriots struck their first blow for freedom by disguising themselves as Native Americans and dumping British tea into Boston Harbor. The gesture was more a blow against corporate tyranny than a protest against big government, for the tea belonged to the East India Company, a ruthless monopoly exempted from the crippling tea tax that every American merchant had to pay. In that sense, the American Revolution began as a protest against excessive corporate power.

Grandma made history come alive, and although she was Irish, she laid claim to America's early history and values as if they were her own. She explained how the Puritans had fled the stifling feudalism of Europe to start new communities based on freedom, equality, and self-determination. She told us how the Indians had saved the Pilgrims during their first bitter winter in New England, and how Pocahontas had rescued John Smith. She loved the dissenters who stood up to religious orthodoxy and government authoritarianism, especially Roger Williams, who founded Rhode Island on the principle of religious liberty.

But her favorite topics were the Revolutionary War and the freedom of our country. She wrote in her autobiography, "I hope my grandchildren and great-grandchildren will realize and remember that the United States of America was one of the few places—perhaps almost the only place—in the world where the saga of the Kennedy family could have happened. I hope they will always feel a deep sense of gratitude toward the country and deep pride in it and an obligation to preserve, protect, and defend it. If they choose to take the means of political life and public office, so much the better, for this has been so much a part of our family tradition, and it would please me to think of it being continued."

The Common Touch

Grandma was a shrewd campaigner and as talented a retail politician as our family ever produced. Beginning with Jack's 1948 congressional race, she campaigned for every Kennedy with all the vigor and savvy she learned at Honey Fitz's knee. She gave speeches in German in Wisconsin, and in French to New England's French-Canadian groups. At eighty-nine she traveled all over Iowa to support Uncle Teddy's 1980 presidential bid, still outshining her swarming progeny. Whenever her relatives gave speeches, she maneuvered quietly to the rear of the venue where she could rub elbows with the crowd and take its temperature, the better to hone her sage advice.

She taught us how to give an interview—"Don't talk about yourself, talk about your issues"—and how to work a room; she visited every table at political dinners, shaking hands with hundreds of people and taking time to chat, her eyes sparkling and full of laughter. "Politics is fun," she'd tell us. "You should always smile. Show them your teeth!" Even after years of communing with kings and queens and world leaders, she never lost the common touch. While being honored at the dedication of an aircraft carrier, she left the officers' table to dance with the enlisted men. She routinely instructed her chauffeur to pick up hitchhikers. And when hoodlums accosted her in Central Park to steal her alligator purse she scolded them for smoking cigarettes, leaving them so thoroughly chastised that they neglected to rob her.

Grandma loved when people recognized her, and she always stopped to chat, whether it was with a famous movie star or the parents of an intellectually handicapped child. "You only get one chance to make an impression," she warned us, "and you should make a good one." On her morning trip to mass, or her daily walks around Hyannisport, she greeted tourists and talked candidly to surprised reporters who gathered every summer morning to cover "the Kennedy beat." When her rambles intersected with a Hyannisport tour bus, Grandma sometimes climbed aboard and introduced herself to each shocked sightseer, then apologized for not inviting them into her house, usually explaining that she hadn't expected quite so many guests. She often reminded us that many of these travelers had come long distances out of love for the Kennedy clan, and she hoped each member of our family would greet that loyalty with respect, gratitude, and kindness. And she told us to always be polite to the throngs of intrusive paparazzi who gathered at the end of our driveway, or at the foot of the Hyannisport pier to snap our pictures when we boarded the boats. "The editor told him to get your picture," she instructed. "And it would be mean to get him in trouble with his boss when he's just trying to do his job."

Grandma had an inquisitive, restless mind. The universal wrench in her self-improvement toolbox was her curiosity about people of all backgrounds. Her style was merciless interrogation. Her victims—anyone with whom she happened to share a moment— encountered a staccato barrage of nosy questions about their lives, their families, their religious beliefs, their circumstances, and their ambitions. Fishermen, actors, cabbies, political leaders, bus drivers, tourists, movie stars, heads of state, strangers in elevators: none were spared these inquisitions. Grandma then reported her findings to the family in enchanting letters, which often also included detailed descriptions of her world travels.

Grandma's correspondence from abroad would infect the most devoted homebody with wanderlust. She never tired of traveling, nor lost her capacity for wonder. Her various missives describe Windsor Castle, her bedroom at the Roosevelt White House, the atmosphere aboard the German ocean liner *Bremen*, or visiting homeless orphans in the streets of Rio, the political views of Italian shopkeepers, and the styles "au courant" in the Paris fashion houses. She conveyed her awe at the vastness of the Pacific seacoast and California's sprawling film studios, her findings about the history of the western missions, the unique foods and architecture of the region, and her views of the great thinkers and literary figures of the day. She traveled alone across the globe until late in life, spending her eightieth birthday in Addis Ababa with Emperor Haile Selassie.

Second only to travel, she loved books and constantly goaded us to read. Her love for reading, she assured us, was the principal reason for Jack's success as a national leader. She told us that Jack read one book a day and that when he departed on an eight-day vacation she had opened his heavy travel bag to find eight books. Each year she wrote current authors requesting a box of signed books for her to distribute as Christmas gifts. In October 1962, she sent such a request to a first-time author, Nikita Khrushchev. The letter arrived on Khrushchev's desk during the height of the Cuban Missile Crisis at a moment when Khrushchev already had his hands full, sorting out letters from other Kennedys. The letter, signed by Rose Kennedy, Hyannisport, Massachusetts, reportedly caused consternation and criticism among both the KGB and CIA.

When I was twenty-one and enjoying a wild stage, during which more cautious family members were reluctant to even loan me their automobiles, Grandma gave me her Palm Beach home for the summer so that I could work on my first book, a biography of Alabama's civil rights hero Judge Frank M. Johnson, Jr. During the early weeks before she left for the Cape, I had the pleasure of sharing the house with her. We dined together, took long walks and swam, attended mass, and mulled her usual wide-ranging interests. We spoke about Henry David Thoreau's stint in the Concord jail protesting the Mexican-American War. She related how Emerson, standing outside the prison window, had asked, "What are you doing in there, David?" Thoreau had replied, "What are you doing out there, Ralph?" Her early embrace of environmentalism gave us a special bond. Her implicit approval would inspire my own acts of civil disobedience, which would land me in a Puerto Rican prison for thirty days.

As far back as the 1930s, Grandma took her children on regular excursions from Hyannisport to Walden Pond, where they would swim and picnic and hear from her about Emerson, Thoreau, and other champions of individual conscience over orthodoxy, and of self-government over authoritarianism. We might best infer the Creator's will, Grandma and her favorite transcendentalists believed, by searching our hearts and by studying Creation. By observing nature one could achieve enlightenment and sense the divine. Her favorite Thoreau quote was: "Heaven is under our feet as well as over our heads." Grandma told me how on a trip with Aunt Kick to the Soviet Union in 1936 she visited the Russian National

Library in Moscow—the country's largest book repository—to determine whether Thoreau had escaped the Soviet censors. She was delighted to find his books still on the shelves.

Her environmentalism aligned with her natural frugality. "Never waste anything" was among her favorite mottoes, and she even squirreled away uneaten fragments of her beloved York Peppermint Patties. She discouraged us from sending flowers, a practice she considered profligate, or riding first class. Her younger children wore hand-me-down blazers and khakis. She was religious about keeping her lights off overnight to save both energy and money, and wrote scolding notes to my mother and Aunt Jackie whenever she looked out from the Big House and spied a lit window after bedtime.

Grandma generally laced her practical advice with tenderness. She sewed on my missing buttons and once helped me dig my jeep off the beach as the incoming tide threatened to submerge it. She counseled me to let some air out of my tires to improve traction in the soft sand. I recall that summer with her as one of the most joyful times of my life. Mostly our conversations wandered among the fascinating range of topics that constantly engaged her interests, but occasionally she would gently say something pointed. Just before she left for the Cape she reminded me how fortunate all of us Kennedys were with the extraordinary lives we'd been given, and of the moral obligation we had to give back. "Never forget that you are a Kennedy. A lot of work went into building that name. Don't disparage it." Sometimes I would prove better at that than others.

An Uncommon Faith

Grandma's vibrant inner life was the wellspring of her reliable joy. She helped instill my father's generation, and my own, with the faith that held our family together and fortified us during calamity. Rosaries were her constant companions; she stashed them all over the house, and carried a string on her daily walks, and also when she traveled by car. When something interrupted her prayers, she clipped a safety pin between the beads to mark her spot. Grandma sometimes brought our afternoon football games to a temporary halt when she emerged from the Big House to walk across the great lawn to pray in her beach cabana, in the years before it flew off in a hurricane.

Prayer was the essence of her life, and she began each day on her knees with morning devotions. Then she was off to church and Holy Communion. In Hyannisport she arrived every morning for seven a.m. mass at St. Francis, peering over the dash of her big blue Chrysler Imperial convertible, golf clubs by her side. She prayed again before breakfast and afterward to thank God for the meal, and bracketed lunch, tea, and supper with similar thanksgivings, ending each day on her knees. Those were the official prayers, but she also prayed continuously throughout the day. She embraced life, yet was undaunted by fear of death, which she saw as a doorway to everlasting peace. Near the end of her life, she told my Uncle Teddy, "I want to go to heaven, because I've seen a lot of you . . . but I haven't seen the other children, or my Joe, in a long time."

Both Grandpa and Grandma were products of an alienated Irish generation that kept itself intact through rigid tribalism embodied in the rituals and mystical cosmologies of medieval Catholicism. When she was younger, Grandma's fundamentalism could be strict, and as with all orthodoxies, occasionally cruel. I didn't know Grandma when she refused to attend Aunt Kick's wedding, or her funeral, presumably because Kick had married outside our faith. By the time I was old enough to appreciate Grandma, she had rid her faith of its dehumanizing elements.

Grandma believed the Church should be a champion of the poor and the powerless, as Christ had instructed in the beatitudes and elsewhere. She spiced her reverence for the Church with skepticism toward the upper clergy, and, like most American Catholics, she winked at the doctrine of papal infallibility, presuming that a refined personal conscience was the ultimate ambition of faith. Her reverence for Emerson and Thoreau was no accident. Both

American philosophers had conspired to level religious and secular hierarchies. Their transcendentalism urged that God resides as much in the individual as in ecclesiastical institutions or divinely appointed monarchs.

Grandma thought deeply about her relationship to God, and she did not allow doctrine to stifle her fiery mind. She often questioned us about inconsistencies in the Scripture. She asked whether we thought Jesus ought to have shown more consideration for his parents when he disappeared from the caravan to Bethlehem, and what it was about the moneychangers in the temple that provoked him to violence and anger. And when Jesus said, "I am the vine, you are the branches," was he really claiming to be a plant? In this way she taught us the distinction between the literal and symbolic truths of Scripture. She believed in evolution and heliocentrism, but implicit in her piety was the acknowledgment that some mysteries are not susceptible to the tools of empiricism. In *Times to Remember*, she wrote, "From faith, and through it, we come to a new understanding of ourselves and all the world about us. It puts everything into a spiritual focus, so that love and joy and happiness, along with worry, sorrow, and loss, become part of a large picture which extends far beyond time and space."

Grandma's Catholic faith allowed her to incorporate the virtues of acceptance and gratitude into every feature of her life, and to dismiss the tribulations and treasures of this world as inconsequential. "If we can truly believe in His presence and goodness, we are never alone or forsaken." She remained content and serene despite tragedies that included the early deaths of her children Joe, Kick, my father, and Jack. She accepted those losses without bitterness, resentment, or despair, focusing instead on the joy of having known them. Her faith was a bulwark against self-pity, which she disdained.

Grandma often spoke of Dorothy Day, cofounder of the Catholic Worker Movement, who, like Saint Francis, campaigned in defense of the poor, the homeless, and the hungry. Day was admired by every generation of Kennedys. In 1940, Joe and Jack visited her for supper and stayed until quite late, chatting with her at the *Catholic Worker* newspaper offices on Mott Street in New York City. Day was imprisoned for protesting for women's suffrage in 1911, for picketing on behalf of César Chávez's farmworkers union in 1973, and many other times in between. After World War II she protested the "warfare state" and nuclear weapons. She was arrested for refusing to participate in civil defense drills, declaring, "We will not be drilled into fear." She abhorred the dangerous pretense that nuclear war was survivable. Day, who will soon be recognized as a saint, believed that we all have a sacred obligation to work for peace and justice.

Above all, Grandma's notion of God's universal love meant that every human being was endowed with dignity and value. Whether our station was humble or noble, rich or poor, our race black or white, our condition genius or cognitively disabled, we were equally treasured by our Creator. Success, her faith told her, should not be measured by accumulation of wealth, power, education, or beauty, but by how each of us uses our assets to serve and benefit others. All material accumulation is a gift from God, entrusted to its beneficiary for some higher purpose. We are each trustees, charged with utilizing our gifts to benefit the whole community, particularly those less fortunate.

My grandmother's eldest daughter, Rosemary, was born intellectually disabled, a handicap later aggravated by a botched lobotomy (which at the time was considered a miracle cure). In those days children with intellectual disabilities were a source of family shame. The remedy was to hide the child in state-run institutions. As a boy I spent time in one of those Dickensian hellholes, volunteering for a hundred hours in the Wassaic State "School" for the Mentally Retarded in Dutchess County, New York, a sprawling stone-and-brick complex that functioned as a holding pen. Inside the giant building there was no evidence of treatment. Children and adults lived together, choking in the putrid odors of urine, excrement, and unwashed bodies. I was scandalized to witness sex acts between intellectually disabled adults, young and old. Violence among inmates was mundane. Robbed of any sense of security, dignity, or opportunities for learning, enjoying almost no stimulus except of their own making, inmates hid in corners and crouched in closets like caged animals, mumbling, masturbating, or mindlessly bobbing their heads. They were mainly forgotten by families exhausted by their needs and shamed by their existence.

Grandma took a different approach. Believing that Rosemary should have all the opportunities for growth and happiness available to their other children, Grandma and Grandpa kept their daughter at home with her siblings. She played sports, attended dances, crewed in sailing races, and took swimming lessons with the other children. The family took Rosemary to dinner at the New York Yacht Club or the Stork Club. When Grandpa went to England as ambassador, Rosemary was presented to the queen at court along with Grandma and Aunt Kick.

Even as she came to terms with the fact that Rosemary's life would be limited, Grandma went public, talking about her daughter in speeches and radio appearances across the country, advocating for people with intellectual disabilities, and urging young women to get good prenatal care. She encouraged families to keep special children at home—a practice now called "mainstreaming"—and urged people with intellectual disabilities to engage in sports. Grandma's stance inspired my Aunt Eunice to found Camp Shriver—a summer and weekend sports camp for children with intellectual disabilities—at her sprawling Bethesda farm. Beginning at age eight I worked there with my cousins and siblings as counselor and coach, giving swimming lessons and organizing running races and baseball games, or simply being a "hugger."

In 1968, the camp blossomed into the Special Olympics, the most widespread athletic program for the intellectually disabled in the history of the world. Today nearly 5.3 million children and adults receive year-round training and compete in 170 countries in over 100,000 events each year. The Special Olympics holds its world games every two years, alternating between summer and winter. After Eunice founded the organization, Grandma traveled the world, urging people to support it, appearing on TV shows like *Merv Griffin* and *David Frost* to speak for people with intellectual disabilities. She devoted all the proceeds from the autobiography she wrote at age eighty-five to that cause.

Thanks to Grandma our whole family became consumed by the cause of civil rights for this most vulnerable population. In 1946, Grandma and Grandpa established the Joseph P. Kennedy Jr. Foundation, named for their deceased son, to "seek the prevention of intellectual disability by identifying its causes, and to improve the methods by which society deals with people suffering intellectual disability, and their families." As president, Jack made intellectual disabilities a new national priority. After his inauguration he founded the National Institute of Child Health and Human Development, which continues to support research on intellectual disabilities to this day. He created a Presidential Panel on Mental Retardation, and eight months later signed the first comprehensive child health legislation—the National Mental Retardation policy amendment to the Social Security Act. Later legislation created research and treatment clinics, diagnostic centers, and special ed. My cousin Tim Shriver continues to run the Special Olympics, his brother Anthony is founder of Best Buddies, and their brother Mark Shriver is president of Save the Children. The creation of the Special Olympics by my aunt Eunice Shriver, and the profound national transformation in the treatment of the mentally disabled, flowed directly from Grandma's deep religious conviction that all of us are created in the image and likeness of God. Their work is Grandma's legacy.

Along with her extraordinary drive, scholarly energy, and religious faith, Grandma had an intensely feminine side. She was fascinated with style. When family members returned from Europe she quizzed them: Were the hems in Paris long or short? Were there any pretty hats? How were people wearing their hair? What art exhibits were on? She always dressed beautifully and thoughtfully. In my memories, I see her peeking over the dash of her blue convertible on her way to mass, impeccably garbed in a wide-brimmed hat, immaculate white gloves, and a chic dress from Givenchy, Pierre Cardin, or Oleg Cassini. Eleanor Lambert's International Best-Dressed List (now *Vanity Fair's*) named Grandma one of the ten best-dressed women in the world so regularly that Lambert finally elevated Grandma to a specially created Hall of Fame, in order to free up space for others on the yearly list.

She shared practical fashion tips in her regular communications to family members. In April 1957 a typical short note to my mother (who was bringing Kathleen, Joe, and me to my father's daily Racket Committee hearing) recommended, "Wear a real boutonniere on your suit at the Senate hearings, either two white carnations crushed together like one large one, or a little lily of the valley." One morning she shared another tip with my cousin Sydney Lawford, after sending her to change out of a revealing halter-top. Presenting Sydney with a stack of Band-Aids to cover her nipples, she suggested delicately, "Darling, you should really wear these under your shirts if you choose not to wear a bra—I've used them myself, and they really do work."

She taught the girls, Caroline, Courtney, Maria Shriver, and Sydney Lawford, to tell real pearls from fake, how to maneuver in and out of a car in the most ladylike configuration, and how to pose for a photograph with their arms at a particular angle that magically makes one look thin and stylish. She showed them the Duchess of York demonstrating the pose in the many pictures of the Royal Family adorning the walls of the Big House. Almost any photo of my girl cousins will attest that they learned the stance. She told us boys not to stand in photos with our hands in our pockets but down at our sides, sending us photos of famous figures in humorous poses to illustrate her point. After viewing a series of such photos of Herbert Hoover comically outfitted in odd haberdashery, including Native American headdress, Uncle Jack vowed never to be photographed in a hat. That resolution is credited with precipitating the collapse of the American hat industry.

Grandma loved style, but, as Teddy wrote, "Mother knew the difference between fashion, which changes, and values, which endure." She kept one foot in the spiritual world and the other in couture. Often she fit both into the same letter. In a typical epistle written to her children in 1966 she said, "They are wearing parchment-shade gloves this year—not white—if you are thinking of wearing them with black dresses, for instance. They can be bought at Bergdorf's. Also say Rosary in month of October. Love, xxx, G'ma."

A continuous cascade of these "G'ma" letters advised us on matters ranging from thumb-sucking remedies and correcting a dog's misbehavior, to how wet towels should be hung and cautions against leaving a bicycle in the driveway. Her notes often included a favorite Bible passage. She never gossiped and always got straight to the point, as she did in this letter to my uncle, written during the 1960 presidential campaign.

Dear Jack,
This is just a note to remind you of Church.

Love,
Mother

EVERY GRANDCHILD REMEMBERS HOW GRANDMA SENT US EACH A letter in her italicized font lamenting the grammatical shortfalls of the current presidential crop. Candidates, she reported in 1965, could be heard nightly on TV murdering English grammar and syntax. "If I was president," she pointed out, "is conditional; proper usage demands the subjunctive verb 'were.'" Yet Grandma could laugh at herself for her perpetual outpouring of missives, recognizing that advice itself could be a vice.

"As you can see, I am busy as always, but I am slowing up a little bit as I read that Socrates gave advice to everybody in Athens, and they finally poisoned him."

Grandma's core values were rooted in her faith and family. We should love generously, and care for our families and the poor without stint. We ought to fight for liberty and the underdog, connect ourselves to principles, commit our lives to some noble purpose, and create homes full of love and laughter. During a life that spanned more than a century, she became a symbol of family in America and all over the world. Polls regularly named her the world's most admired woman. "Whatever any of us has done, whatever contribution we have made," Teddy remarked at her funeral, "begins and ends with Rose and Joseph Kennedy. For all of us, Dad was the spark. Mother was the light of our lives. He was our greatest fan. She was our greatest teacher." Teddy then wept, saying that he hoped, in heaven, she would, someday, welcome us all home.

Grandma died of pneumonia on January 22, 1995, at 5:30 p.m. She was 104 years old. All the cousins gathered in her house with Teddy, Jean, Eunice, and Pat, and we spent the next day sharing stories about her. The following day, we took her from Cape Cod back to East Boston and buried her next to Grandpa at our family plot in Holyhood Cemetery in Brookline. There was a light snow falling, but my brothers and I swam in the ocean on Grandma's favorite beach before joining the funeral cortege. Thousands of people lined the seventy-six-mile road from Hyannis to Boston, many of them holding their children on their shoulders. It was a frigid morning and hundreds of people waved from bridges as we brought her from the Cape to St. Stephen's Church in the North End, where in 1890—when Benjamin Harrison was president— she'd been christened. Her father, Honey Fitz, was baptized there in 1863.

As with everything about her life, Grandma's death and burial seemed intertwined with the story of America. America's great architect, Charles Bulfinch, had designed St. Stephen's Church for Congregationalists in 1802. Paul Revere, an early member, had cast its bell. The Boston Brahmins abandoned it when they fled to Beacon Hill before the wave of Italian immigrants flooding the North

End. The Italians adorned the interior with colorful Mediterranean iconography, and in 1965 Cardinal Cushing restored the church to Bulfinch's design.

People all over the world loved Grandma. Decades later they still stop me in airports, on sidewalks, and at public events to say how much she inspired them with her faith that sustained her through even the greatest sorrow, along with her lifelong love affair with America and her optimism about our country and its role in the world as an exemplar of humanity's loftiest ideals.

● ● ● ● ●

The Skakels

The unity of freedom has never relied on the uniformity of
opinion.
                              —JOHN F. KENNEDY

THERE WAS LITTLE IN MY MOTHER'S PEDIGREE TO PREDICT HER YEARS
as an activist for justice and tolerance. Her paternal great-
grandfather was a scoundrel named German Jordan, who married
above his station and, when his fifteen-year-old wife, Mary Roach,
died in childbirth in 1853, at age twenty-five inherited a large plantation, Tallulah, and its substantial contingent of slaves, in Yazoo
County, Mississippi. German, a calculating playboy, had eloped
with Mary after assisting her midnight escape from the Sacred
Heart Convent School in Georgetown and spiriting her across the
Potomac in a rowboat.

Family legend has it that the Jordans, whose ancestral name was
Branch, hailed from an estate near England's border with Scot-
land, and had procured their biblical name in 1096 from Robert,
Duke of Normandy, son of William the Conqueror, in gratitude for
services performed by one of German's ancestors during the First
Crusade. German's direct progenitor, Samuel Sylvester Jourdain, of

Stratford, England, published an account in 1610 of his shipwreck and marooning during a hurricane in Bermuda, on his way to Jamestown, Virginia. That yarn later inspired Shakespeare's play *The Tempest,* when Samuel Jourdain returned, briefly, to the bard's neighborhood before successfully sailing again for Jamestown in 1620.

German hailed from the shallow end of the Jordan gene pool. On a Mississippi paddle-wheeler he met his second wife, another teenager, Sarah Virginia Hendricks, who attended Emma Willard's school for girls in the Upper Hudson Valley town of Troy, New York. They married on July 2, 1856, and had two daughters and a son. Their first girl, Grace Mary Jordan ("Gracie"), was my mother's paternal grandmother. In 1863, the entire family dodged both Sherman's bloody march and the Yazoo County sheriff (German had killed his slave overseer, Arthur Bennett, with a pistol) by fleeing Mississippi to Jordan White Sulphur Springs, a Virginia health resort owned by German's family. Ironically, Sarah sickened there and died of pneumonia when Grace was eleven. Grace abided at the spa until after Appomattox, then migrated north to Chicago, where she met and married James "Curt" Skakel, in 1885, despite the fact that her marriage to a Yankee would make her a pariah to her family.

Descended from Scottish stock, Curt Skakel was born on November 8, 1850, near Montreal, the youngest child of George Skakel and Sarah Sawyer, an Englishwoman. Curt would grow up to become a pious Dutch Reform Protestant and a bigot who directed his formidable hatreds equally against blacks, Catholics, and Jews. By all accounts, alcoholism was a prominent family feature back to Neanderthal times, and Curt's booze-fueled violence frightened Gracie, but his bigotries provided them common ground. Gracie boasted of owning slaves, and of her charity work sewing costumes for Virginia Ku Klux Klansmen after the Civil War.

In Chicago, Curt and Gracie produced three children, Bill, George, and Ann Margaret. George, born July 16, 1892, would be my grandfather. By then, Curt's drinking had landed the family in crushing poverty, so Curt turned for help to his elder brother, Billy "the Clock" Skakel. Billy had fought for the Union Army's Massachusetts Regiment during the Civil War and then worked as a bricklayer in Baltimore, accumulating enough scratch to move back to Chicago, where he opened a string of gambling houses, bordellos, and bucket shops. His success in these enterprises made him a power broker in Chicago's Democratic machine. He presided over the city's rotten First Ward, one of the Windy City's dirtiest precincts. With his profits from politics and vice, Billy purchased three great ranches from South Dakota homesteaders during the 1880s and nineties.

To extract his brother Curt from Chicago's alehouses, Billy sent him and his family up the Missouri River to one of their prairie spreads near Tyndall, South Dakota, where Grace bore her fourth child, James. Unfortunately, there were eleven saloons in Tyndall, built to service the Irish railroad workers, and Curt rapidly located them all. With Curt drunk and Grace broken and exhausted, their older children, George and Bill, took to hunting and fishing as an alternative to school. Neither boy would ever graduate from high school. On New Year's Day 1917, Curt Skakel ended his morning bar-hopping by passing out in a frozen roadside ditch. Five days later he died of pneumonia, leaving his family penniless.

Grandpa George, who loathed his father for his drinking and violence, had long since run away from home. At fourteen he rode the rails back to Chicago, while his brother Billy stayed out West, avoiding school but learning to speak fluent Ute, Crow, and Lakota. According to my mother, Billy was one of the few white men ever invited to participate in the Sun Dance, and he sported scars on his breast and back to prove it. Despite his gift for languages, Billy was parsimonious with his words. "He only said 'yep' and 'nope,'" mother recalls. "He was very tall, and thin as a rail. He always wore his six-shooters, even at the dinner table." Grandpa George, similarly laconic, never lost his love for the American West. Much later, after World War II, Grandpa bought a ranch in Moab, Utah, in hopes of finding uranium. That search came up empty, but like nearly every bet he made, the venture hit the jackpot: he discovered oil. Astounded by the easy money, Grandpa spoke one of the longer sentences anyone recalls of him: "Why didn't I get into this racket earlier?"

From his youth, Grandpa George Skakel had ambition, determination, and an indomitable entrepreneurial spirit. These qualities, along with a gift for numbers, a photographic memory, and a near reckless appetite for risk, would make him one of the wealthiest men of his generation. My mother remembers him easily multiplying seven-digit numbers in his head. His first regular job was as a freight rate clerk on the Sioux City Line of the Chicago, Milwaukee, and St. Paul Railroad. The year was 1913. George was twenty-one and working for $4.50 a week. Six months later, the railroad promoted him to traffic manager, the youngest in the company's history, and rewarded him with a two-dollar-a-week raise for recognizing the serial numbers on a string of stolen railcars as they passed below his station on a switch tower.

A teetotaler for most of his life, Grandpa George had a kind and judicious disposition. free of his parents' bigotry. He quit the railroad, furious when the company fired another worker two weeks before the man's pension vested, and quickly found work with the William Howe Coal Company, a distributor. In 1917 he married Ann Brannack, a tall Irish Catholic from Wabash Avenue on Chicago's tough South Side. Like the Kennedys, her grandparents on both sides had sailed from Ireland in 1848, at the height of "the starvation." Grandma Ann was loud, brash, and half a foot taller than her husband. She blasted up prayers to the Almighty for hours each day and cheerfully catechized everyone within earshot. She attended daily mass with a fervor my mother would inherit.

Ann's dad, Joseph Brannack, a jut-jawed Irish cop, was a colossus, with appetites to match, including a thirst for whiskey that spooked off his Irish-born wife, Margaret Brannack. She fled her husband soon after her daughter's nuptials and spent the rest of her days with Ann and her son-in-law, George Skakel. In her dotage, afflicted by a genial dementia, Grandma Ann would stuff her stockings with fruit and convince herself that she was Admiral Horatio Nelson. The first time my father met her, she shouted from the second story, "Who goes there?" My dad, not long out of the Navy, answered, "Lieutenant Kennedy. Permission to come aboard?" She slid down the banister to formally greet him with peaches and bananas tumbling from her undergarments.

Out of high school, Grandma Ann attended a secretarial academy she found advertised on a matchbook. Poor health soon drove her from the febrile Chicago slums to work on a South Dakota Sioux reservation, teaching English to tribal children under Catholic patronage. She returned a year later "fit as a fiddle" and met George Skakel, who, though initially repulsed by her muscular Catholicism, ultimately surrendered to her garrulous personality and bulldog determination. They married on November 25, 1917, at St. Mary's Church in Chicago. A decade later they had seven children, three sons and four daughters, all living at Fifty-seventh and Woodlawn, in Chicago's most Catholic neighborhood, together with George's fervent mother-in-law. My mother, their fifth child, still recalls both the poverty and the happiness of her early youth.

Grandpa George enlisted in the Naval Reserve in 1918 and became an officer, even though he lacked a high school diploma. He trained in Cleveland and served eight months as a Navy ensign during World War I, docking tugboats in New York Harbor. He came home on Christmas Day 1918, so destitute that his only suit was his Navy uniform, in which he returned to work at the coal company. Using his gift for numbers, George recovered $50,000 for William Howe after discovering that the railroads had been cheating the coal company on hauling fees during the war. When the corporate brass refused him his commission for the recovered fortune, he quit, swearing to never work for a boss again.

Grandpa persuaded two fellow employees, Walter "Wally" Graham and Rushton Fordice, to cast their lots with his. Each contributed $1,000 to a joint pool and the three men vacated William Howe on the same fateful day to launch their own coal brokerage. They named their venture the Great Lakes Coal and Coke

Company (later, Great Lakes Carbon). Great Lakes adopted the perilous course of buying and selling only extremely large quantities of the highest-quality coal, a strategy for which both the coal sellers and their buyers, the large oil refiners, rewarded them with preferred pricing. Wally Graham's father-in-law was a senior executive at Standard Oil Company, which became their first customer.

The trio were canny salesmen, though effectively paupers at the outset. "He didn't have two pennies to rub together," my mother remembers, pointing to a silver-framed photo of her father clowning with Wally Graham in the front seat of a giant white Cadillac convertible. Graham and Grandpa had hitchhiked to Houston and spent their last dollars to purchase the ten-gallon Stetsons they wore in the picture. Then, claiming they wanted to "test drive her," they borrowed the Caddy from a GM dealership. Thus stylishly decked, the duo called on a Houston tug company, making sure the owner saw their Caddy. Using bluff and honey-tongued palaver, they succeeded in persuading the barge carrier to haul their first load of coal from Chicago to a Galveston refinery COD.

Grandpa spent his career in the world's most polluting industries, leaving, in F. Scott Fitzgerald's words, "foul dust that floated in the wake of his dreams." While pollution probably didn't trouble him as a moral transgression, it turned out that his genius lay in reducing it by devising schemes to monetize the industry's abundant waste. Mining companies customarily abandoned coal dust in mountainous heaps around their excavations. When the piles of these "fines" grew inconveniently large, the companies plowed them into the nearest river. Grandpa foresaw that during the labor strikes that periodically roiled the industry, this coal dust might be profitably marketed to individuals and industries desperate to fire empty furnaces. He contacted mine owners across coal country, proposing to buy their nuisance fines for five cents a ton. Great Lakes Carbon collected thousands of tons of the stuff, and when the United Mine Workers struck, Grandpa sold the fines for over six dollars a ton, a bargain for oil refineries desperate for fuel. The oil companies showed their gratitude during the threadbare years of the early Depression, when Grandpa's appetite for leverage and risk nearly drove Great Lakes Carbon into bankruptcy; all the majors suspended billing, allowing him to fend off a hostile takeover by Peabody Coal.

Even though George Skakel never made it past grade school, he devoured books on metallurgy and chemistry, geology and business. Browsing through *Business Journal* he learned of an exploding worldwide appetite for the pure carbon needed to create aluminum for the burgeoning aviation industry. He knew that carbon could be refined to purity using petroleum coke (petcoke), a waste product of turning oil into gasoline. As with coal fines, the oil industry's solution for ridding itself of alpine heaps of this refinery waste was a bulldozer and a local river. Grandpa signed ninety-nine-year contracts with Standard Oil and its competitors to purchase all those companies' petcoke for pennies per ton, gaining a virtual monopoly just as commercial aviation was leaving the runway.

Over the next decade, Great Lakes Carbon sold millions of tons of petcoke in the eastern United States as a substitute for coal in domestic heating, and to the electrochemical and metallurgical industries. In 1935, Grandpa and his partners spent $50,000 constructing a giant furnace to refine petcoke in Port Arthur, Texas. They had the notion to sell the purified version themselves, thereby relieving aluminum companies of an extra step in their manufacturing process and enabling the production of aluminum of unrivaled strength. They quickly persuaded Alcoa, Reynolds, Kaiser, and the other major aluminum manufacturers to purchase their product, bragging later that following its construction, their oven paid back its capital costs every six weeks. The Port Arthur plant was among the first such ventures for a company that would soon have manufacturing facilities across the globe. By 1941, Great Lakes Carbon had additional plants in Illinois, California, Wyoming, Texas, and New York, with others in Canada, Portugal, Spain, England, Sweden, and India, and the company was selling millions of tons of refined coke to the aluminum, graphite, and steel industries. During the war years, Great Lakes Carbon also obtained lucrative military contracts for lubricating oils and diatomaceous earth.

As Grandpa's wealth multiplied, he remained a devoted family man, an eclectic reader, and a lover of the outdoors. Although he used it sparingly, Grandpa George loved language, was precise with his grammar, and measured in his self-expression. Like Grandpa Joe, he didn't swear; coarse language or off-color stories mortified him. My mother inherited his puritan streak, and the absence of biblical precepts forbidding vulgar language did not stop her from religiously beating the tar out of any child who swore. I tasted soap and endured thrashings for minor maledictions like "butt" and "damn."

Grandpa rarely spoke disparagingly, except concerning Democrats, labor unions, and Franklin Roosevelt, all of whom he loathed. According to my mother, he had "beautiful table manners," learned by imitation, but "he never frowned on bad manners in others." She inherited his deportment, but not his restraint in the latter regard. She interrupted many family dinners to pitch a fit at elbows on the table, or bad posture, or improper management of cutlery, whether by her children, friends, or complete strangers.

Despite his lack of formal education, Grandpa studied Shakespeare at night. At the family breakfast table he commonly read aloud the *New York Herald Tribune*'s human interest stories, or tales about animals from the columns of Thornton Burgess. My mother recollects him as "a wonderful dad, always sweet, sensitive, thoughtful, and loving toward his children," and a competent naturalist. He took his kids for long weekend walks and countryside drives, often stopping to identify a bird or a rare tree, or to explain a bit of natural history, his mastery of which was encyclopedic. My mother describes Grandpa George as "not religious," but "Christian in the best sense of that word." As she put it, "He didn't mistake religion for a belief in God." A policeman once stopped her in New York and told her, "Your father paid to educate my family, and all I did was help him across Park Avenue." That personal touch was typical of his approach to charity. As my mother recalls, "Grandpa was nice, kind, and generous, and never forgot his humble origins.

Even after accumulating great wealth, he often cautioned the family, 'We could all be thrown out on the street tomorrow.'"

Big Ann

While Grandpa was a workaholic with a genius for turning muck into gold, Grandma had a knack for spending moolah nearly as fast as Grandpa made it. By the time my mother was born, Grandma was already known to everybody as "Big Ann." Her weight was then well north of 250 pounds, and it never stopped climbing. She accumulated ballast with each of her seven pregnancies and never relinquished an ounce.

While Grandpa was taciturn and laconic, Ann was loud, garrulous, and capricious. My mother describes her as "outgoing." Like all Skakels, she was extraordinarily generous, and prone to extravagant gestures and impulsive gifts. If you admired something she was wearing, she would give it to you. Every meal at her home was a lavish feast. As openhanded as she was with friends and charitable causes—principally the Catholic Church—Ann was a skinflint with vendors and merchants. A haggler who recoiled at retail pricing like a cat from water, she toted a fat purse packed with business cards identifying her by diverse professions that might garner a discount. She was variously an official of Chicago's public schools, a matron of the orphanages of Chicago's Catholic Archdiocese, a lingerie retailer, an interior designer, and an architect, depending on the circumstance. Other cards identified her as Ann B. Skakel, School for Young Girls; or Ann B. Skakel, Garage Parts and Accessories; or Ann B. Skakel, Hardware and Tools. These little frauds, and her large-quantity purchases, enabled her to buy commodities on the cheap; milk, for example, cost her twenty-three cents per bottle instead of twenty-seven cents, and such meager savings were her purest delight. "She thought that was great," my mother recalls. "She must have had a hundred cards. The cards probably cost more than whatever she saved. But she got such a kick out of it."

My mother lived in Chicago until she was eight, when she, her parents, and her older siblings—Georgeanne, Jimmy, George Jr., Rush, and Patricia—moved east, following the family business to New York. The Skakels spent their first East Coast summer in Red Bank, on the Jersey shore. My mother recalls a tidal wave flooding their home and Grandpa rowing a dinghy to work each morning on the first leg of his commute to Manhattan. Later that year they moved to Larchmont, which offered a quicker commute to Great Lakes' headquarters in Midtown.

My mother and her brothers learned to swim and sail on Long Island Sound. The sea's stormy chaos, the neat simplicity of the vessel, and the laser focus required to maintain a compass point were apt metaphors for her tempestuous life, and may explain sailing's enduring appeal to my mother. "I like the discipline of it. You need absolute concentration to stay in tune with the wind and currents and avoid getting distracted from your bow heading. And I love the openness and the exhilaration of sailing and the way it pits you against the elements—and I love the surprises." She won the Larchmont Yacht Club championship at age eleven against a field of fifty boats, mostly piloted by older skippers. Winning made her happy.

The money was also rolling in. By the time my mother turned eleven, Grandpa's company was on its way to becoming one of the largest privately held businesses in the United States. During the Depression, there were only twenty-four known millionaires in the country, and among them were Joe Kennedy and George Skakel. But unlike Grandpa Kennedy, George Skakel always kept his profile low. "You can't quote silence," was a favorite mantra. Two years later he and his family moved east on the Sound to Greenwich, Connecticut.

At that time, Greenwich was rural horse country, with fewer than six thousand residents, many working or living on large estates. The Skakel house on Lake Avenue was a hybrid of *The Philadelphia Story* and *The Beverly Hillbillies*. One of the most enormous mansions in one of America's wealthiest burgs, the Skakel home was elegant, graceful, and roomy, with an eight-car garage, a village of outbuildings, an Olympic-size swimming pool, and a fountain rivaling that at Buckingham Palace. The manse, which easily accommodated thirty houseguests in addition to the family, crowned a ten-acre lawn surrounded by a hundred acres of woodland and fields. "Even though it was large, it just felt like home," recalls my mother. "There was plenty of land for the kids to hunt and ride horses." Swans preened while the children skated, fished, and swam on a big lake.

"We had three hundred chickens. I would collect the eggs and milk the cows, which I loathed," my mother recalls. Caring for the horses was another matter. "I rushed home from school to curry them, clean the tack, and muck out the stables. I loved it!" On weekends, my mother competed in horse shows and rode with the Greenwich Hounds. The foxhunt routinely convened in the Skakel yard, exciting her obsession for horses—hundreds of blue ribbons and trophy cups, from local horse shows to the prestigious International in Madison Square Garden, adorned the walls and bookshelves of her room. She was the star student of Stabern Stable's famous trainer Teddy Wahl, and she knew that when she grew up, she wanted to be a veterinarian.

My mother's first horse was a smoky black pony with white socks named Guatemala, or "Gwamy," followed by a chestnut mare, and afterward by a dark-bay quarter horse called Beau Mischief, upon which she bested the women's world high-jumping record. Riding Beau Mischief, she cleared a seven-foot-nine-inch rail at a jumping competition in Old Westbury, Long Island, in 1947. "She was a superb horsewoman. She was like the horse whisperer before horse whispering was invented," recalls her friend Lady Dot Tubridy, a fellow equestrian. "If women had been allowed to compete in equestrian events at the Olympics, Ethel would have been on the U.S. Olympic team." (Olympic show jumping was not opened to women until 1952.)

Her daring horseplay, galloping bareback through trackless woods, vaulting stone fences, jumping automobiles or a tractor by moonlight, became as notorious as her brothers' perilous antics.

On my mantel I have a photo of my mother in Chester, Vermont, flying bareback over a five-foot fence on a giant gelding with her girlfriend Anne Morningstar clinging to her waist. The gelding has a white smear on its forehead, striped with pink lipstick—a fashion the girls pioneered.

Horsemanship was not my mom's only skill. Under Big Ann's watchful eye she unwillingly endured piano lessons for ten years with Dr. Carlos La Diabanco, who had somewhere lost three fingers on his left hand and two on his right, accounting for my mother's odd style of hitting all the keys with just five digits. At the Academy of the Sacred Heart in the Bronx, Sister Garafalo taught her the art of tap, which she mastered with sufficient dexterity to hoof her way to parochial stardom in the school's senior-class musical. In later years those talents served her, and disquieted her children, whenever she trotted them out during those mercifully few occasions when circumstances invited. Even into her late seventies, despite her mediocre voice, it was an easy prompt to get her to sing and tap.

My mother remembers the Greenwich house as a happy, cheerful place, alive with the same species of boisterous chaos over which she would later preside. "We had sixteen dogs," recalls my mom, "mainly Irish Setters, all named Rusty. They slept in our beds." Grandma was a great cook, preparing daily meals for seven children and ten household staff members. "She did it with such good cheer and fun. She turned her kitchen into a community center where everybody would come and hang out, trade gossip, and enjoy each other's company. She was Julia Child before Julia Child. She just had a love of cooking, and the numbers did not throw her. It didn't matter how many she cooked for," my mother recalls. "And it wasn't like the meals we have now; there would be five vegetables, and two or three different kinds of meat; and cakes, cookies, and pies homemade in the butler's pantry, which served as her bakery." After dinner the seven children retired with their father and mother to the living room, where they would talk or play cards, or backgammon, at which my mother excels (and which remains a favorite sedentary pastime of hers). They rarely turned on the television.

The Skakels played ice hockey on the gargantuan fountain each winter and would drain it to bicycle or roller skate in the spring. The fountain could spray water fifty or sixty feet in the air. When the empty pool filled with skaters, the routine called for pushing the button in the breakfast room and soaking the crowd. "That was great fun," my mom recalls.

My mother attended Greenwich Academy as a high school freshman and sophomore before transferring to the Academy of the Sacred Heart in the Bronx for her junior and senior years. Sacred Heart's militaristic Catholic regimen had the girls praying in rote and marching silently in rank from class to class. She matriculated to Manhattanville College—another rigorously structured parochial institution. My mother was not a scholar, and neither studied nor read much beyond the Belmont tout sheets. Throughout her scholastic career she largely limited her academic efforts to last-minute cramming. She couldn't sit still in class; her Sacred Heart French teacher, Mademoiselle Figuet, diagnosed her as *"un paquet de nerfs."*

My mother's irreverence for authority and habitual evasion of petty rules started early and would remain a lifelong trait. Along with riding and prayer, she filled her days with tiny acts of disobedience. While enrolled at Manhattanville, she engaged a bookie, studied the racing forms daily, and played the horses. "Every morning at college, from eight-thirty to nine I read the odds, and I went to the track several afternoons a week when Belmont was open. I loved the track." Since car ownership violated college rules, she parked her 1917 Pierce Arrow convertible off campus.

Manhattanville's disciplinary records show that my mother earned demerits for "chewing gum," "tardiness," "clowning," and "combing her hair in public." She had twenty-eight demerits, compared to my aunt Jean Kennedy's twenty-one. "I beat Jean," she says with satisfaction. When, in her junior year, my mother racked up too many demerits to go to the Harvard–Yale game, she and her roommate, my Aunt Jean, burned the demerits book in the school incinerator. "Why would they keep it right next to the furnace?"

she asks. "It was an invitation!" My mother loved the nuns, but gravitated toward the renegades, like Mother Muldoon, who had largely abandoned strictures of any kind. Fortunately, Muldoon was too advanced in years for any earnest mischief. Eastern European folk dancing was the best she could muster. "She had been retired as a teacher, and she kept to herself on the third floor," my mother recalls. "So I used to run up to say hi, and then we'd sing and do the polka."

When my mother was twenty and attending the International Horse Show in Madison Square Garden, she hatched a scheme to pose as a *New York Times* reporter and arranged an interview with the show's champion—the dashing Irish rider Captain Michael Tubridy. My mother and her best pal, an Irish girl named Dot Lawlor, had major crushes on the renowned Gaelic footballer and show jumper who had captured world attention for his equestrian skills and movie-star good looks. My mother arranged for the "interview" to take place at Schrafft's ice cream parlor, but when she got there it took Tubridy only moments to recognize the hoax. Pegging my mother as a stalker, he fled abruptly. In revenge, she and Dot Lawlor snuck into the Madison Square Garden stable and painted Tubridy's prized white stallion with indelible green food coloring. Tubridy rode the championship round on an emerald stud. As it turned out, he had a sense of humor. A year later Dot married Lord Tubridy, who, sadly, died in a riding accident in April 1954, just a few months after I was born. Dot remains among my mother's closest friends.

Grandma Ann's devotion to the Catholic Church was legendary among her friends, both in Chicago and in Greenwich. My mother recalls that her mom kept the house thick with clergy, frequently hosted meetings by religious societies, and assembled a whirl of church groups for teas. Opus Dei made its American debut at the Skakels' Lake Avenue house, in keeping with the society's strategy of aligning itself with wealthy, powerful, and politically conservative Catholics. "At the time I didn't realize," recalls my mother, "how vigorously the organization opposed the Church's mandate for social justice." Grandma Ann worked for years on her husband's conversion, with merciless catechizing and organized invocations by his family and the clerical team imploring the Creator's blessings on the enterprise. It was the single request from her that he resisted, at least until shortly before his death.

Grandma befriended the influential Trappist monk Thomas Merton, whom she served as personal secretary and stenographer. Ann met Merton through her friendship with Father Abbott Fox, the charismatic monk who ran the Abbey of Gethsemani in Louisville, Kentucky. In the early fifties Merton inspired a great Catholic revival in the United States with the publication of his book *The Seven Storey Mountain,* an account of Merton's transformation from beatnik to Catholic monk. He believed that human pride and materialism lay at the root of most of the world's woes. To rein in those impulses, the Trappists practiced disciplined silence, prayer, and obedience. Merton was an eloquent and influential pacifist and an opponent of nuclear arms and civil defense. He wrote my mother regularly during the early 1960s, praising and encouraging President Kennedy's resistance to the military-industrial complex. His ecumenical writings made him a hero, not just to Catholics but to contemplatives in every religion. He died, from accidental electrocution at a Buddhist conference in Bangkok, six months after my father. Grandma and Grandpa Skakel supported the Gethsemani monastery and built its chapel, which bears their name.

Grandma also had many other outside interests. She organized a card club every Thursday and ran salons and seminars with hundreds of attendees. She would, of course, feed all her guests. In addition to her various charitable and culinary enterprises, Grandma was an entrepreneur. Among her many business ventures, she owned and operated St. Paul's Guild bookstore in New York and a consignment store under the elevated train in New York City called Lots o' Little. Among the many items she accepted for sale during the war years were three dolls of Dominican nuns with tiny rosaries dangling from the habits, two of which Grandma sold, giving the third to my mother. Unbeknownst to Grandma, an Irish-born

German spy named Emily Dickenson had used one of the dolls to transport a coded message. The FBI tailed Grandma for six months until agents arrested Dickenson and recovered the codes from one of the dolls Grandma had sold, only closing the investigation once they were satisfied that Grandma was innocent of espionage.

The story came full circle years later, when my mother and father, recently married, were living in a small house in Georgetown, my dad fresh out of law school. At the end of each workday, my mom would meet my father walking home from his job at the Justice Department. Sometimes they rendezvoused in a local bar, where my mother would order a rum and Coke, leaving the drink untouched while she waited. (She never tasted liquor until she was forty-four years old.) One day a large man took a seat beside her and introduced himself as James McInerney, an FBI agent who had tailed Grandma for six months without her ever being the wiser. My mother was stunned and amused to hear the story. McInerney later worked with my father at the Justice Department, where my dad led the criminal division and forced his old boss, J. Edgar Hoover, kicking and screaming, to investigate the Ku Klux Klan. My parents loved McInerney. Hoover despised him. Emily Dickenson, the spy McInerney had helped collar, also had a continuing involvement with my family. She was one of my Aunt Eunice's cellmates during Eunice's undercover assignment in a West Virginia penitentiary. Eunice conscientiously employed former convicts and people with intellectual disabilities as her household staff, and Dickenson worked as Aunt Eunice's private secretary following her release from prison.

———————

MY FATHER AND MOTHER MET DURING A WINTER SKI TRIP TO MONT-Tremblant, in Québec, introduced by Jean Kennedy, my mother's Manhattanville College roommate, who had conspired with my mom to unite their families. After that first encounter, Kennedys and Skakels dated each other furiously. My mother remembers Jimmy and George Jr.—a notorious lady-killer—dating most of the

Kennedy sisters at one time or another. Uncertain what to make
of this feisty hoyden, my father first dated my mother's studious
elder sister, Pat, before falling in love with my mom. The two families seemed ideally matched, with their shared Catholic piety, their
proficiency in sports, particularly sailing, riding, tennis, skiing, and
golf, and their love for outdoor adventure. Both Kennedys and Skakels were similarly smitten with a passion for football. George Jr.
had played on the varsity squad at Amherst. And, like the Kennedys, all the Skakel brothers served in the Navy during World War II.

In addition to everything else they had in common, these were
the two wealthiest Catholic families in America. But fault lines
soon emerged. The differences were partly stylistic. Grandpa Kennedy played golf with his oldest friends, and did his business by
phone from the seclusion of his Palm Beach bullpen. He operated
a tight ship, demanding decorum, discipline, and modesty from
his children. The Skakels, by contrast, seemed unwilling even to
acknowledge the rules of civil society. "When my brothers took
the train to New York, they never rode on the inside," recalled my
mother. "They always rode on the top of the train." The Skakel
boys were big and tough, prone to gunplay and fistfights and a kind
of monumental recklessness, exacerbated, at times, by an inherited
weakness for alcohol.

Starting at six in the evening, Grandpa Kennedy would listen to
classical music while sipping a daiquiri, and then eat a semiformal
dinner with his family punctually at seven fifteen. He dressed for
dinner each night in a smoking jacket and slippers. By contrast,
neither formality nor punctuality were considered virtues in the
Skakel home. Grandpa Skakel never considered getting formally
dressed under his own roof. "You didn't know whether you were
going to have dinner at five or ten at night," my mother recollects.
"You had no idea. You'd get so hungry." Far from participating in
organized dinner-table discussions, the Skakel children often ate
ice cream in the pantry while their parents entertained adult guests
in the dining room.

Where the Kennedys were comparatively tightfisted and modest, the Skakels reveled in immoderate consumption. They squandered their riches on Cadillacs and aircraft, and during the month of August retreated with the swanks to bet the horses at Saratoga. While Joe Kennedy forbade his children from playing polo, Uncle George Skakel was a founder and outstanding player at the Blind Brook Polo Club in Purchase, New York, and at the Bethpage, Long Island, club; in 1962 he organized and promoted an indoor version of the sport. As the Skakels were fond of saying, the only way to make a Skakel a millionaire was to leave him $100 million. "Whatever they've got," Georgeann Skakel told me, "they're going to give away or spend." My mother inherited that profligacy, whereas my father was among the most parsimonious of Grandpa Kennedy's frugal children.

Both families shared a love of nature that might have provided the genetic antecedents to my predilection for the outdoors, but while the Kennedys embraced the challenges of sea, river, and wilderness, the Skakels gloried in capturing, collecting, or subduing. Grandpa Skakel took his customers on extravagant hunting expeditions to the many private ranches he had acquired in Idaho, Canada, Mexico, and Cuba. All the boys shared his love of guns and were expert marksmen, as were the girls, including my mother. The Skakel romance with armaments went beyond firearms to include more primitive weaponry like bows, knives, throwing spears, and harpoons. They pretty much captured, shot, stabbed, hooked, or speared anything that moved, including each other. The three sons—tall, large-boned, and brawny—spent their leisure time hunting and fishing at Grandfather's far-flung ranches, to which they sometimes disappeared for months at a time. Graceful athletes, they loved backcountry skiing in the Bugaboos, shooting dove at Cabo San Lucas on the Baja, canoeing at their salmon camp in Newfoundland, or trout fishing at Lake Certo in Ontario. All were accomplished with spinning and fly rods.

A grizzly once mauled Uncle Rush during a bear hunt, snagging him by his boot and crushing his foot with a bite, but more often the Skakel boys got the better of nature. In 1961 Uncle Jimmy harpooned a gray whale from the bow of a Portuguese dory off the Azores, and after an eight-mile "Nantucket sleigh ride," slew the cetacean amid a spray of blood, then dragged it to the beach to be butchered. *Life* magazine chronicled that adventure in an eight-page photo-spread in December 1962. On another occasion Jimmy killed a caribou Inuit style by climbing on its back as it swam the icy Tatshenshini River in the Yukon Territory, and slicing its carotid with a hand blade.

Jimmy's house in Bel-Air was filled with giant aquariums populated by sharks, abalone, octopus, and diverse marine creatures he bagged while diving in the kelp beds of the Humboldt Current. Once each day, Jimmy donned his scuba gear and entered the tank to hand-feed his maritime friends. A pair of mountain lions roamed Uncle George's house. He had treed and lassoed them as cubs in Western Canada. As they grew, he packed a sidearm, because they never really got tame. Facing an ultimatum from the Greenwich Police following their repeated escapes, George drugged his big cats with Miltown pills from the Greenwich Pharmacy and flew them north in the family's plane, a twin-engine Convair 580. When his pilot refused to fly them across the border, George had him land in Burlington, Vermont; from there George smuggled the snoozing felines past immigration in the trunk of a rented Cadillac, and released them on the family salmon ranch. Prime Minister Pierre Trudeau would later tell my mother that the experiment had been successful and that, as a result, a small population of mountain lions inhabited the woods of Newfoundland—though this detail might be apocryphal.

At Christmas, Uncle Jimmy customarily gave us extravagant gifts. These included animals he collected on his wilderness expeditions, among them rhesus and spider monkeys, marsupials, a pair of bush babies, and a Central American honey bear. Following an expedition in British Columbia, Jimmy had promised us "Christmas seals," which we gloomily assumed were the prosaic holiday coupons that earned discounts at Safeway. But when the crate came off the truck on Christmas Day, out came a young California sea lion, Sandy, who took up residence in our swimming pool and joined the growing menagerie I had collected from the countryside around Hickory Hill, the McLean, Virginia, home where my parents moved in 1957. Sandy ate mackerel by the barrel, devouring everything but the eyeballs, which we found scattered like marbles across the pool, patio, and lawn. Sandy rode in the car, accompanying my mother when she picked us up after school, and my sister Kathleen taught him to jump off the diving board through a hula hoop.

The sea lion soon fell in with the dogs and started roving. As a pack, they crossed the metal cattle guards—pipe-covered trenches across both driveways intended to keep the horses and livestock from escaping—and wandered the countryside, venturing occasionally into the nearby mercantile district, Salona Village. One winter morning as I spooned down Cream of Wheat, a news flash on the kitchen radio reported a live seal in McLean causing a rare rush-hour traffic jam on the Georgetown Pike. A few hours later the milkman brought him back to our house in his truck. After about a year, Sandy's expanding peregrinations, the complaining neighbors, the great surplus of fish eyes, and a mishap that landed Kathleen and her Hula-Hoop in an icy swimming pool finally persuaded my mother to donate Sandy to the National Zoo in Washington.

Despite their shared loved of exotic creatures, the Kennedys regarded wealth as a trust from God that they should use to serve others, whereas the Skakels were more inclined to see their wealth as the just reward for their industry, with no strings attached. They balanced their Catholic piety and natural generosity with a love for booze and general devilment. "They were plenty dangerous," recalls my mother. Young George, my mother's brother, once shot his brother Jimmy while they were working as roustabouts in the Texas oil patch. And Jimmy, in turn, shot my mother with a .22 rifle, from a great distance but with sufficient accuracy to shatter the neck button on her blouse. Grandpa George slept with a gun beside his bed, and the Skakels carried guns nearly everywhere, gleefully discharging them at street signs and traffic lights as they drove their convertibles, or at chandeliers if they happened to be inside.

On a visit to Hickory Hill when I was four, Uncle Jimmy offered to teach me to swim; as I contemplated the proposition, he threw me into the deep end of the pool and then drolly observed my thrashings, hoping, I suppose, that this bare-bones lesson would take. To their Greenwich neighbors, the Skakel boys seemed to be wanting in the head, but my mother loved the chaos, her brothers' rowdy irreverence, and their reckless sense of fun. I was with my mother one day when Tim Hagan, her friend, asked her if she had ever been frightened of her brothers. "Well, I adored them," she replied evasively, while acknowledging that "there was always an element of risk." Then she added, "It was amazing they survived so long." Still, when she left home to attend Manhattanville College, my mother remembers impatiently waiting for the weekend, eager to return to Greenwich to be with her brothers.

When my mother was eleven, her gentlest brother, Rucky, five years her elder, solicited her help testing a knot of his creation. After wrapping her in a coil of rope, the other end of which he had fastened to an oak limb, he pushed her from a second-story window so that she swung wildly through the branches and back, shattering two plate glass doors and careening into a crystal chandelier. Still entangled, she swung back out again. On her second backswing into the house she managed to free herself and fall to the ground, then hid behind a woodpile, fearful she would take the heat from her mother for busting the chandelier. "She would never blame her precious boys," my mother added, laughing.

All my mother's brothers were pilots, and their aircraft stories are legendary. George once flew the company DC-3 under the George Washington Bridge at rush hour. Rush Skakel had to surrender his license—probably saving his skin—because on one occasion he followed a commercial airliner's descent through a cloud bank and onto an O'Hare runway, after wandering lost above a dense cloud bank for an hour with no instruments and dwindling fuel, in a float-plane on his way home from a Canadian hunting expedition.

Cars ended up in the Skakel swimming pool so often that a winch stood mounted on a nearby tree ready to pull them out. That tradition began when Uncle George drove Granddad's Cadillac into the deep end to douse the flames that had erupted after he lost a round of a game he and his siblings had devised: while one brother pursued the other two around the lawn in the convertible, they endeavored to throw lit matches into his open gas tank.

Lem Billings's first encounter with George Skakel, Jr., was at my parents' wedding. Lem was a formidable man, almost six foot four, and a champion college wrestler and crew captain. My mother and father were characteristically late to the affair, having lost their car keys. George, seeking relief from the tedium, flung a handful of change into the aisle. When Lem helpfully stepped from his pew and crouched to retrieve the coins, Uncle George kicked him so hard from behind that Lem slid, spread-eagled down the aisle, almost to the altar, with agonizing pain in his coccyx. It was the first time Uncle George had set eyes on him.

In 1998, I stopped in Moab after paddling down the Colorado River with four of my children, Bobby, Kick, Kyra, and Conor. We found Grandpa George Skakel's old homestead, the Taylor Ranch, a thousand-acre spread that once included much of today's downtown Moab. Grandpa had bought the ranch after falling in love with a picture postcard of boulder-strewn Moab Valley. After striking oil, he installed Uncle Curt as the spread's manager. In their teens, Uncle Jimmy and Uncle George worked on the wells as roustabouts. Sam Taylor, editor of Moab's *Times-Independent* and a great-grandson of the Taylor the ranch was named after, recalled the Skakel boys partying nonstop and "taking potshots with their deer rifles at the Irish statuary" their father had installed in the yard. "I was in junior high then, and they were known as the biggest hell raisers among a breed of hell raisers. I idolized them." The fun ended when Uncle Jimmy, working on the derrick, was blown sky high by a rogue gas bubble. He spent a year in the hospital.

The Fault Lines Open Wider

The Skakels were Republicans, but so little interested in politics that my mother marrying into a prominent Democratic family did not set off alarm bells, at least not at first. "When I married Daddy, it wasn't a big deal," she recalls. "They thought I was a little Communist, but they all had good humor about it. It wasn't taken seriously." But when she started actively campaigning, my mother crossed a line. The Skakels were oil and coal Republicans, and, like most folks in the extractive industries, they leaned conservative. While Grandpa Kennedy was FDR's top donor in 1932, Grandpa Skakel was piping his carbon lucre to the GOP, and only rejected an offer to become the Republican Party's national treasurer due to his antipathy for the spotlight. "When I first went to dinner at Grandpa Joe's house and everyone was talking about what a great fellow FDR was . . . well, that was something entirely new in my experience," my mother recalls. While they were comfortable with people of every station and donated millions of dollars to Catholic charities, the Skakels were not particularly sympathetic to the role of government in protecting the rights of the downtrodden, or in fostering an equal playing field.

Although Grandpa Skakel had shaken the bigotry that ran so prominently in his parents' lineage, and was unfailingly kind and considerate to everyone he encountered, rich or poor, black or white, the broader issues of civil rights did not concern him. Oil being thicker than blood, my mom's brothers and their carbon-sector cronies labored to derail Jack's electoral ambitions—and later my dad's; they supported Nixon in 1960, a wound my mother bore in silence, while continuing to love her family. During the inauguration ceremonies in 1961, Uncle George distributed a pile of coveted Kennedy family tickets, which my mother had acquired for the Skakels, to some hardboiled hobos from Washington's skid row. Uncle Jack's friend and fellow PT boat commander Paul "Red" Fay, who would become undersecretary of the Navy, found himself sitting among a dozen pickled winos in the reviewing stand. That, of course, was very funny, but the Skakel brothers became more and more vocal about opposing my father in his political endeavors, and contributed heavily to his opponents. By sheer coincidence, my dad's first case at the Department of Justice was an antitrust suit against my mother's brother Rush and his partner, Tommy Reynolds, when they tried to move the Milwaukee Braves baseball team they owned to Atlanta. The Justice Department lost the case. As these fault lines opened wider, my mother increasingly adopted my father's values, and the distance lengthened between her and her siblings. As a result, we children didn't see much of the Skakels while we were growing up.

In September 1955, Grandpa Skakel made the momentous announcement that he wanted to convert. It was the answer to years of supplication by every member of the Skakel family and its diverse posse of clergy. Grandpa told Father Abbott Fox at Gethsemani that he meant to begin the conversion process when he returned from a business trip to the West Coast. He and Grandma took off from Idlewild in Great Lakes' converted Air Force B-26 bomber on October 3, 1955, spent a day in Tulsa, and left at nine forty-five the following evening for Los Angeles. Thirty minutes into that flight both aircraft engines exploded in midair. The bomber lit up the sky like a comet and plunged to the ground near Union City, Oklahoma. Grandpa and Grandma were both sixty-three years old. Police identified her body by the rosary wrapped around her hand.

Great Lakes Takes a Nosedive

Uncle George Skakel, Jr., had joined his father's business in 1947, and two years before Grandpa's death he had become president of Great Lakes Carbon Corporation. By this time Great Lakes had three thousand employees, and was manufacturing and selling petroleum, coke, carbon, and graphite products including electrodes and fiber, charcoal briquettes, crude petroleum, and natural gas filters, as well as building materials, from its headquarters at 18 East Forty-eighth Street in New York. George ran the family's far-flung corporate enterprise, including the newest factory in Cuba, where Great Lakes also sold filters widely used in Cuba's sugarcane industry. The company also owned several sugar plantations and mills that Fidel Castro eventually nationalized, but even after the revolution Uncle George went regularly to the family home at Varadero Beach (about eighty-five miles east of Havana) to hunt wild boar and quail, or troll for marlin, often accompanied by his pal, CIA officer Lewis Werner II, who, no doubt, used the trips as a cover for government business.

As with many other large oil, coal, and mineral companies, Great Lakes enjoyed a symbiotic relationship with the CIA, an agency that could always be counted on to protect their interests. The Skakel men made a habit of informing the Agency about political developments in countries where it maintained offices. A few months after Castro's takeover, Uncle George was vacationing in Cuba with friends when a group of armed soldiers tried to expropriate his brother's forty-two-foot, high-speed Chris-Craft, the *Virginia*, as George and his friends prepared to go marlin fishing. The former guerrillas pointed tommy guns at the three Americans and ordered them off the boat. One of them recalled to me his terror when George responded coolly, "We're going fishing. You can go to hell," and eased the vessel determinedly out of its slip. A month later, CIA friends of Jimmy borrowed the boat, mounting a machine gun in the bow for some piece of mischief. Shortly afterward, the Agency permanently expropriated the *Virginia* for use by a Cuban exile group to raid the Cuban coast. The Skakels shrugged off the vessel's loss. Overall, the CIA had been a great business partner.

The Skakel boys knew that, sooner or later, they would have to pay a price for their devilments—and they were right. Uncle George Skakel died in an Idaho air crash on September 24, 1966, flying into a remote wilderness airstrip in the Salmon River Valley near Riggins, where he and some friends were staging for a ten-day pack trip to hunt for elk in the wilderness. My father's close friend

Dean Markham, CIA agent Lew Werner, and two other buddies of Uncle George's died with him. The plane, a single-engine Cessna, became trapped in the nearly vertical walls of Crooked Creek's box canyon, and crashed while the pilot was making a desperate, last-ditch attempt to turn. An eyewitness to the crash told me that as the plane headed for the wall, George waved goodbye from the copilot's seat to his friends who had come in on earlier flights and stood in a throng, awaiting his arrival at the end of the tiny dirt airstrip. He flashed them the broad smile he reserved for moments of reckless mischief, and raised his cup as the plane nicked some pine trees, hit the canyon wall, and tumbled into the river. George was forty-four years old. Eight months later his widow, Pat Skakel, choked to death on a piece of meat during dinner at her home in Greenwich, leaving my five cousins orphaned.

George had inherited his father's business acumen, but after his death, Uncles Jimmy and Rucky, who had little interest in running the company, took Great Lakes on a rollicking joy ride till the wheels fell off, then sold it for pennies on the dollar. Under new, competent management, the company quickly regained its value, ending up as a shining star in the Koch brothers' coal and oil portfolio. Bill Koch told my mother that GLC might be the most profitable acquisition in the Koch family conglomerate's history.

My mother lost her brother Jimmy just five months after my brother Michael died in December 1997. In his dotage, Uncle Jimmy slept with a loaded shotgun, and his friends considered it a peril to visit him at night. He was both intensely paranoid and blind from diabetes and alcohol. When I attended his funeral on April 27, 1998, Jimmy's eulogists talked about his love for life, for conflict, struggle, and challenge, his roaring laughter, and how he squeezed pleasure from every moment. They mentioned his extraordinary generosity—something all the Skakels shared—and his abiding faith. Between bouts at the dog track, Jimmy attended three masses each day at St. Patrick's, St. Joseph's, and St. Martin's in Biscayne, Florida, praying so fervently that, even when his eyesight was good, he did not recognize friends at church as they passed by. His

relationship with Aunt Virginia was legendary for its tempestuous-
ness and ardor; Virginia divorced Uncle Jimmy twice and married
him three times.

My mother's youngest brother, Rucky, was confident that he
was immune to the alcoholic dementia that ran through the Skakel family, despite abundant evidence to the contrary. "I already
had that," he once told me, thoughtfully. "And I don't think you
can catch it twice." His confidence was misplaced. He died of the
disease in 2003, at age seventy-nine. Fortunately, he was *non compos
mentis* by the time his son, Michael, was convicted during a media
lynching in 2001 of a murder he did not commit, and sentenced to
serve twenty years to life in prison.

My mother's siblings generally inherited their parents' virtues,
including Grandma Ann's piety, loyalty, and boundless generosity,
and Grandpa George's driving curiosity about history and the natural world. They espoused highly personalized brands of honor,
but despite their genuine efforts at culture and charity, the Skakels
never tried to pass themselves off as proper folks. They were carbon nabobs, and their roughness, vitality, and irreverence showed
through the generations. When Uncle George died in 1966, his
lifelong friend, conservative icon William F. Buckley, Jr.—another
CIA officer—bemoaned his death in a *New York Times* obituary that
could have applied to any of my Skakel uncles. Buckley recalled
George Jr. as a young tycoon and sportsman of "enormous competence, curiosity, charm" who was "impulsive in mischievous and
irresistible ways . . . in the tradition of the total American man.
The range of his interests was phenomenal, his capacity to live and
exploit the pleasures of the earth extraordinary—they call it a 'lust
for life,' the kind of thing great artists and writers and statesmen
have had—Michelangelo and Victor Hugo, Thomas Jefferson and
Winston Churchill."

CHAPTER 4

The White House

I WAS ONLY NINE WHEN MY UNCLE JACK DIED, BUT UNLIKE MY SKAKEL uncles, whom I saw only rarely, his gravities ordered my early life. He was the deep central channel that gathered the collected currents of our clan—family, country, and faith—and from which the braided and winding tributaries of our lives have flowed ever since. His high standards and successes cemented the sense that we were part of a gallant effort to help our country achieve its ideals as an exemplary nation.

As an adult I've studied—I hope with dispassion—the history of that era in an effort to understand my family, my culture, my country, and my own place in the world. I've profited from the observations and stories of the friends and family members who best knew Uncle Jack and my father. I've read the most significant of the multitude of histories and biographies that have dissected their lives down to the granular level. I've come to know their human frailties and to admire how consistently they set themselves to a higher purpose. And with those insights I've come to respect, more than ever, the wisdom and judgment with which they served our nation. Camelot was not, after all, an illusion of my youth.

And today JFK's great concerns seem more relevant than ever: the dangers of nuclear proliferation; his notion that empire is inconsistent with a republic, and that corporate domination of our democracy at home is the unholy partner of imperial policies abroad; his vigilance for the pitfalls of power abused; his sense of the importance of the rule of law and the sacredness of the Bill of Rights; his deep understanding of the perils to our Constitution from a national security state; his mistrust of zealots and ideologues, political or religious; his curiosity and tolerance for other perspectives; his notion that democracy should be enlarged to accommodate all people; his belief that nations ought to fight their own civil wars and choose their own governments—and that America has no business doing those things for them.

I believe that it was Jack's capacity to imagine a different future for America—a future consistent with our ideals—that made him, despite his short time in office, one of the most popular U.S. presidents in our history. A 2009 poll of sixty-five historians ranked him sixth in presidential leadership, just ahead of Thomas Jefferson. He is the only one-term president whom historians consistently include in their list of our top ten presidents.

The 1960 Campaign

I was just shy of seven years old during the 1960 campaign, but I still have vivid memories of Uncle Jack's presidential race, which seemed to us a long celebration. A 45-rpm single of Frank Sinatra's campaign song, "High Hopes," blared ceaselessly from the living room Victrola, while Kenny O'Donnell, Dave Powers, Ted Sorensen, Arthur Schlesinger, Pierre Salinger, and my uncles Teddy Kennedy, Steve Smith, and Sarge Shriver huddled with my father in the TV room at Hickory Hill, in their dark suits, white shirts, and skinny ties, or out on the patio in shirtsleeves.

Our pride in Uncle Jack magnified our exuberance for the campaign. He was a Pulitzer Prize–winning author and historian. Hollywood was making a film of his World War II heroics, starring heartthrob Cliff Robertson. When a Japanese destroyer rammed and sank his fogbound PT-109 off the Solomon Islands in 1943, Jack, despite his own injuries, led his eleven surviving crewmen through perilous nighttime waters to safety. A Harvard swim team veteran, he dragged his badly burned mechanic by a lanyard clenched in his teeth for five hours to a tiny island. Marooned for days, he and his crew dodged Japanese patrols during the day. Jack searched for help during long nocturnal swims in the perilous currents of the Ferguson Passage. On Nauro Island he finally encountered a pair of Solomon Islanders, Biuku Gasa and Eroni Bumana. Paddling their dugout, the two Melanesians carried a coconut, on which Jack had etched his coordinates, thirty-five miles to an Australian outpost. Jack invited the two men to his inauguration. Only after his death did we learn that a British colonial official, embarrassed by their poor mastery of English, prevented their travel. Four decades later, in 2002, my little brother Max traveled with underwater explorer Robert Ballard to search for PT-109 in the Blackett Strait. They found very little of Jack's boat, but they did find Gasa and Bumana, who had carried the coconut. Bumana was wearing a T-shirt proclaiming, in large block letters: I SAVED JOHN F. KENNEDY. The two men told Max that President Kennedy was their "chief." They wept piteously and hugged him like a long-lost brother.

After Jack's rescue and discharge, he ran for Congress in 1946, capturing the seat abandoned by Honey Fitz's rival, James Michael Curley, who was, by then, in the Danbury Penitentiary, and served three terms before winning the 1952 U.S. Senate race in Massachusetts against incumbent Henry Cabot Lodge—another Honey Fitz rival. During the 1960 presidential primaries, we made campaign trips with him on Grandpa's Convair, the *Caroline*. We wore PT-boat tie clips and sported Kennedy buttons on our school blazers. After school, we shinnied up the neighborhood stop signs, adding Nixon stickers below the word STOP. Over Christmas in

Palm Beach, Uncle Jack, indulging my interest in animals, showed me his bedroom bathtub, where he'd formerly kept a pet alligator. The grown-ups, shielded from the trade winds in the lee of a cedar hedge, laughed and lunched with Franklin D. Roosevelt, Jr., who that day agreed to campaign for Jack in West Virginia.

My father managed his big brother's campaign and gave Uncle Teddy, then twenty-seven, and Jack's former Choate roommate, Lem Billings, both green campaigners, the toughest districts, to prove themselves. Teddy's states included impossible-to-win Wyoming. Piloting a single-engine Piper Cub, Teddy scoured the wide-open spaces of Big Sky Country for tiny gatherings, landing on two lanes outside remote towns to speak to coffee klatches and breakfast diners. He was so hard up to address a crowd that, in one instance, he agreed to ride a bucking bronco for the privilege of speaking at a rodeo, and, in another, to take flight off a sixty-meter ski jump for the chance to stump the audience at an Alpine competition. That summer, Lem and Teddy serenaded us one evening at Grandpa's house with a song they'd composed to commemorate their experiences with a crooked political boss from La Crosse, Wisconsin; they called it "The Al DiPiazza Rhumba."

> We went down to La Crosse
> To meet our new boss:
> Al DiPiazza!
> Everything is OK—
> Why don't you boys go away?
> Al DiPiazza!
> His wife likes beautiful things.
> She wears great big diamond rings.
> Al DiPiazza!

And so on.

They belted it out, at high volume, in the clipped, gravelly style of Jimmy Durante, with Teddy on the piano. Even as a six-year-old, I had an overwhelming sense that politics was joyful and fun.

A Religious Crusade

Uncle Jack's quest for the presidency seemed more than a political campaign: it was a historic revolution to make Catholics full partners in the American experience. It's difficult to imagine now, but when John Kennedy ran for president in 1960, his Irish Catholicism was the campaign's central burning controversy. Block-voting Irish still dominated America's urban political landscape, but every politician remembered how anti-Catholic bigotry immolated Al Smith's 1928 presidential campaign. Flames of hatred fueled national cross-burning ceremonies by the Ku Klux Klan. Conventional wisdom held that widespread prejudice against Papists still burned hot enough to keep any Catholic out of the White House.

This meant that for Irish, Italian, Polish, French-Canadian, and Eastern European immigrants, the 1960 campaign was a battle for equality and enfranchisement. I was a second-grader in Washington, D.C., at Our Lady of Victory during the campaign, where Monsignor Hesse and the Sisters of Notre Dame saw Jack's presidential race as a religious crusade. I recall the droves of sisters in their habits, along with black-robed priests, seminarians, and lay brothers who marched fervently from the rectories, convents, and priories for Uncle Jack. Nuns followed him like teenyboppers, campaigning in every primary. While anti-Catholic rhetoric was a mainstay of American Protestantism, Southern Baptist fundamentalists, in particular, openly loathed Catholics. Before the West Virginia primary, Reverend Billy Graham convened Southern Baptist preachers to warn congregants against "the Roman menace." Conservative newspapers, radio evangelists, and dark right-wing manifestos terrorized the Protestant majority with ominous prophecies of a White House gone Papist. JFK's Roman minions would rename the Statue of Liberty "Our Lady of the Harbor," they warned, and string together bowling balls to bedeck her with rosary beads. While Protestant leaders like Norman Vincent Peale fired off bigoted broadsides, mainstream journalists debated whether Jack's religion would doom the Democratic ticket.

Together with my school chums I collected some of the quarters defaced by bigots with nail polish to depict a cardinal's skullcap covering Washington's head, an ominous warning against a "Roman" in the Oval Office. Eight years later my Protestant schoolmates at Millbrook School in the Hudson Valley still teased me as a "mackerel snatcher," a slur derived from the Catholic tradition of eating fish on Fridays and during Lent. Among my earliest childhood friends were Fays, Markhams, O'Donnells, and McNamaras, all children of frontline troops in the political battalions who would march for America's first Irish Catholic American president.

Jack's first real test on the religious issue came in the mountains of West Virginia, the state with the smallest Catholic population. Jack had already licked his most formidable opponent, Hubert Humphrey, in the Wisconsin primary, right next door to Humphrey's home state of Minnesota. But Humphrey was a labor champion and a Protestant, and now Jack had to beat him in a union state that was only 3 percent Catholic, and notoriously sectarian. Early polls in West Virginia showed JFK trailing Humphrey by eighteen points, as venomous preachers and hate-radio jocks blanketed the airwaves with apocalyptic portents.

Paul Corbin, a labor leader and early JFK supporter who became a kind of mischievous alter ego to my father, told me he'd spoken to a pair of nuns sporting Kennedy buttons debarking from a Greyhound at the Charleston bus depot. Apparently, two young crackers, seated behind the sisters on the bus and annoyed at the tall habits that obscured their views, had complained loudly, "I thought the whole point of living in West Virginia was that there was only 3 percent Catholics here." One of the nuns told Corbin she had turned to face the men, and advised them, "Why don't you go to Hell; there are no Catholics there!" Corbin was a gravel-throated rogue of abundant charm and elastic scruples who famously fooled an FBI polygraph and lied under oath to Joe McCarthy about his activities as an organizer among the West Coast longshoremen. In later life, Corbin, probably out of affection for my parents rather than genuine yearning for redemption, converted to a sort of lapsed Catholicism, with my father and mother serving as his godparents at his baptism.

During the spring of 1960, both my parents disappeared for six weeks to West Virginia, part of the vanguard of Kennedy clan members who would ultimately shake hands with nearly every voter in the Mountain State. Kenny O'Donnell later recalled Jack persuading West Virginians to support him, in a series of speeches "marked by a fire and dash that I had seldom seen in him." O'Donnell and Jack's other advisers had previously urged him to approach the religious issue obliquely, arguing that a direct approach would offend public sensibilities. But the West Virginia crowds were universally sullen and my aunts and mother found that no one would even host a tea party. "We like him," the ladies would explain, "but he's a Catholic."

Finally, Jack decided to buck his advisers and confront the issue head-on. In a spontaneous, off-the-cuff address to a morose and indifferent Morgantown audience, he asked the Mountaineers to consider whether 40 million Americans had forever given up their right to run for president on the day they were baptized as Catholic. Jack appealed to the patriotism of West Virginians, who had the highest enlistment rate of any state. "Nobody asked my brother if he was Catholic or Protestant before he climbed into an American bomber plane to fly his last mission." From then on, he employed similar refrains in every stump speech across the state. Franklin Roosevelt, Jr., told Lem Billings that Jack hammered the issue so hard that West Virginians felt they *had* to vote for him to prove their tolerance. He won the state by an astonishing 65 percent.

Jack's resounding West Virginia victory allayed the most serious doubts about whether a Catholic could win the presidency. Overwhelming wins in Oregon, Maryland, Nebraska, Indiana, Pennsylvania, Illinois, and New Hampshire and a series of hard-won

gubernatorial endorsements cemented a margin that my father—
who was in charge of counting delegates—knew at the outset of the
Democratic Convention would give Jack a first-ballot nomination.

In mid-July, Kathleen, Joe, David, and I flew in the *Caroline*,
along with other family members, from Cape Cod to California
for the convention. We stayed at the home of Marion Davies, William Randolph Hearst's consort and muse. Her house had a long
cobblestone driveway topped by a swimming-pool-sized fountain,
where we waded and played among statues of peeing angels and
cupids. Grandpa wore his scarlet bathrobe and black velvet slippers. He kept a low profile during the campaign and convention,
but he strategized with Jack and my father inside Davies's mansion.
My mother loved Davies, with whom she had developed a lasting
friendship during my parents' honeymoon. Davies was loud, garrulous, and funny. "She made everyone laugh!"

We visited my cousin, Chris Lawford, at his parents' Malibu
home and spent afternoons playing on the beach and swimming in
the pool. Chris's mother, Aunt Pat, was the most glamorous of the
Kennedy sisters. She was tall and slender, with thick, long chestnut
hair, square shoulders, a perfect figure, and a beautiful laugh. We
were thrilled to meet Peter Lawford's Rat Pack buddies, Sammy
Davis, Jr., Frank Sinatra, and Dean Martin, who joked, mixed mar-
tinis, and bantered with actresses Angie Dickinson, Kim Novak,
and Lauren Bacall. One evening, Frank Sinatra hosted a party for
Uncle Jack at the Davies manse. He greeted every guest at the door,
made the rounds shaking hands and chatting, and then sang a duet
with Dean Martin and a trio with Uncle Teddy.

On the night of Jack's nomination, Joe, David, and I, clad in our
gray flannel shorts and knee socks, made our way with Kathleen to
the gallery high above the Los Angeles Memorial Coliseum floor.
My dad had a command center in the cottage outside the arena and
forty men on the floor keeping the delegates in line. Joe, Kathleen,
David, and I knew most of them and made a game of spotting them
from the gallery. We knew their job was a first-ballot victory, which
would take 761 votes. It was like watching a football game. We kept

asking for the updated score. Grandma's proximity kept us mostly on our best behavior. In her red dress, she seemed to dominate the convention center. My mother also looked beautiful, in a simple green dress and diamond brooch given to her by her father. I played with a gopher snake I'd captured a few days before, and studied the sea of delegates waving banners and sporting white Styrofoam voter hats emblazoned with the name Kennedy in red, white, and blue. I wondered where they got those costumes, and whether they dressed like that year round.

When we saw the chance to escape, I stowed the snake in my blazer pocket and scrambled with my brothers through the cigar-smoke-filled corridors, collecting buttons, ribbons, and placards from the vendors and delegates. We returned to the balcony for the final vote count.

We saw Uncle Teddy standing in the middle of the Wyoming delegation, smiling when that state put Jack over the line. Then applause convulsed the stadium as confetti and balloons dropped from on high and the band played "Toora Loora Loora" and "Happy Days Are Here Again." The following night we gathered again to hear Jack's acceptance speech, in which he spontaneously coined the phrase "New Frontier" for the first time. The Coliseum was a football stadium open on its west side, and from our lofty aerie we could see the Pacific Ocean behind Jack as he spoke, creating a majestic spectacle. The next day we all flew back to Cape Cod on the *Caroline.* Uncle Jack sat next to me and I briefed him on my animal collection, a subject that, despite his other preoccupations, he seemed to find of great interest. He later presented me with a picture of the two of us chatting, inscribed with the note: "A President gets his advice from many sources."

Throughout the general election campaign against Nixon and the GOP's vice presidential nominee, Henry Cabot Lodge, conservatives, particularly evangelicals, continued to dog Jack with questions as to whether a Roman Catholic could respect the constitutional guarantee separating church and state. On September 12, Jack finally settled those questions satisfactorily for most

Americans in a speech in Houston, Texas, to Reverend Norman Vincent Peale's Protestant Minister Association, and his deceptively named National Conference of Citizens for Religious Freedom, which Peale had newly founded to oppose Jack's election. Peale publicly warned that a Papist in the Oval Office would be the slave of the Roman hierarchy. Jack's advisers had unanimously warned him against accepting Peale's invitation but Jack was anxious to confront the issue head-on. He told Dave Powers and Kenny O'Donnell, "I'm getting tired of these people who think I want to replace the gold in Fort Knox with holy water." That evening he began by telling the resolutely hostile assemblage that the nation had far more vital concerns than "the so-called religious issue." At stake in the present election, he said, was the welfare of "the hungry children I saw in West Virginia; the old people who cannot pay their doctor bills; the families forced to give up their farms; an America with too many slums, with too few schools, and too late to the moon and outer space." Unfortunately, he lamented pointedly, "the real issues in this campaign have been obscured—perhaps deliberately—in some quarters less responsible than this, by questions about my religious faith." He reminded the predominantly Baptist preachers that it was the harassment of Baptist ministers in Virginia that prompted Jefferson to draft the state's statute of religious freedom. He urged them to work with him to abolish religious intolerance "and promote instead the American ideal of brotherhood." He then described what would become for decades America's branded vision of church-state separation.

"I am not the Catholic candidate for president. I am the Democratic Party's candidate for president, who happens also to be a Catholic. I do not speak for my church on public matters, and the church does not speak for me. Whatever issue may come before me as president—on birth control, divorce, censorship, gambling, or any other subject—I will make my decision in accordance with these views, in accordance with what my conscience tells me to be the national interest, and without regard to outside religious pressures or dictates. And no power, or threat of punishment, could cause me to decide otherwise. But if the time should ever come . . . when my office would require me to either violate my conscience or violate the national interest, then I would resign the office; and I hope any conscientious public servant would do the same."

Later that autumn we were among seventy million Americans who watched the first televised presidential debate—the largest audience to date in television history. Even at my tender age, seeing Nixon's darting eyes and sweaty upper lip, I had the solid impression that this "Tricky Dick" was shifty and dark. If my life were a Superman comic, Richard Nixon would be Lex Luthor, the pertinacious supervillain. He seemed always to be waiting in the wings with some new perverse scheme for world conquest.

On Election Day we watched the vote tallies with fingernail-biting anxiety late into the night. The grown-ups had all gathered at Cape Cod for the returns. Because my dad was campaign manager, our house served as the command center, with fourteen "boiler room girls" working twenty-four phones installed on the enclosed front porch and pollster Lou Harris encamped in my bedroom with his team of vote analysts. My father stayed up all night fretting, with Grandpa restlessly sallying between the makeshift campaign headquarters and the Big House. The election proved much tighter than Jack's in-house polls had predicted. Despite the enthusiasm for Jack among lay Catholics and lower clergy, the Catholic hierarchy had been lukewarm toward Jack's candidacy. Grandpa considered it a devastating personal betrayal when Cardinal Spellman, to whom he had donated millions of dollars in charitable contributions over many decades, supported Nixon. As a result, the Catholic turnout was much lighter for Jack than the Kennedy campaign had anticipated, and my father and Grandpa spent the night on pins and needles. With the outcome still up in the air, Jack went to bed at three a.m., walking across the compound to his own house. In the morning, his daughter, my three-year-old cousin Caroline, prompted by Jackie, entered his room and greeted him, "Good morning, Mr. President." In this way John Fitzgerald Kennedy learned that he would soon become the first Irish Catholic president of the United States—and, at age forty-three, the youngest president ever elected. My mom told us that we must never again call him "Uncle Jack." From now on it would be "President Kennedy," even among his family.

And my father got a new job, too! Uncle Jack named my father attorney general. Jack believed he would profit from my father's loyalty and judgment in his Cabinet. "I want the best man for the job and they don't come any better than Bobby." The press didn't love the idea. The *New York Times*, which had warned Jack against bringing his brother into the White House, now howled that my father had never practiced law. Thanks to those critiques, I learned the words "nepotism" and "ruthless," just as I would learn the term "carpetbagger" four years later, when my dad ran for senator in New York. Sensitive to the optics, my father repeatedly declined the attorney general appointment. On January 16, he went to Jack's Georgetown house on N Street to give his elder brother a final, definitive no. After hearing my father out over breakfast, Jack instructed him to go upstairs and comb his hair; there were reporters waiting outside. Press Secretary Pierre Salinger had already posted news of my father's appointment. Everyone in our family except my father thought it funny when Jack justified his choice to the press by explaining that he'd made the appointment because he felt my dad "needed some legal experience before going into private practice."

Nineteen sixty-one marked a time of great optimism for America. Our country was at the apex of its power. Following the Allied victory in World War II, Europe had yielded global leadership to the United States. Our nation had emerged from the war with half the world's wealth. Our industries and agriculture dominated the global economy. "Made in America" was the gold standard for quality. Everyone wanted American products: our automobiles, radios, food, appliances, electronics, blue jeans, and even our democracy. Hollywood and rock and roll were infiltrating world culture, and we had just invented color TV. The lingua franca was American English, the global currency the dollar. Jack planned to reassure the world that we would use our new leadership and moral authority to stand up for justice and resist tyranny in all its guises—communism, fascism, and imperialism. America had not sought world leadership. The vast majority of Americans had shared Grandpa Joseph Kennedy's prewar sentiment that America should avoid foreign entanglements. But World War II had thrust leadership upon us, and Jack was determined that ours should be the exemplary democratic leadership of Athens, and not the martial plutocracy of Rome. History, it seemed, had prepared us to be an exemplary nation. In 1789, America was the world's lone democracy. By 1866, there were six. By 1960, the entire world was poised to imitate the American experiment with self-governance. Jack was determined that our role as an exemplary nation should be just that—leadership by example. We should perfect our union and *model* democracy for the nations of the world, not force it upon them.

Jack was hyperconscious, in the formulation attributed to Lincoln, that we had "become a great nation because we were a good nation" and that we would fast lose our greatness if we stopped behaving well. "We shall nobly save," Lincoln put it, "or meanly lose the last best hope of earth." We were off to a good start. Jack's generation had saved the world from the murderous dictatorships of Hitler, Mussolini, and Tojo. It had taken up the cause of humanity by making America a bulwark against the homicidal regimes of Mao and Stalin now festering behind the Iron Curtain. American idealism was literally reconfiguring the globe: as a condition for America entering the war, FDR had forced Churchill to sign the Atlantic Charter, an agreement that Europe would relinquish its empires and lift the burden of colonialism from the oppressed peoples of Africa and Asia. American generosity, embodied in the Marshall Plan, was reconstructing the ruined civilizations of Europe. Instead of punishing our vanquished foes in Germany and Japan, we were helping them to rebuild as thriving, prosperous democracies. America was the unchallenged emblem of optimism, democracy, and freedom. We were living up to our nation's promise to be, in Thomas Paine's words, "the last great hope for humanity."

On January 20, 1961, with blinding sunlight reflecting from freshly fallen snow, we watched from frigid bleachers as Jack paraded, hatless, to the reviewing stand for his inauguration. The high point of the day for me was shaking hands with Commander Kohei Hanami, the Japanese captain of the destroyer *Amagiri*, which had sunk Uncle Jack's PT boat. Jack had graciously invited Hanami to the celebration that his seamanship had so nearly precluded.

We all began the day at mass in Holy Trinity Church with Cardinal Cushing presiding, and it was Cushing who gave the invocation in his gravelly voice and swore Jack in as president. Jack's inauguration was rife with symbolism of his determination to perfect the union and make America a worthy model for democracy and freedom. Mahalia Jackson, whom five years earlier the Daughters of the American Revolution had banned from performing at Constitution Hall because she was black, sang the national anthem at Jack's request. Later that afternoon, Jack called his special assistant, Dick Goodwin, to ask why there had been no black faces in the Coast Guard contingent that had marched past his reviewing stand in the inaugural parade. Goodwin's heart was bursting with excitement as he phoned the Coast Guard commander to order the immediate and aggressive recruitment of African-American cadets. "It's really happening; we are about to change the world," he thought to himself.

Uncle Jack took the unusual step of inviting America's greatest living poet, eighty-five-year-old Robert Frost, to read a poem. That short recital, and the presence on the reviewing stand of W. H. Auden and John Steinbeck as Jack's special guests, symbolized the elevation of art and culture that would become a hallmark of Camelot. (Jack's favorite author, Ernest Hemingway, was too sick to attend.) Grandma Rose taught two generations of Kennedys that culture was the soul of a vital nation. She reminded us that not long after their victorious battles, the generals of ancient Greece were forgotten, but two millennia later people around the globe still treasure Athenian philosophy, art, poetry, and plays.

In that spirit, Uncle Jack was determined that U.S. leadership would mean more to the world than materialism and military power. He and Jackie made art a central preoccupation of his administration; their dream was to make Washington a world cultural center. The White House became a stage for classical music concerts and for gatherings of great artists, poets, writers, and musicians, including Spain's virtuoso cellist Pablo Casals, who, following American recognition of Franco's fascist regime, had sworn to never set foot in the United States, but reconsidered his vow as Jack steered the nation back to a course of idealism. Jack drafted artist William Walton and his best friend Lem Billings to begin planning a cultural center that would make Washington a global mecca for opera, ballet, art, and poetry. After his death, that edifice would become the Kennedy Center.

Jack's Inaugural Address summoned Americans to a period of national engagement and sacrifice, to lead the world in a new direction, away from war, imperialism, and injustice. The stakes, he reminded Americans, had never been higher: "For man holds in his mortal hands the power to abolish all forms of human poverty or all forms of human life." He reached out to the Soviet Union to end the ruinous arms race that had left "both sides overburdened by the costs of modern weapons and both rightly alarmed by the steady spread of the deadly atom, yet both racing to alter that uncertain balance of terror that stays the hand of mankind's final war." And the vitality, hope, and energy of the high-quality men and women who flocked to work in his administration infused the country with a sense of high purpose. At the time of his death, a record 85 percent of Americans trusted that government by the people could be made to work for the people.

Pledging America's support "to those peoples in the huts and villages of half the globe, struggling to break the bonds of mass misery," he put our country squarely on the side of liberation struggles by developing nations against their former colonial masters. He promised to make our nation a beacon of justice around the globe, embodied in the Alliance for Progress and the Peace Corps. These programs would showcase America's strength, her can-do attitude, vigor, humanity, and idealism. Jack was determined that the face of America in Africa and Latin America would no longer be a soldier, but a Peace Corps volunteer.

At home JFK launched a physical-fitness program to invigorate and toughen Americans for the tasks of global leadership. He mobilized our nation to reach the moon and stars through the space program. Responding to the gauntlet tossed by the Soviets when they launched the first man-made satellite, *Sputnik*, he committed our country not just to explore the solar system but also to a public education program that would make young Americans the world's leaders in math, science, and engineering.

And Americans answered his call with idealism, optimism, and energy. Eight years after Jack challenged the nation to put a man on the moon, we accomplished that lofty goal. Sixty-five thousand Americans flocked to join the Peace Corps within three months to serve overseas for less than an Army private's pay, "to go anywhere, do anything needed. Its first direction," my Uncle Sarge Shriver recalled, "to serve the poor. Not to Americanize individuals or nations. Not for money or glory. But just to serve their fellow human beings." Uncle Jack shifted America's two-century relationship with Latin America, placing democracy and justice for the poor ahead of serving local oligarchs and U.S. corporations. Before his death he laid plans for a massive assault within our own country against poverty and racism. And his economic programs launched the country on its longest sustained expansion since World War II.

The nobility of these national efforts recalled the stirring images of Camelot, which would become synonymous with the Kennedy administration. I read the heroic tales of Arthur and his knights in T. H. White's *The Once and Future King* and *The Sword in the Stone*, in *Sir Gawain and the Green Knight*, and in Sir Thomas Malory's *Le Morte d'Arthur*. Our parents took us to see the Washington premiere of *Camelot*, the musical play written by Jack's Choate and Harvard classmate Alan Lerner, and starring Jack's childhood chum Robert Goulet. From my mother's record player the soundtrack pursued us into every nook of our Virginia home, Hickory Hill.

In the afternoon following the inauguration, we went to the White House to watch my father be sworn in as attorney general. We toured the presidential mansion to look at paintings and historical artifacts. I took particular interest in FDR's "Fish Room." I was disappointed that his aquariums were long gone, but Joe, David, and I watched men hanging a mounted swordfish that Jack had caught during his Acapulco honeymoon. My dad would hang a slightly larger specimen from his own Acapulco trip in his Justice Department office. Unleashed for the afternoon in the sanctum sanctorum of American democracy, we dashed through the tunnels in the basement, dived in the pool, and threw wild gutter balls in the bowling alley. We tore through the upstairs bedrooms where our cousin Caroline and her new baby brother, John—born sixteen days after his father was elected—would live. Following the ceremony, our dad slid down the brass banister with us. Our Costa Rican nanny, Ena Bernard, scolded him, "Mr. Kennedy, you'll never grow up."

Foreign Affairs Challenges

Uncle Jack's inaugural address combined the era's obligatory Cold War saber-rattling palaver with conciliatory words that revealed his central preoccupation with avoiding nuclear annihilation. He asked "that both sides begin anew the quest for peace, before the dark powers of destruction unleashed by science engulf all humanity in planned or accidental self-destruction." He believed that nuclear war would mark a grave moral failure by the leaders of the nuclear powers and would end civilization. "We have the power to make this the best generation of mankind in the history of the world, or to make it the last." His commitment to eradicate war was the defining feature of his administration. Speaking with his friend Ben Bradlee (later the executive editor of the *Washington Post*), Jack said that his highest ambition for his presidency was to be remembered with the epitaph, "He kept the peace." Elaborating, he told

Bradlee, "I think that is the primary function of the president of the United States—to keep the country out of war." His national security adviser Bob McNamara would later say that JFK succeeded in this goal under the most adverse circumstances, surrounded as he was by warmongering generals, congressional hawks, and a highly militarized Washington establishment, all convinced that war with the Soviets was both inevitable and desirable.

Hugh Sidey, a journalist and friend, wrote this about Jack after his death: "If I had to single out one element in Kennedy's life that more than anything else influenced his later leadership it would be a horror of war, a total revulsion over the terrible toll that modern war had taken on individuals, nations, and societies, and the even worse prospects in the nuclear age. . . . It ran even deeper than his considerable public rhetoric on the issue." But to the conservative jihadists who viewed the struggle against communism as a religious crusade, even questioning the Cold War boilerplate was cowardice—or worse. Barry Goldwater, the conservative senator from Arizona, characterized the idea of coexistence with Communists as treason. In his best-selling diatribe *Why Not Victory?* he proclaimed, "Our objective must be the destruction of the enemy as an ideological force. . . . We will never reconcile ourselves to the Communists' possession of power of any kind in any part of the world." Jack viewed Goldwater's words as reckless, irresponsible, and idiotic.

President Eisenhower had predicted that it would not be a warrior who got us into the next war. "Men acquainted with the battlefield will not be among the numbers that glibly talk of another war." Subsequent events have affirmed that sentiment. Despite his use of Cold War boilerplate, Uncle Jack was a war-weary pacifist. The only president awarded the Purple Heart, he had risked his life in battle and seen men die. He had lost his brother Joe in the war, along with his brother-in-law Billy Hartington, three members of his PT boat crew, and many, many friends. He had witnessed firsthand the waste and stupidity of armed conflict. He had proven his courage in the Solomon Islands and didn't have to prove it again.

During his three-year presidency, John Kennedy didn't send a single military unit to fight a war.

While it's true that JFK presided over an intense military buildup, his goal was to bring the Soviets to the negotiating table and avoid a nuclear showdown. Jack often quoted Churchill's 1949 remark: "We arm to parlay." It was preferable, Jack explained in 1959, to meet the Soviets "at the summit than at the brink." And despite his own blustery rhetoric, Nikita Khrushchev, an experienced combat veteran of some of World War II's bloodiest battles, would find common ground with Jack on the road to peace. "I had no cause for regret once Kennedy became president," Khrushchev wrote. "It quickly became clear he understood better than Eisenhower that an improvement in relations was the only rational course . . . He seemed to have a better grasp of the idea of peaceful coexistence than Eisenhower had. . . . From the beginning he tried to establish closer contacts with the Soviet Union with an eye to reaching an agreement on disarmament and to avoiding any incidents which might set off a military conflict."

The Environment and Physical Fitness

Along with avoiding nuclear annihilation and ending colonization, the grave threats of overpopulation and air and water pollution were also central preoccupations of the New Frontier that I watched Jack describe in such inspiring terms when he accepted the Democratic nomination in Los Angeles. He campaigned on his concerns about the "population explosion," even though the conservative Catholic hierarchy had condemned that term as "a recently coined terror technique phrase." Uncle Jack also frequently quoted Teddy Roosevelt's insistence that "The nation behaves well if it treats [its] natural resources as assets which it must turn over to the next generation increased, and not impaired, in value." He created the National Seashore, which included preserving 43,500 acres of Cape Cod shoreline, dunes and beaches that today draw more than four million visitors each year. He saved Georgetown from developers who wanted to plow under its historic townhouses to build office parks. As his secretary of the interior he appointed Stewart Udall, a committed outdoorsman from New Mexico and one of the nation's leading conservationists. The Udall cousins Tom and Mark, who both became U.S. senators, were our occasional playmates at Hickory Hill. At the time of my father's assassination in 1968, Stewart Udall was his first choice for vice president.

When Rachel Carson wrote her groundbreaking 1962 runaway best-seller *Silent Spring*, she came under savage attack by a corporate coalition led by Monsanto and its PR firm, Hill and Knowlton. In a meticulously orchestrated smear campaign that would lay the strategic groundwork for later battles by the tobacco and carbon industries, the chemical industry branded Carson's meticulously fact-checked book a collection of lies, and claimed that more study was needed. They vilified Carson herself as a "bitter spinster"— the contemporary euphemism for lesbian. The American Farm Bureau, the American Medical Association, the American Garden Club Association, and powerful journalistic voices including *Time* and *Life* all joined in the orchestrated assault against Carson. Dying of cancer, she was hardly able to defend herself, but Uncle Jack went to bat against her powerful detractors. He created an independent commission of scientists that quickly vindicated Carson, verifying all of her controversial claims about the lethal dangers of pesticides. That report, highly critical of conflicts of interest within his own USDA, led to the eventual ban on DDT and laid the groundwork for the creation of the EPA and modern environmental protection laws. In 1962, my parents hosted Carson for a seminar at Hickory Hill—one of the treasured moments of my youth.

In 1963, urged on by my father's close friend Wisconsin senator Gaylord Nelson, Jack embarked on a national whistle-stop tour to alert Americans to the looming environmental threats from pollution and overpopulation, and to garner public support for land conservation. He was frustrated that at each stop reporters refused to engage him on those subjects, asking him instead about topics ranging from the threat of Soviet expansion to the upcoming elections. Like leaders today, he found it challenging to keep media attention on the world's most pressing issues. After Jack's death, Senator Nelson and his energetic aide Denis Hayes turned that frustration into action and organized the first Earth Day in 1970. I marched that day in New York: it was the largest public demonstration, to date, in American history.

I had my own strong thoughts on these issues. I was particularly agitated, at age seven, to learn that entire species of animals were going extinct, and I was angry at the men who had forever doomed the passenger pigeon and the dodo. Killing off the last members of a species seemed to me to have the desolate finality of a mortal sin. I announced that I would write a book about pollution, wildlife, and the environment and, with my dad's encouragement, began drafting the opus. Using embossed stationery that I'd received for Christmas, I wrote Uncle Jack requesting an appointment to air my concerns. He invited me for a private audience in the Oval Office. As a gift, I brought him a seven-inch spotted salamander I'd captured the previous afternoon and given the biblical name Shadrach. It was Shadrach's misfortune that Hickory Hill had recently switched from well to town water, and, not understanding the implications, I'd unwittingly killed the amphibian in a chlorine bath. However, I remained in sturdy denial about its demise as I entered the Oval Office. Uncle Jack and I spent a good deal of our meeting discussing the inert salamander's health, and I still have a photograph of myself seated across from him, studying the large crystal vase between my legs. President Kennedy is reaching across his desk and probing the salamander with the dull end of a pen. I recall his concern. "He looks ill to me. Are you sure he's well?" I assured him the salamander was resting, but the president wasn't convinced. "I think he might be dead," he said, gravely, although he told the press corps he was delighted with the gift. Afterward, we went outside and released Shadrach in the White House fountain, out of sight of nosy journalists. In private, I admitted to him that the salamander's lack of animation was striking. But otherwise the meeting was a success.

Jack agreed to arrange for me to interview key figures in his administration, including Interior Secretary Stewart Udall, the nation's leading environmentalist. I met Udall in his cavernous offices at the Department of the Interior, where I marveled at the murals of western mammals and Native Americans. I brought my bulky reel-to-reel tape recorder and asked him questions about air and water pollution and extinction. I was grateful for the attention when the story of my impending publication made *Time* magazine and the *Saturday Evening Post*, though the actual writing eluded me for some thirty-three years. I published my first environmental book, *The Riverkeepers,* in 1996.

Given the steady disappearance of America's wilderness and the migration from farms to the cities, JFK fretted that America was becoming soft. The remedy was his national campaign for physical fitness. Inspired by a 1903 letter from Teddy Roosevelt to the commandant of the U.S. Marines, Uncle Jack also encouraged Americans to meet the challenge of a fifty-mile hike. Years later I would accept this summons myself, walking from Boston to Cape Cod with Lem Billings, who, after my dad's death, became a surrogate father to me. I had invited three of my Harvard rugby teammates to accompany us on the stroll. Lem, then sixty, worried that this group of fit, young athletes would quickly leave him eating dust and badly demoralized, so we agreed that he should begin with a one-hour head start. According to the plan, we would jog at first to catch him, and then be satisfied to continue at his slower pace for the remainder of the hike. But Lem was a sprightly walker, and we never caught up. My friends dropped out with various injuries, the last of them after thirty miles, and I walked the final twenty alone, though I was heartened by occasional signs of Lem's progress, including an empty asthma inhaler and his familiar handkerchief soaked with sweat.

I recall my father undertaking his own fifty-mile hike a decade earlier, without the same exquisite planning. On a frigid Friday afternoon he impulsively rounded up four of his "band of brothers" who had the misfortune to be working late at the Justice

Department—Ed Guthman, Dave Hackett, Lou Oberdorfer, and Jim Symington—for the challenge. At five a.m. the next morning, wearing a pair of tennis sneakers, he started down the snowbound C&O Canal towpath in the dead of winter, headed for Camp David, where we awaited him. The temperature never rose above 26° F and dropped precipitously in the late afternoon. After thirty-five miles the last of his companions, Ed Guthman, the Pulitzer Prize–winning crime reporter for the *Seattle Post-Intelligencer,* whom my father had tapped to join his staff in a mafia investigation, dropped out. Guthman, an avid outdoorsman, had walked across most of North Africa and Italy during World War II, as an infantry platoon commander, winning a Silver Star and a Purple Heart. It killed him to quit. "You're lucky; your brother isn't President of the United States," my father kidded him, grimly. Guthman later wrote, "Bob would have completed the fifty miles if he had had to crawl." At Camp David, I waited anxiously for my father late into the night and saw him come in and collapse on the bed with my mother massaging his blistered and bleeding feet.

The following summer my brother David and I went on a bird-watching expedition, walking a twenty-mile stretch of the old C&O Canal towpath between Washington and Great Falls, Maryland. We accompanied my dad and his friend, Supreme Court justice William O. Douglas, who was leading the ultimately successful battle to preserve the towpath as a national park. During the Great Depression my grandfather had hired Douglas to work for him at the SEC, and Grandpa later persuaded Roosevelt to appoint his protégé SEC chairman and, later, to the Supreme Court. Douglas, an avid hiker and naturalist, went on to become America's greatest environmental jurist; Douglas's dissenting opinion in *Sierra Club v. Morton,* arguing that trees should have standing to sue lumber companies, is required reading for every student of environmental law. The towpath tour was my first environmental direct action. I was beat by the end of it.

In 1962, Ed Guthman and William O. Douglas took us kids, along with our parents and my Shriver cousins and their mother,

Eunice, on a ten-day trip on horseback through the most remote reaches of Washington State's Olympic peninsula. Douglas, a robust and virile woodsman, developed an unrequited ardor for Eunice, and spent an afternoon vigorously pursuing her around her tent. I witnessed the chase, with my brother David, from a hideout beneath thick cedars in the rainforest, and I can attest to the earnestness of her evasive maneuvers—and to her successful escape.

It was Justice Douglas who taught me how to fly-fish. He had a long, graceful cast that I've spent a lifetime emulating. On that trip we all ate trout for breakfast, lunch, and dinner, and drank from the streams. I remember how excited I was to catch wild fish using dry flies and salmon eggs, and to toss them into a frying pan. A photo depicts me standing in a stream, in my cowboy boots, holding a creel with a dozen brookies in my left hand and a respectable rainbow trout in my right. Afterward we fished for salmon in Puget Sound and caught more fish than I'd ever seen.

The CIA Unleashed

On my seventh birthday, January 17, 1961, three days before my Uncle Jack took the oath of office, outgoing president Dwight Eisenhower delivered the most important and memorable speech of his political career. In his farewell address to the American people, Eisenhower decried the dangerous rise of "the military-industrial complex," which he described as "the conjunction of an immense military establishment and a large arms industry . . . new in the American experience." Eisenhower feared that increased defense spending could destroy America's free enterprise system and transform our country into an imperialist power abroad and a police state at home. "The potential for the disastrous rise of misplaced power exists and will persist," he cautioned. "We must never let the weight of this combination endanger our liberties or democratic processes."

That same day, the Belgian government, with the support of

the CIA, assassinated Congo's independence leader and first freely elected premier, Patrice Lumumba. Lumumba was Congo's George Washington, the only figure popular enough to peacefully unite all its disparate tribes and regions. But Lumumba had promised to nationalize foreign holdings within his country, and to use Congo's vast mineral and natural resource wealth to benefit its people. Jack's public support for Lumumba had created alarm within the CIA, adding a sense of urgency to Allen Dulles's clandestine efforts to assassinate the African leader before Jack took office. The CIA sent an assassin to spike Lumumba's toothpaste with poison, but the Belgians beat him to the punch, kidnapping and shooting Lumumba three days before Jack's inauguration. Lumumba's assassination elevated the man who led the coup, Colonel Joseph Mobutu, to premier. For fifty years Mobutu and his thieving cronies would loot their resource-rich country, relegating the Congolese people to destitution and embroiling the nation in endless civil wars.

Deeply disturbed by Lumumba's murder, Jack repeatedly rebuffed Pentagon, CIA, and State Department requests for military occupation of Congo. He worked instead to encourage an independent and united Congo by supporting UN peacekeeping efforts. As his UN ambassador Adlai Stevenson explained, "The only way to keep the Cold War out of the Congo was to keep the United Nations in the Congo." That objective clashed with the ambitions of the CIA and the multinational corporations that the Agency treated as clientele: oil, mineral, and resource companies hoping to carve the former Belgian colony into bite-size morsels in order to more easily devour its natural riches. Behind Jack's back, the CIA was busy subverting his policy of nonintervention by secretly arming secessionist forces in the Katanga region who were in thrall to international mining companies.

President Truman had originally created the CIA in 1947 as an intelligence-gathering agency, and would come to rue the cloak-and-dagger powers it later appropriated. Congress had terminated the wartime intelligence agency, the OSS (Office of Strategic Services), following the Axis surrender, and both political parties questioned the creation of a peacetime spy agency. Republicans and Democrats alike considered intelligence services an artifact of tyranny, and condemned proposals to create the CIA, which they characterized as an "American Gestapo." Consequently, the initial statute empowered the new agency to collect information but not to engage in clandestine activities. Nevertheless, the Agency's first civilian director, Allen Dulles, vastly expanded the CIA's mission. In 1949, without presidential, congressional, or constitutional authorization, he issued a top-secret National Security Council directive that authorized the Agency to conduct covert activities, including assassinating national leaders and overthrowing governments. Dulles's doctrine of "plausible deniability" decreed that his army of spooks could lawfully lie under oath about their activities—even to Congress—in order to protect the "national interest," which, presumably, only Dulles and his minions possessed the genius to discern. To make matters worse, Dulles and his protégés reliably interpreted the CIA's mission as advancing the mercantile ambitions of U.S. corporations, the expanding power of the military-industrial complex, and the cause of U.S. imperialism. As his biographer David Talbot observed, in Allen Dulles's mind, America's "national interest was seamlessly intertwined with the profit interests of its largest corporations—many of whom Dulles had represented as attorney for the white shoe law firm of Sullivan and Cromwell."

In 1949, at Dulles's urging, Congress freed the CIA from its final restraint, budgetary review, and the Agency quickly became a clandestine government within our government, with its own enormous budget, legions of bureaucrats, and rogue armies, all completely unaccountable to American democracy. As it expanded power, the Agency metastasized like a cancer, to threaten the very democracy and national security that it was commissioned to safeguard.

While he remained president, Truman largely succeeded in keeping the spooks under tight rein, forbidding them from engaging in the business of overthrowing foreign governments. But the moment Truman left office, the Agency began deploying its vast resources in nefarious ways antithetical to American values, including coups and assassinations, destabilizing foreign governments, and rigging elections. Eisenhower, with his veteran warrior's loathing of committing ground troops, used the CIA as his instrument of first choice, inadvertently enabling its dangerous expansion. He largely abdicated his foreign policy to the Dulles brothers—Secretary of State John Foster Dulles and his brother, Allen, the CIA director.

As historian Stephen Kinzer has observed, "Democracy for the Dulles brothers was cosmetic, and they revoked it when it produced leaders who threatened their business friends." Days after Eisenhower's inauguration (in 1953), the CIA, in cahoots with the Anglo-Arabian Oil Company (now BP), overthrew the popular, democratically elected prime minister of Iran, Mohammad Mosaddegh, a piece of mischief Truman had steadfastly resisted. *Time* had named Mosaddegh its "Man of the Year" because of the hope he had inspired among national movements across the world, but Mosaddegh made the lethal mistake of moving to nationalize Anglo-Arabian's Iranian operations. Mosaddegh expelled the British when he discovered Churchill's government was planning a coup, but continued to trust the United States, which he regarded as the paradigm of democracy and anticolonial resistance. Exploiting that trust, Dulles fermented a faux revolution, using paid mercenaries, gangsters, and street thugs, to overthrow Mosaddegh and install Shah Mohammad Reza Pahlavi as head of state. We—and much of the world—are still paying the costs of that caper.

The following year, at the behest of former Dulles law client United Fruit, the CIA overthrew Guatemala's enormously popular, and democratically elected, president, Jacobo Árbenz, and put a brutal military dictator in command of the country. Following the CIA coup, its agents assembled a "murder list" of leftists and liberals, whom its new dictator, Castillo Armas, dutifully liquidated. Guatemala has never healed, and continues to be one of the most violent and oppressive nations on the planet. During this period the CIA also installed corrupt puppets in Laos and Vietnam, fixed myriad elections in Italy and Greece, organized coups or supported tyrants in Iraq, Syria, Lebanon, Egypt, and Jordan, all resulting in catastrophic blowback that bedevils U.S. relations across the Mideast to this day. In 1957, Agency spooks serving private mining interests orchestrated a failed coup against Indonesian prime minister Sukarno. Eisenhower publicly denied any U.S. involvement until after the Indonesian Army shot down and captured a CIA pilot after he bombed a school. The incident marginalized the United States, pushed Sukarno to the left, and strengthened the indigenous Communist movement. All these shenanigans would come back to haunt us.

Between 1947 and 1989, the CIA would initiate seventy-two coup d'états—the equivalent of one-third of the world's governments. The CIA seemed determined that the American experiment in self-government would not spread to other nations. Instead of becoming an exemplary nation and leading the world by example, the CIA would push America toward imperialism.

In 1956, my grandfather Joseph Kennedy sat on a panel that first sounded the alarm about the growing power of the CIA and its clandestine activities. The Hoover Commission urged Eisenhower to limit the CIA's powers, "which find us involved covertly in the internal affairs of practically every country to which we have access." Two years later, when Uncle Jack was a U.S. senator, he publicly castigated the Dulles brothers for their secretiveness, for their "condescending disdain of those who challenge their omniscience," and for their reluctance to "give up the luxury of their exclusive control over our role in world affairs."

Today, the CIA has sprawled beyond recognition or control by the democracy it is supposed to serve. Secret budgets now support an astonishing 1,271 additional intelligence agencies that function under CIA control and 1,931 private companies engaged in spying, murder, kidnapping, espionage, sabotage, and torture. Over 30,000 spy agency employees, engaged in eavesdropping, intercept over 1.5 billion telephone conversations daily. The history of CIA activities during the sixties and seventies, when it had relatively insignificant budgets and power, gives us abundant reason to fear the awesome and unaccountable power of an agency that now poses the most daunting threat to American democracy and national security, a threat that dwarfs the peril to America from Islamic fundamentalism. My family's grappling to bring that agency and its Pentagon collaborators under the control of our democracy was the central defining battle of the Cold War era.

The Bay of Pigs

On New Year's Day 1959, Fidel Castro's revolutionary forces marched into Havana and overthrew Cuban dictator Fulgencio Batista. With the support of his army, a corrupt oligarchy, the United Fruit Company, and the American Mafia, Batista had ruled the island nation with fear and torture for twenty-five years. Castro led his followers as they tore down the casinos, sex clubs, and bordellos, while Batista and his Mafia overlord Meyer Lansky boarded planes and fled to Miami.

In 2014, I took my wife, Cheryl Hines, and our children to meet Fidel during a scuba trip to Cuba. In answer to a question from my thirteen-year-old son Aidan, Castro said that he had learned antimaterialism from his Jesuit teachers while growing up in the western province of Santiago. Castro's first wife, Mirta Diaz-Balart, was the daughter of a United Fruit attorney, so Fidel had the opportunity to see firsthand the gated communities in which the company's American employees lived in opulence while fencing out the poor. His father, he told Aidan, had come to Cuba from Spain in 1898, as an idealistic adventurer, intent on joining the Cuban revolution against Spain. The United States had, according to Fidel, first supported, and then hijacked, Cuba's war for independence, in order to steal Cuba's strategically important coal port at Guantánamo Bay. Fidel believed that U.S. corporations and organized crime gangsters had then ransacked Cuba of its political sovereignty, its economic independence, and its dignity. But he was most disgusted by the Cosa Nostra's role in running Havana. During an earlier meeting with me, in 1996, he had expressed his continuing bitterness about the corruption and humiliating control of his country once exerted by American Mafiosi. It was just after the collapse of the Soviet Union, and although Fidel had predicted that Cuba would soon move away from a strict Communist model and adopt some species of a marketplace system, he pounded his fist on the table at the mention of casinos, telling me in Spanish, "The casinos, no! Never again in Cuba!"

The Eisenhower administration, which had supported Batista's bloody regime, reluctantly recognized Castro's new government. Eisenhower, however, snubbed Fidel during the Cuban leader's visit to New York. Explaining that he had scheduled a prior golf game, Eisenhower delegated the meeting to Vice President Richard Nixon, who subsequently issued this verdict on the Cuban leader in a secret memo: "I was convinced that Castro was either incredibly naïve about Communism or under Communist discipline." Nixon added, ominously, "We would have to treat and deal with him accordingly." He pressed President Eisenhower to arm Cuban exiles to overthrow Castro, and the CIA began training 1,500 paid Cuban guerrillas, a unit known as "Cuba Brigade 2506," on secret bases in the "made-in-America" dictatorships of Nicaragua and Guatemala.

United Fruit owned most of Cuba's arable land and kept the bulk of its holdings uncultivated, guaranteeing high prices for sugar and bananas and low labor costs, and keeping Cuban peasants in a permanent state of serfdom. When Castro began expropriating United Fruit's properties, he offered fair compensation, based on the company's own property tax valuations, which United Fruit now complained it had kept artificially low. The company had powerful friends in the Eisenhower administration: Secretary of State John Foster Dulles had been United Fruit's lawyer, and CIA director Allen Dulles had served on its board of directors. The two brothers had already orchestrated a coup in Guatemala to protect the company's interests there by overthrowing the democratically elected government of Jacobo Árbenz. Ironically, Castro's Argentinian protégé, Dr. Ernesto "Che" Guevara, had been radicalized while working as a visiting doctor among Guatemala's poor during that U.S.-backed coup. Castro and Che now anticipated a similar intervention by the CIA in Cuba, and took appropriate precautions.

So when Castro announced his expropriations in May 1960, and the United States retaliated by cutting imports of Cuban sugar, Fidel; his brother, Raul; and Che quickly negotiated a deal with the U.S.S.R. to step into the vacuum. Khrushchev agreed to trade Russian oil for Cuba's sugar crop. As a gesture of solidarity with United Fruit, Texaco announced in June that its Cuba-based refinery would not process Russian oil, a move designed to strangle the Cuban economy. In response Castro nationalized the refinery on June 29, offering fair compensation to Texaco. For this affront to Texaco—another former Dulles client—the United States moved to dissolve diplomatic relations. Castro's government reacted by shuttering the remaining gambling casinos operated by American Mafiosi and nationalizing one billion dollars in U.S. assets, including a private home on Varadero Beach owned by my grandfather George Skakel, along with several dozen sugar mills and plantations also owned by the Skakel family.

Under Nixon's patronage, Brigade 2506, the CIA's guerrilla army of anti-Castro Cuban exiles had trained in Central American jungles for over a year before Uncle Jack's inauguration. Although Nixon considered the Bay of Pigs caper his "brainchild," the Republican administration left its execution to the incoming Kennedy team. Eisenhower's senior advisers, Allen Dulles and Richard M. Bissell, Jr., promised a reticent Uncle Jack that there was zero chance of failure. The Cuban people, they assured him, would greet the "Brigade" as liberators, join the revolt, and happily overthrow Castro. Dulles later acknowledged that Jack, only three months into his presidency, had recoiled at the project from the outset. However, Dulles and Bissell goaded him with the prospect of being portrayed as less vigorous than Eisenhower in resisting Communist expansion in the hemisphere, and warned of the danger of calling off the invasion, thus bringing dangerous Brigade members back to the

United States with nothing to do. Jack reluctantly moved forward only after receiving ironclad guarantees from Dulles and Bissell and the operation's Pentagon planners; General Lyman Lemnitzer, chair of the Joint Chiefs of Staff; and other military brass that there would be no need for direct U.S. military involvement. Lemnitzer gave his guarantee. Jack was adamant on this point because he didn't want the United States to look like a bully as the Soviets had when they invaded Hungary in 1956. He wanted the New Frontier to represent a clean break with what he considered our reprehensible history of intervention, which had particularly caused such consternation and resentment among our Latin American neighbors. To nail down this caveat, he went off script to announce publicly at his weekly press conference on April 12, 1961, that there would never be, "under any condition, an intervention in Cuba by United States forces." That statement, largely ignored by the American press, met widespread praise across Latin America.

Dulles and Bissell knew they were lying about the easy success of their caper and about the promised uprisings; Castro was immensely popular with the majority of the Cuban people and had already assembled a formidable intelligence apparatus. The CIA's secret 1960 national intelligence estimate, which Dulles never disclosed to Jack, predicted that there was no action "likely to bring about a critical shift of popular opinion away from Castro." Fidel's army was disciplined, effective, and outnumbered the insurgents a hundred to one. Knowing this, the CIA nevertheless assured Jack that the Cuban military was demoralized and would quickly break. The Agency's backup plan was likewise a sham: if the prime objective failed, the Brigade would melt into the wilderness and become an ongoing guerrilla movement. CIA planners had picked the Bay of Pigs, a flat swamp utterly unsuited for guerrilla action, expecting the Brigade's assault would falter, and hoping to force the young president into ordering a full-scale military invasion by U.S. forces to remove Castro—the CIA's and the Pentagon brass's preferred option all along.

During the night of April 17, 1961, 1,500 members of the Cuban

Brigade 2506 who had sailed from Nicaragua, on ships owned by United Fruit, landed at the Bay of Pigs on Cuba's south coast. The operation was a disaster. Within hours Fidel and his army had the Brigade pinned down. Three days later, 200 were dead and 1,197 were surrounded and soon to be captured. Then the Pentagon and CIA got a shock: Jack refused to order U.S. air strikes against Castro's air force and ground troops. While swearing to Jack that no U.S. military assistance would be necessary, the CIA was also lying to the Brigade 2506 commander, telling him that Kennedy had promised air and naval support. That lie planted bitter seeds that would later bear the poison fruit of abiding hatred by many Cuban exiles against my family. E. Howard Hunt, the CIA's chief of political action for the invasion, who went on to become operations director for Nixon's Watergate break-in, later acknowledged, bitterly, that he and his Agency colleagues had dismissed as posturing Jack's insistence that there would be no bailout by the U.S. military. "I did not take him seriously," said Hunt.

Jack, in turn, realized he had been drawn into a trap. He told Dave Powers, "They were sure I'd give in to them and send the go-ahead order to the *Essex* [the U.S. Navy aircraft carrier]. They couldn't believe that a new president like me wouldn't panic and try to save his own face. Well, they had me figured all wrong." Jack realized, as his father had predicted, that the CIA had become a monumental problem for American democracy. "I know that bunch," Grandpa told him. "I wouldn't give a nickel for the lot." While the Cuban Brigade was still battling to gain a beachhead, Jack angrily told Arthur Schlesinger that he wanted to "splinter the CIA into a thousand pieces and scatter it to the winds." He added, "It's a hell of a way to learn things, but I have learned one thing from this business—that is, that we will have to deal with the CIA. No one has dealt with the CIA. I made a mistake in putting Bobby in the Justice Department. . . . Bobby should be in the CIA."

Although my father resisted Jack's requests that he run the Agency, he did agree to oversee an inquiry into its Bay of Pigs actions. As a result, late in 1961 Jack fired the Agency's top three officials—Director Allen Dulles, Deputy Director Richard M. Bissell, Jr., and Deputy Director Charles Cabell, the Bay of Pigs architects. He brought in retired general Maxwell Taylor to regain control of the Joint Chiefs. Taylor was a maverick whose career had been derailed for his philosophical revulsion against nuclear weapons and for his championing of flexible response—the notion that we could contain Soviet expansionist ambitions with political strategy and measured, small-scale military action. Taylor would become one of my dad's closest advisers and most trusted friends. My brother Maxwell Taylor Kennedy is his namesake.

After my father and General Taylor oversaw the postmortem on the Cuba operation, Jack tried to clip the CIA's wings. He slashed its budget, removed its advisory powers on military matters, and gave U.S. ambassadors at every U.S. embassy supervisory control of local CIA activities. He created a President's Foreign Intelligence Advisory Board, appointed James Killian of MIT to run it, and put Washington attorney and veteran presidential adviser Clark Clifford on the Board, charging it with finding out what the CIA was up to. My dad thought a Republican who was not the president's brother needed to be running the nation's spy agency, so they chose John McCone, a deeply pious Catholic (with whom my mother developed a particularly close friendship during his wife's struggle with terminal cancer). My father would come to deeply regret their decision to appoint McCone, but for a time they were friends. CIA headquarters in Langley was only a half mile down a wooded path from Hickory Hill, and McCone made a habit of eating lunch at our house daily and coming to swim in our pool after work in the spring and autumn. I remember him taking easy laps as we played "sharks and minnows" in the deep end after school.

My father and Jack remained at odds with the military brass and intelligence apparatus for the duration of Uncle Jack's presidency. The Bay of Pigs episode set the course for the next one thousand days, as Jack's own military and intelligence agencies engaged in incessant schemes to trap him into escalating the Cold War into a hot one—a circumstance that would, in even its worst-case scenario, increase their power.

The Bay of Pigs action turned into a potent illustration of the CIA's chronic curse, the law of unintended consequences. At the Punta del Este conference in Uruguay that summer, Che Guevara told JFK aide Dick Goodwin, "It has been a great political victory, for it enabled [the Castro] regime to consolidate, and transform Cuba from an aggrieved little country into an equal." For the next forty years the ill-conceived boycott of Cuba would play the same role—allowing Castro to consolidate power and escape accountability by blaming his nation's economic woes and his many other failures on the United States. Fidel and Raúl Castro pointed to the embargo to justify their continued oppression of the Cuban people. The embargo was, in fact, the biggest gift we could have conceived for the Castro brothers. After his firing, Dulles mainly refrained from openly expressing his anger and disgust at Jack's core foreign policy programs, the Peace Corps and the Alliance for Progress. To his friends he faulted Jack for the "weakness" of "yearning to be loved by the rest of the world." Dulles remarked, "I should much prefer to have people respect us than to try to make them love us."

In a departure from such tight-lipped assessments of President Kennedy while Jack lived, Allen Dulles told a young writer in 1965, "That little Kennedy, he thought he was a god." LBJ would later appoint Dulles to the Warren Commission investigating Jack's assassination, a curious choice at a time when some Americans, including my father, suspected the CIA's involvement in JFK's murder.

Support for National Independence

The Cold War, which began almost immediately after the surrenders of Germany and Japan, was for many Americans a religious crusade against godless communism, but Jack saw it as a purely political conflict. Cold Warriors viewed communism as monolithic and every indigenous anti-colonial revolution in Asia, Africa, and

Latin America as hatched in Moscow. Although his administration began with a U.S.-sponsored invasion of Cuba, cooked up before he became president, Jack's core belief was that America should not worry if emerging nations decided to experiment with Marxist economic systems. He went out of his way to buttress movements for national independence—even those that were left wing and overtly Marxist. He gave both blessings and financial assistance to leftist revolutionary leaders Eduardo Mondlane of Mozambique, Ahmed Sékou Touré in Guinea, and Holden Roberto of Angola. Instead of trying to subvert those systems with military or covert intervention, he arranged to bring their university students to the United States to study democracy. He even supported President Kwame Nkrumah of Ghana, alarming the State Department, his own advisers, and even my father, who disapproved of Nkrumah's romance with the Kremlin.

Jack told British Guiana's Marxist president, Cheddi Jagan, "We are not engaged in a crusade to force private enterprise on parts of the world where it is not relevant. If we are engaged in a crusade for anything, it is national independence. That is the primary purpose of our aid. The secondary purpose is to encourage individual freedom and political freedom. But we can't always get that, and we have often helped [Communist] countries which have no personal freedom, like Yugoslavia, if they maintain their national independence [freedom from Kremlin control]. That is the basic thing. So long as you do that, we don't care whether you are socialist, capitalist, Marxist or whatever. We regard ourselves as pragmatists."

Like his father, JFK believed that open dialogue and peaceful competition in markets for goods and for ideas were the best antigens against the spread of communism. He thought Communist China should be admitted to the UN, and that we should expand our ties with the Communist governments of Yugoslavia and Poland. Although Jack was adamant about opposing Soviet expansionism, he was equally determined that we should negotiate with our adversaries. He said in his inaugural address, "Let us never negotiate out of fear. But let us never fear to negotiate."

While Jack often disagreed with Grandpa over foreign policy, they shared a skepticism of the Cold War theology that tasked America as the global police force against Communist proliferation. Grandpa knew the Dulles brothers, and lunched with Allen at the home of his Palm Beach neighbor and mutual friend, right-wing Texas oilman Jack Reisman. Grandpa regarded Allen Dulles as oversecretive and silly. The notion that America could dictate the destinies of other sovereign nations was, in his view, dangerous and antidemocratic. He believed that effort would weaken America. "Grandpa thought Dulles was a donkey," my mother recalls. "Dulles believed that everybody who wanted to overthrow a colonial government or nationalize an American corporation was a Communist. He was unable to discuss world events except from his narrow perspective of power politics. He made no distinction between what was good for our country and what was good for oil companies and other corporations." Most ironically for an intelligence officer, Dulles seemed coldly uncurious about perspectives that differed from his own and that might be held in the rest of the world. "He seemed disinterested about anything, or any fact, that didn't serve his ideology," said my mother. After one luncheon, Grandpa was astonished that Dulles didn't have a clue who Marilyn Monroe was, even though her picture had been on the cover of *Life* magazine. "Dulles's view was constricted and dark and Grandpa doubted, with those limitations, that his judgment could be trusted."

In August 1961, my parents left us on Cape Cod while they visited West Africa, to carry Uncle Jack's message of support for independence movements. Their destination was the Ivory Coast, which was celebrating the first anniversary of its liberation from France. Both Jack and my father considered anticolonial nationalism to be the most powerful and exciting impulse in the developing world. Africa was no longer a European garden. Everywhere my parents traveled they handed out PT Boat tie clips as souvenirs to the African celebrants, often to people who had never seen a necktie. They returned with a treasure trove of zebra-skin drums, spears, buffalo-skin shields, seven pairs of African stilts upon which we all learned to race, and a scrapbook of photos of their meetings with tiny pygmies and giant Watusis—the world's tallest people— dancing stupendously in ostrich feather headdresses. Both my mom and my dad had read Alan Moorehead's *The White Nile* and enthralled us at dinner with the book's strange and wonderful descriptions of Africa—its wildlife, history, and extraordinary people. Those recitations prompted me to spend hours reading in my bedroom, mesmerized by Moorehead's books.

I'd been obsessed with Africa ever since second grade, when my father brought home a movie called *Africa Speaks*, which we watched with an old 16mm projector on a small portable movie screen in the basement. My fascination eventually led me to read all twenty-seven of Edgar Rice Burroughs's Tarzan novels and Frank Buck's *Bring 'Em Back Alive*. I spent hours studying photos of mountain gorillas, Serengeti wildlife, and African villagers in *National Geographic*. I combed *Life* magazine for African stories, and read everything I could about Albert Schweitzer, Dr. Livingstone, and the great explorers, including Burton, Speke, and Stanley. I wrote my third-grade research paper on Niger River explorer Mungo Park.

I was desperate to see this exotic continent for myself, and my dream came true in the summer of 1964 when I accompanied my cousin Bobby Shriver and his father, Sarge, the first director of the Peace Corps, on a tour of East Africa. We went on safari with a famous hunter, Leon Lynn, touring the Serengeti, marveling at the vast herds of wildebeest, impalas, and elephants. We visited the Maasai Mara reserve and the Samburu on the northern frontier, flew over the Rift Valley, and slept in the Ngorongoro crater, where lions and hyenas serenaded our tent with their barks and coughs, and a vexed rhino charged our Jeep. It was beyond my boyhood dreams. I met African photographer Peter Beard and began a decade-long correspondence with him, which included many letters he penned in his own blood. He regularly sent me his most gruesome photos, including one of a Peace Corps volunteer stacked in pieces in an aluminum beer tub after Beard had removed his remains from the belly of a Lake Victoria crocodile.

I returned from that trip with a sixteen-pound leopard tortoise I named Carruthers. I'd brought him home under the diplomatic protection of Uncle Sarge's passport in a Gucci suitcase my mother had loaned me (which could not be used thereafter). Carruthers wandered our house and yard and became a fixture at Hickory Hill for the next twenty-one years, growing to thirty-five pounds. I also brought back an iron spear, a gift to me from Masai villagers with whom we visited. The weapon would play a pivotal part several years later, when an intruder with malicious intent made his way into our home.

Mongoose

On November 30, 1961, shortly after the Bay of Pigs, JFK authorized Operation Mongoose, a mainly covert program of sabotage and infiltration designed to encourage popular revolt against Castro. My father played a key role in supervising the operation, persistently badgering the CIA to support an indigenous counter-revolution among the island's discontented citizenry whom the agency deceptively portrayed as aflame with revolutionary fever. While Jack was cool and detached, my father ran hot. The daylight between their views—often seamless in other areas—is illustrated by two notes by my father to Uncle Jack in 1962, recommending to Jack an escalation of Mongoose. Jack ignored the first memo. My father wrote an almost sulking follow-up indicating that he took Jack's failure to reply as disagreement with the project. My father's contempt for communism was ideological, and his distaste for Fidel was personal. The Cuban leader was exporting violent revolution to the democracies of Latin America—particularly targeting left-wing democracies that the Alliance for Progress was supporting. In addition, my dad felt a genuine strategic imperative to prevent Fidel from making Cuba a platform for Soviet expansion in the hemisphere. Finally, there were political considerations.

National Republicans, like Nixon and Goldwater, and Southern Democrats were denouncing Jack for losing Laos and Cuba and badgering the administration to "do something" about Castro. My father, who had been left out of the Bay of Pigs deliberations, was furious at the CIA for its orchestration of the fiasco that had stranded courageous Cuban fighters on the Bay of Pigs beaches and wounded his brother's presidency. He probably saw Mongoose as his chance to extract penance from the Agency and show the incompetent spooks how to manage the project properly. For Cuba hands inside the Agency, Robert Kennedy's attention felt punitive, naive, and overbearing. To aggravate the tension, our house was less than a mile, on a country two-lane, from CIA headquarters, and my father would drop by daily to demand reports and to spy on the spies. For the Agency, this was oppressive, humiliating, and unbearable.

An ideological rift between the White House and the CIA hobbled Mongoose from its conception. The State Department and my father wanted to support an anti-Castro movement within Cuba. They wanted the revolution to be Cuban. The Kennedys did not believe that the United States should be in the business of overthrowing governments. While it was morally acceptable to give limited aid to assist indigenous rebellion against an oppressive dictator—after all, the French had assisted American patriots during our own revolution—they considered direct intervention by American forces waging, in essence, undeclared wars to be immoral, unconstitutional, and antithetical to every essential American value. The Kennedy brothers' objective, according to Major General Edward Lansdale, the Pentagon's Operation Mongoose director, was to depose Castro the way Castro had ousted Batista, to have "the people themselves" overthrow the Castro regime rather than United States–engineered efforts from outside Cuba.

The CIA, knowing there was little Cuban support for counterrevolution (a fact it tried to hide from the Kennedys), favored direct CIA action to depose Castro. Spies like William Harvey, Sam Halpern, and David Atlee Phillips bridled at my father's oppressive, nagging demands that they "do something" about Castro. They disdained the Kennedy brothers for being too squeamish to assassinate Castro and for lacking the mettle to launch a U.S. invasion, options that the CIA knew were the only realistic roads to regime change.

Furthermore, my father and the State Department specifically wanted to avoid entanglements with former Batista army and intelligence agents, whom they regarded as gangsters, goons, and oligarchs. The Agency's favorite allies, however, were precisely those cohorts; Batista's former lieutenants, Cuban oligarchs (now living in Miami) and American mobsters who had forfeited million in casino revenues. Neither democracy nor social justice were priorities for these villains; their concern was the restoration of their political power, their nationalized properties, and their Havana gambling empires. My father's favorite exiles were idealistic Cuban nationalists like Enrique "Harry" Ruiz-Williams, Miro Cardona, Manuel Artime, Manuel Ray, José San Roman, and Encidio Olivia, who had fought alongside Fidel against Batista, and then turned against Castro when his revolution became Communist. These men became familiars at Hickory Hill. The CIA was not interested in working with such Cubans who, like most true nationalists, were independent, politically unreliable, and difficult to control. The CIA and its Batista allies scorned these men as "Bobby's Cubans" and the White House strategy as *fidelismo sin Fidel.* "The CIA wants to subordinate everything else to tidy manageable operations." Arthur Schlesinger wrote in his daybook, in reference to the schisms over Cuba policy between the Kennedy brothers and the CIA's operational branch, "hence it prefers people like [Joaquin] Sanjenish [head of the sinister and secretive CIA death squad, Operation 40] to proud and independent people like Miro [Cardona, head of the Cuban Revolutionary Council and Castro's first prime minister]."

Howard Hunt, the dirty-tricks spook who would later gain notoriety as the architect of the Watergate burglary, abandoned the

Cuban project soon after my father took over. His disgust typified Agency reaction to my father's strictures. "It was obvious there was no serious intent in overthrowing Castro," Hunt explained contemptuously. The entire purpose of Mongoose, he insisted, was "to give the appearance of activity."

Operation Mongoose died soon after the Cuban Missile Crisis. In the end, the operation amounted to little more than an ineffective series of sabotages of bridges, roads, power supplies, and sugar mills. Arthur Schlesinger called Mongoose "Robert Kennedy's most conspicuous folly," an assessment with which I would agree. Meanwhile, the CIA had been conducting its own plans to rid the world of Castro, out of presidential purview.

Assassinating Castro

During the Eisenhower administration Allen Dulles and Richard Bissell, the CIA's director of plans, assembled a team of Cuban expatriates, assassins, and terrorists and launched a covert program known as Operation 40, presided over by Vice President Richard Nixon and CIA deputy director Richard Helms. The team ultimately grew from forty to seventy men. One member, Frank Sturgis, another future Watergate burglar, explained: "This assassination group [Operation 40] would, upon orders, naturally assassinate either members of the military or the political parties of the foreign country that you were going to infiltrate, and if necessary some of your own members who were suspected of being foreign agents. . . . We were concentrating in Cuba at that particular time." According to Fabian Escalante, the former chief of Cuba's intelligence service, among its dubious successes, Operation 40 is credited with blowing up a Belgian munitions ship in Havana Bay on March 4, 1961, killing seventy-five people and injuring more than two hundred.

Operation 40 was the CIA's club for murderers. It assembled the Agency's most homicidal scoundrels, almost all of them alumni of the CIA's 1954 Guatemala coup and/or the Bay of Pigs operation. Its primary function was to have been the liquidation of Cuba's Communist officials and its influential leftist and liberal leaders following a successful coup or U.S. invasion. According to researcher Jim DiEugenio, its murder lists would almost certainly have included influential liberals like Manuel Ray, Harry Williams, and the others favored by the Kennedy brothers and derisively known as "Kennedy's Cubans." Its rosters included the CIA murderous alcoholic spymaster William Harvey; propaganda wizard David Atlee Phillips, who among many other villainies orchestrated the numerous coups against legitimate Western Hemisphere governments, including the Chilean coup against Allende in 1973 and the Washington, D.C., car-bombing murder of former Chilean foreign minister Orlando Letelier in 1976; David Sanchez Morales, the feared CIA hitman who often bragged of his part in both the JFK and RFK assassinations; Orlando Bosch and Luis Posada, the "unrepentant terrorists," in Attorney General Dick Thornburgh's assessment, who blew up a civilian Cuban airliner in 1976, killing all seventy-seven passengers; Cuban contract killers Eladio del Valle and Hermino Diaz; drug-smuggling pilots; Barry Seal, later murdered by the Medellín Cartel; future Watergate burglars Bernard Barker, Eugenio Martinez, Frank Sturgis, and E. Howard Hunt. Robert Blakey, the chief counsel to the House Select Committee on Assassinations, told me he is nearly certain that two Operation 40 members, Eladio del Valle and Herminio Diaz, were involved in the JFK assassination. Several Operation 40 members later played key roles in the Iran-Contra scandal. Along with Morales and E. Howard Hunt, del Valle and Diaz also boasted of playing roles in the JFK assassination before their deaths. In a series of deathbed confessions for his sons, E. Howard Hunt admitted to participating in the planning of President Kennedy's murder in a plot that he said included Morales, Sturgis, Harvey, and David Atlee Phillips.

Operation 40 was headquartered in the CIA's Miami station, code-named JM-Wave. It was the CIA's largest station and Miami's largest employer. The influx of CIA operatives to the station created a real estate boom in South Florida during the 1960s. The unit continued to operate until 1970, when the CIA disbanded it after the crash in California of one of its airplanes carrying large quantities of cocaine and heroin.

During the Eisenhower, Kennedy, and Johnson administrations, members of this cabal, under CIA Deputy Director Richard Helms and Cuba project overseer William Harvey, conducted a super-secret project, known as ZR/RIFLE, to assassinate Fidel Castro. Helms's group sponsored dozens of attempts on Castro's life, using an arsenal that included poison pills, a lethal pen, exploding cigars, a contaminated wetsuit, booby-trapped conch shells, and a murderous girlfriend. Harvey recruited to this enterprise a former FBI and current CIA agent named Robert Maheu, who was then working as a security and operations officer for Las Vegas billionaire-recluse Howard Hughes, who owned the world's largest defense contractor. Maheu's job was to enlist the support of mobster Johnny Roselli and Mafia capos Santo Trafficante and Salvatore "Sam" Giancana to murder Fidel.

My father stumbled across the caper in early 1962 after Maheu bungled a plot to bug the Las Vegas hotel room of comedian Dan Rowan, costar of the top-rated TV comedy show *Rowan & Martin's Laugh-In*. The bugging escapade was a CIA favor to mob godfather Sam Giancana, who believed Rowan was secretly dating his girlfriend, singer Phyllis McGuire. When Las Vegas Police caught Maheu's operative planting the bug, the CIA went to bat for its stooge, asking the Justice Department to drop the prosecution of Maheu to avoid the potentially embarrassing fallout from this CIA-sanctioned crime. FBI Director J. Edgar Hoover informed my father of the CIA's request to his department.

According to Hoover, my father was horrified when he learned of the CIA's romance with mobsters, and of the Agency's involvement in the assassination racket. Worst of all, Giancana and Trafficante were my father's top targets for prosecution. My father ordered an end to all clandestine operations to murder Castro, and redoubled his efforts to put Giancana behind bars. CIA officers Sheffield Edwards and Larry Houston assured my father that the plots had ceased. They were lying. The Agency and its mob partners secretly continued their efforts to assassinate Fidel.

In 1975, Senator Frank Church held hearings on the assassination plots before the Senate Intelligence Committee. Church called the CIA "a rogue elephant charging out of control." The Senators were shocked to learn that the Agency was not only developing exotic murder weapons and mind-control (*Manchurian Candidate*) drugs, but was also conducting assassination attempts. Castro himself gave the Church Committee a list delineating twenty-four attempts by the CIA to kill him between 1960 and 1965, and the Church Committee confirmed at least eight of these. When the committee subpoenaed CIA officials to testify, Richard Helms, who, alongside counterterrorism chief James Angleton, had masterminded and orchestrated the murder program, swore that Presidents Eisenhower, JFK, or Lyndon Johnson were unaware of the plots. "Nobody wants to embarrass a President of the United States by discussing the assassination of foreign leaders in his presence." My father had disliked Helms from the outset. My dad admired candor and considered Helms dark and furtive. My mother told me, "Helms was sinister and haughty. Daddy never trusted him. It puzzled him how Bob McNamara had a blind spot toward Helms. Helms also kept the plots secret from CIA Director John McCone, anticipating that McCone, a pious Catholic, would have torpedoed the operation. McCone confirmed that fear. When Helms delicately broached the subject of assassination, McCone told him, "If I get myself involved in something like this, I might get myself excommunicated." When, in August 1963, McCone read a newspaper account of the CIA's engagement with Giancana during the Eisenhower era, McCone forcefully demanded assurances that the Agency had terminated the relationship, which Bill Harvey obligingly gave him. Harvey was lying to his boss.

In 1967, as my father was considering a primary run against LBJ, Helms's aide-de-camp, Sam Halpern, began leaking disinformation suggesting that the Kennedys were aware of the assassination attempts on Castro. Such false reports caused LBJ to exclaim, "We had been operating a damned Murder Inc[orporated] in the Caribbean." My father was enraged by the accusations and furiously denied them. President Johnson turned to Richard Helms to confirm the Kennedy's knowledge of the CIA's murder plots. Following LBJ's request, Helms asked the CIA's Inspector General Supervisor J. S. Earman, Inspector General Scott Breckenridge, and Deputy Inspector General K. E. Greer to research whether the Agency could implicate the Kennedys in the Agency's assassination plots. Researcher Lisa Pease has obtained a copy of that report and provided it to me. The report concludes that the Kennedys had no prior knowledge of any of the CIA assassination programs and that there was no way for the Agency to credibly blame them. Helms, however, did not want to see the Kennedy brothers exonerated. When the inspector general submitted the report to Helms, the CIA Director ordered all draft copies destroyed and all of the coauthors' notes to be shredded. He kept the single surviving original draft in his safe.

Inspector General Scott Breckenridge was the principal coauthor of the CIA's report on the Kennedy brothers' involvement in the assassination. Breckenridge conducted exhaustive interviews with CIA personnel and extensive document reviews. On June 2, 1975, the Church Committee specifically questioned Breckenridge as to the Kennedy brothers' knowledge of the CIA's murder plots against Fidel. Under intense interrogation (Breckenridge's transcribed testimony, first obtained by JFK historian and researcher William Davy, runs more than 110 pages), Breckenridge testified there was no evidence of presidential approval for the assassination plots. Again and again, he reaffirmed that neither John nor Robert F. Kennedy had prior knowledge of the plots. He said that CIA officials Sheffield Edwards and Larry Houston had briefed RFK on the plots after he discovered them when CIA operative Robert Maheu's bagman was caught trying to bug Phyllis McGuire's hotel room for Giancana in 1972. Breckenridge stated that Sheffield and

Houston further lied to RFK when they told him that the CIA had terminated its assassination program. Breckenridge testified that Edwards knew that he was lying to my father when he said the Castro murder project was terminated. Breckenridge said that Bill Harvey and Richard Helms both knew that RFK had been lied to by their agents Sheffield and Edwards. Breckenridge said that Harvey and Helms also ordered that John McCone be kept in the dark. He explained that Harvey and Helms knew that McCone would order the assassination attempts stopped. As the dark wizard behind the assassination program, Helms, of course, knew, without having to read the inspector general's report, that the Kennedys had no knowledge of the plots. Helms testified to this fact himself before the Church Committee and two years before his death, he acknowledged to a Vincent Bugliosi that John and Robert Kennedy were unaware of the CIA's assassination plots. Despite the fact that Helms absolutely knew that the Kennedys had no prior knowledge of the assassination schemes, Helms's deputy, Sam Halpern, who was deep in the network himself, devoted enormous professional energies to besmirching the Kennedy family. For more than thirty years Halpern continued to spread slanderous stories, painting JFK and RFK as the masterminds of the CIA's assassination intrigues. This, despite the fact that Halpern told the CIA inspector general the opposite during his interviews for their report. Halpern was the wry and diminutive deputy to Desmund and William Harvey, first in Vietnam and later in the Western Hemisphere Division. Halpern himself was in the thick of the assassination program and bore the brunt of Robert Kennedy's bruising and berating harangues during the Mongoose era and came away scarred by the bruising my dad gave his team. "He was arrogant," Halpern said of my father. "He knew it all. He knew the answer to everything. He sat there, tie down, chewing gum, his feet up on the desk. His threats were transparent. It was 'If you don't do it, I'll use my brother.'" He later acknowledged that he would have considered murdering my father had he confirmed the CIA suspicion that operation Mongoose was just "busy work."

Halpern was one of journalist Seymour Hersh's key sources for much of the misinformation about my family in Hersh's 1997 Kennedy hatchet job, *The Dark Side of Camelot*. Historian David Talbot, who has written a book about the Kennedy presidency (*Brothers*) and the CIA during the Cold War (*The Devil's Chessboard*), interviewed Halpern before his death in 2007 "Halpern had made a career of contaminating the Kennedy name. He was doing it even after he retired from the CIA," Talbot told me. "He rolled out all this nonsense that was demonstrably untrue. He clearly had an agenda." Hersh is a brilliant, Pulitzer Prize–winning reporter whose best journalism emerged from his inside access to military and intelligence sources. His inside track to the CIA helped him break the My Lai massacre story in 1968, and to reveal the George W. Bush administration's machinations to "fix the intelligence" to justify the Iraq War in 2003. But there was a downside to his exclusive relationships, which Hersh readily acknowledged. "I'm a prisoner to my sources," he mused. Mixing tantalizing confidences with malicious falsehoods against the Kennedys allowed Halpern to manipulate writers—even seasoned, brilliant journalists like Hersh—like sock puppets to besmirch the Kennedys. Talbot meticulously obliterates Hersh's key evidence supporting his premise that the Kennedys orchestrated the CIA murder plots on Castro. Hersh bases his thesis on Halpern's claim that my father deployed the late Charles Ford as his liaison with Mafia assassins that he was sending to kill Castro. However, Talbot dug up Ford's memo to the Church Committee. Ford specifically rebuts Halpern, telling the committee that the subjects of his meeting with my father in Cuba were "the efforts of a Cuban exile group to foment an anti-Castro uprising, *not on Mafia assassination plots*" (italics added). Talbot, correctly concludes that Halpern "fabricated this story about Bobby Kennedy and the Mafia. . . . Officials like Helms and Halpern tried to deflect public outrage over their unseemly collusion by pinning the blame on the late Attorney General." As Jim DiEugenio has observed, "Halpern should have already been suspect to Hersh because he is listed as a witness in the CIA Inspector General report (which exonerates RFK and JFK in any involvement or pre-knowledge of the assassination plots)." Halpern was charged by the CIA with producing a report on the activities and history of the Agency's subsequent Operation 40 Death Squad. The CIA has, to date, refused to make public the report despite the JFK Assassination Record Review Act (ARRA), which mandates its release. The man who appointed Halpern to that post was Richard Helms—"the man who kept the secrets." It wasn't just Hersh who appears to have been bamboozled. Halpern had automatic credibility with writers as a former *New York Times* journalist before joining the CIA. Halpern was the *Times'* Havana bureau chief. Using his control over previously secret CIA files as a "dangle," Halpern helped spawn a three-decade publishing boon of "CIA insider" books, all suggesting Kennedy involvement with all sorts of dark shenanigans. (The "tell" is usually some mention by the author of Sam Halpern in the acknowledgments. The reader may then expect malicious slanders about the Kennedys within.)

Halpern is a source for many widespread slanders about my family now attached to the American national consciousness like barnacles on steroids, having infiltrated both liberal and conservative thought. JPK's bootlegging, his involvement with mobsters; Giancana's fixing of the 1960 election; the Castro assassination plots were all the pap of Halpern's fabrications. According to presidential researcher and author Lisa Pease, Halpern was the spear tip of a broader CIA agenda to besmirch the Kennedys. "The CIA ran a well-organized disinformation campaign against the Kennedys, by seeding its documents with hints that RFK knew of the assassination plots. But on investigation, none of the documents show anything of the sort!"

Of course, the CIA campaign has not been a good thing for my family. It's bad for our country as well. That "fake history" has the effect of lowering our moral standards. It's a natural step to rationalize that if liberal icons like the Kennedys were murdering foreign leaders and overthrowing governments, then liberals are in no position to object.

In an era when the CIA's power and technical reach have

multiplied, it is worth reflecting here just how far the CIA was willing to stray in its bloodlust for Castro. The Agency had lost all its moral bearings, all sense that our value system and honor are the only essentials that make America, America.

It was not just the CIA that was willing to trample American values to murder or oust Castro. The Joint Chiefs were equally unhinged. And the military submitted a competing proposal that, while less surgical than the CIA's, included its own portfolio of cloak-and-dagger shenanigans—including murder.

On March 13, 1962, the chairman of the Joint Chiefs of Staff, General Lyman Lemnitzer, unveiled a blueprint for ousting Fidel to be managed by the Defense Department intelligence agencies. The brass intended the escapade, code-named Operation Northwoods, to create a provocation to justify a U.S. invasion of Cuba. The plan, which proposed all kinds of immoral monkeyshines, unmistakably criminal under U.S. and international law, would be risible had it not been offered by the highest officials in the U.S. military with lethal zealotry. The Joint Chiefs' proposal illustrates the perilous waters through which Jack was then navigating, and how badly the American military leadership had lost its moral bearings in the national security madhouse of the Cold War.

The generals offered a menu of "well-coordinated incidents . . . to take place in and around [the U.S. Marine base at] Guantanamo to give genuine appearance of being done by hostile Cuban forces." These provocations would give the United States a credible excuse for invading Cuba.

"A 'Remember the Maine' incident could be arranged in several forms: We could blow up a U.S. ship in Guantanamo Bay and blame Cuba. We could blow up a drone (unmanned) vessel anywhere in the Cuban waters. We could arrange to cause such incident in the vicinity of Havana or Santiago as a spectacular result of Cuban attack from the air or sea, or both."

The generals' backup bid was that the U.S. government gin up a terrorist campaign on American soil to create popular clamor for military action against Cuba. The top-secret memo suggests that

"We could develop a Communist Cuban terror campaign in the Miami area, in other Florida cities and even in Washington. The terror campaign could be pointed at Cuban refugees seeking haven in the United States. We could sink a boatload of Cubans en route to Florida (real or simulated). We could foster attempts on lives of Cuban refugees in the United States, even to the extent of wounding in instances to be widely publicized. Exploding a few plastic bombs in carefully chosen spots, the arrest of Cuban agents and the release of prepared documents substantiating Cuban involvement also would be helpful in projecting the idea of an irresponsible government."

Lemnitzer and his staff wanted to make sure that they, and not their competitors at the CIA, would get all the credit for this piece of genius. He impressed upon Secretary McNamara that "a single agency will be given the primary responsibility for developing military and para-military aspects of the operation." Lemnitzer assured President Kennedy that his proposal had the unanimous support of the Joint Chiefs. Three days after the submission, JFK told Lemnitzer he could not foresee any circumstances "that would justify and make desirable the use of American forces for overt military action" against Castro. Despite this rebuke Lemnitzer, on behalf of the Joint Chiefs, continued pressing for a military invasion. JFK finally dismissed him in September 1962.

———

THAT OPERATION NORTHWOODS MEMO SHOULD SERVE AS A WARNING to the American people about the dangers of allowing the military to set goals or standards for our country.

CHAPTER 5

● ● ● ● ●

Hickory Hill

AFTER MY FATHER FELL FOR MY MOTHER, THEY EMBARKED ON WHAT Arthur Schlesinger described as "one of the great love stories of all time." My dad loved Ethel Skakel's fiery spirit, and was intensely proud of her reckless competitiveness and athleticism, her self-confidence, her humor, and her peculiar blend of deep religious faith and mischievous irreverence. Her fearless, fun-loving, outgoing personality perfectly complemented my father, providing encouragement to a man who was inherently quiet, vulnerable, and shy. Her devotion to him became a platform for his growth as a public leader. Where Jack was detached and deliberative, my father burned with passion. She fueled those flames.

My parents were so affectionate with each other that they often appeared pasted together, arms draped over each other's shoulders, kissing and calling each other "sweetheart," "darling," and "honey." I had an early allergy to corniness, and these antics would have put me in anaphylactic shock had they not been so casual and adorable. They held hands when they ice-skated or walked on the beach. On river trips they lay against each other beside the campfire. He proudly introduced her at his speeches, and she always took the front row and hung on every word, even after hearing the same stump speech a hundred times. Her disciplined attention to his talks always impressed me. I thought, "She must have learned that in Catholic school, where they rapped your knuckles for wavering attention." She sat behind him in the motorcades, and he would look around whenever he lost track of her. He told reporters that his greatest achievement was "marrying Ethel."

Journalist Andrew Glass, a close friend of my father, once said: "Bobby had a Calvinistic moral sense. He really believed in absolute right and wrong, and this strict code guided his moral life." He and my mother shared that Manichean worldview. Jack, by contrast, was dispassionate and judicious; he always assumed that the other guy had a point of view, and maybe a good reason for holding it. My father was doctrinaire. He saw the world as a place where good battled evil, yet next to my mother, he seemed equivocal. While my father might grant the benefit of the doubt to his enemies and critics, my mother would be out on the battlefield, slaying the wounded. Reporters knew to steer clear and dodge her calls after writing something uncomplimentary about my dad. Even when my dad refused to say bad things about J. Edgar Hoover or Lyndon Johnson, my mom was quick to the barricades and unforgiving. She deked when Lyndon tried to kiss her cheek at my father's burial. When Eugene McCarthy tried to console her after my father's death, she brushed him aside, still hot from his earlier slights.

From their first encounter on Mont-Tremblant, my mother fit in with the Kennedys. Lem Billings described her as "more Kennedy than the Kennedys," and my father's siblings embraced her like no other in-law. Coming from a family that favored boys, my mother was impressed by both the strong, independent Kennedy women, and the respect their brothers and father accorded them. "It was remarkable to me that in this family, the girls' opinions counted as much as the boys'," she recalls. Unlike other Catholic fathers of his time, Grandpa Joe insisted that all his daughters attend college and find jobs, rather than wait for a husband. My mother says, "They were way ahead of their era when it came to expecting a lot from the girls. Jack always said Eunice would have made a better president than he. The truth is that he honestly believed it!"

My mother and father married at a lavish wedding in Greenwich in June 1950. My father had twenty-eight ushers, including his brothers and teammates from his Harvard football squad. My parents honeymooned in Hawaii, making a special pilgrimage to the leper colony on Molokai, for mass at the chapel of Father Damien, the Catholic saint who had contracted the crippling disease himself while ministering to his flock. Afterward, my parents flew to California and drove cross-country in a convertible they purchased from Pat and Peter Lawford, settling down in Charlottesville, Virginia, where my father attended the University of Virginia Law School. My mother tried cooking, but fortunately for their union, her mastery of the epicurean arts was not the reason he fell for her. Her cuisine must have been inedible, because my spartan and penurious father quickly agreed to hire a cook.

At UVA my father founded the Student Legal Forum to recruit outside speakers each month. In 1950 he hosted the liberal Supreme Court justice William O. Douglas, Republican leader Thurman Arnold, Franklin Roosevelt's antitrust czar, and Seth Richardson, the ardent anticommunist who chaired the Subversive Activities Control Board. My grandfather gave his "Fortress America" speech as my father's guest of the Legal Forum. Then my father set off a bombshell by inviting UN ambassador Ralph Bunche, a black man, to speak. Segregationists at UVA denounced my father for "bringing a nigger" to campus in violation of state laws, which forbade racial mixing in public gatherings. Bunche, a Nobel Peace Prize laureate was insisting his university audience be integrated. The university pressured my dad to withdraw the invitation.

In response, my father drafted a legal brief, citing a Texas case in which the U.S. Supreme Court had exempted college and university audiences from state segregation laws. Then he organized a petition drive and browbeat his fellow law students into signing. Many of his classmates who had supported Bunche's appearance in principle, refused to sign; for aspiring southern barristers, endorsing the appeal would have amounted to professional suicide. Their reticence infuriated my father, who branded it moral cowardice. In the end, Bunche spoke to the first integrated audience in the university's history. He slept in my parents' home. "It was the safest place for him to stay," my mother told me. "We had just one bedroom, so I made a room for him in the attic. He was charming and noncomplaining and wonderful." Rednecks surrounded the cottage, chucking rocks, bricks, pottery, fruit, eggs, and more malevolent missiles, including a Molotov cocktail. For the first time in her life, my mother saw the ugliness of bigotry. "I got my first real taste of what black people in our country had to go through."

Having stirred up campus conservatives, my father then alienated liberal students by inviting Joseph McCarthy, the powerful red-baiting Wisconsin senator, to speak. McCarthy was one of the few Irish Catholics on the top rungs of political power in Washington. He was high-spirited and fun-loving, and the Kennedys admired his physical courage. A boxer at Marquette University, Joe McCarthy had guts and a willingness to stand up against the most powerful opponents. He knew Jack from the South Pacific, where he had served as a bomber squadron tail-gunner and had visited the Cape, where he played touch football with respectable recklessness. He most impressed the young Kennedys by gamely joining them in diving from their sailboat onto a towline, even though he couldn't swim; he nearly drowned.

It has been widely reported that McCarthy was godfather to my eldest sister, Kathleen, who was born in 1952. That rumor is not true. Her godfather was Father Daniel Walsh, a Manhattanville College theology professor and mentor to the Trappist monk Thomas Merton.

After my father graduated from UVA Law School, he took a job at the Department of Justice for an annual salary of $4,200, and the couple rented a small Colonial in Georgetown, a few blocks from Uncle Jack. In my father's very first case, he prosecuted a close circle of President Harry Truman's cronies for skimming money from a freezer chest business. The DOJ sent my dad to New York for the trial, where he worked for the Eastern District in Brooklyn under U.S. attorney Frank Parker. During this case my father first learned how organized crime ran protection rackets, bribed judges and political officials, and infiltrated labor unions. Uncovering the existence of a well-organized criminal consortium infiltrating legitimate business and perverting the justice system fortified his sense that life was a clash between good and evil. My mother was proud when my father won a conviction against Truman's pals; other politically ambitious young DOJ lawyers had dodged that political hot potato.

During the trial the young couple lived at Grandpa Kennedy's apartment on Park Avenue, with its big sunny rooms beloved by my mother. She attended mass daily. "I'd just walk a couple of blocks to St. Patrick's, then I'd go have lunch at Howard Johnson's. I sat at the counter, reading the *New York Times* and the *Daily News*. I felt very grown-up." Having previously limited her reading to religious works and racetrack tout sheets, now she adopted a lifelong habit of studying two or three newspapers a day. When we were growing up she read the *Washington Post*, the *Daily News*, and the *New York Times*, finishing the *Times* crossword at warp speed as she waited in carpool lines. She read *Time, Newsweek,* and three to four books every week—mainly mysteries, spy novels, and biographies.

When Uncle Jack ran for James Michael Curley's vacated congressional seat in 1946 (Curley was in the Danbury Penitentiary), my mother campaigned for him in Boston. Unlike my father, she was a natural at retail politics, provided she herself was not in the spotlight; microphones made her stupid. She campaigned door-to-door in the blue-collar, minority, and ethnic neighborhoods of Cambridge, Brighton, Brookline, and Chelsea. Then, as now, she loved to talk to people. She was naturally curious and had the Kennedy habit of interrogating strangers, carpet-bombing them with nosy questions about their lives and their opinions. By the end of the campaign she had banged on nearly every door in Brookline and climbed every three-decker in Charlestown. "We would drive up to Boston and canvass door-to-door, then lick stamps. I thought, *This is so exciting!* I was rubbing elbows with people I'd never encountered. A lot of them were minorities and financially challenged people. Best of all, we won!" The poverty and prejudice she encountered challenged assumptions gleaned from her revered father about the evils of government intervention. "For the first time I started thinking, during this campaign, 'These are people who face a lot of hardship and struggle, and if the government can help, then that's a good thing.'" And she loved the diversity: "It was fun mixing it up and all being on the same team."

In 1952, my mother took a much greater role in Uncle Jack's statewide senatorial campaign against the seemingly invulnerable twelve-year incumbent Yankee blue-blood pillar, Henry Cabot Lodge. She, Aunt Pat, Eunice, and Jean hosted their famous Kennedy ladies' tea parties, often with my grandmother, in towns across Massachusetts. Ladies from blue-collar families in Fitchburg, Leominster, Fall River, and Amherst would put on their Sunday best for tea with the Kennedy girls. They each hosted up to nine tea receptions every day, serving a total of seventy-four thousand ladies over the course of the campaign. My mother went into labor while she was giving a speech in Fall River. Undeterred, she drove to Boston, delivered my brother Joe, and returned to the campaign later that week. With the exception of another quick break to christen Joe in Hyannis, she worked until election day. Republican senator Joe McCarthy also contributed. Due to his friendship with Grandpa and the kids, McCarthy agreed to stay out of the state at a time when his immense popularity gave him the power to flip elections, particularly in a heavily Catholic state like Massachusetts. Boston political boss Paul Dever commented that McCarthy was the only man alive who could have whipped the popular Cardinal Cushing in a head-to-head race in Cushing's home neighborhood in Southie. Jack's Senate win was one of the few Democratic victories during Eisenhower's 1952 Republican national landslide. Jack won by a margin of 75,000 votes and Lodge later claimed that the Kennedys drowned him in "75,000 cups of tea."

After Jack won his Senate seat, McCarthy, a gifted soapbox orator, became chairman of the Senate Permanent Subcommittee on Investigations of the Committee on Government Operations, a platform he would use to destroy thousands of lives with reckless accusations and poisonous fearmongering. Before McCarthy began his reign of terror, my father, against Jack's advice, applied to work for him as chief counsel, but McCarthy had already given the job to Roy Cohn, the former New York prosecutor who had sent the Rosenbergs to the electric chair for handing atomic bomb plans to the Soviets.

My father accepted the job as deputy counsel instead, but quickly found he could not work with Cohn. The two men detested each other on sight. While Cohn, an anti-Semitic Jew and closeted gay homophobe, was ginning up a witch hunt against gays and supposed Communists at the State Department, the Pentagon, and the Voice of America, my father kept his distance, successfully prosecuting U.S. shipping magnates for illegal profiteering during the Korean War. Those victories confirmed his innate gifts as a skilled, meticulous, and tough investigator. Prior to the commencement of the notorious Army–McCarthy Hearings, my dad resigned in disgust. He later explained that he had lasted a mere six months because "most of the investigations were instituted on the basis of some preconceived notion by [Cohn] or his staff members, and not on the basis of any information that had been developed."

My father returned to McCarthy's Senate committee in February 1954, a month after my birth, this time as counsel to the Democratic minority, where he would help orchestrate McCarthy's downfall. The carnival that followed—the thirty-six-day Army–McCarthy Hearings—glued twenty million Americans to their televisions, where they witnessed the epic rise and precipitous fall of an American populist firebrand. The Army countered McCarthy's allegations that the Pentagon had come under Communist influence by charging McCarthy and Cohn with abusing their power to procure cushy treatment of Cohn's intimate friend, G. David Schine, a Korean wartime draftee. My mother was in the front row of the Caucus Room every day with Kathleen and Joe, and occasionally me. My dad handed his questions to Senators John McClellan, Henry "Scoop" Jackson, and Stuart Symington, in order to bait a seething Cohn, who channeled McCarthy's gift for incendiary demagoguery to his own wicked purposes.

Cohn would later complain about my father's prodding of Jackson, who relentlessly dissected David Schine's plan for reorganizing the American military, finding something hilarious in each recommendation: "Every time Kennedy handed him something, Senator Jackson would go into fits of laughter. I became angrier and angrier as the burlesque continued." Finally Cohn snapped. In the hearing room lobby, during a break, he threw a wild punch at my father, who dodged the blow before two spectators grabbed Cohn and pulled him away. To his credit, Cohn later acknowledged that he would not have fared well in a fistfight with my dad. According to observers, my mom needed to be restrained from getting her own licks in. The following day, the New York *Daily News* ran the headline "Cohn, Kennedy: Near Blows in 'Hate' Clash."

In December 1954, a special Senate committee censured McCarthy—my father helped draft the resolution—for conduct unbecoming a senator, and stripped him of his chairmanship. McCarthy left the Senate and died a pariah in 1957, of alcoholism and despair. My father described him as an "ironic mixture of extraordinary kindness and generosity combined with cold cruelty." Despite the universal revulsion for McCarthy among Democrats, my father nevertheless flew to Appleton, Wisconsin, for McCarthy's funeral. Some liberals mistrusted my dad until the day he died for his association with McCarthy. When asked about the relationship, my father usually avoided the longer explanation and said simply, "I was wrong."

In 1955, a year after my birth, Aunt Pat, Aunt Jean, and my mother traveled to Europe, rendezvousing with my father in Leningrad. He had gone ahead on a trek across Soviet Asia with William O. Douglas. There my father's armor of orthodox anticommunism would be breached for the first time when he spoke with a proud woman soldier who had battled the czar, fought Hitler's armies, and now ran a two-thousand-person communal farm. She was tough, ideal-istic, and immensely proud of the opportunity that communism now promised to former Russian serfs who had suffered unspeak-able deprivation under both the medieval feudalism of the czars and the occupation by Germany. By an odd serendipity, my brother Douglas met the same woman forty years later, during his own hike across Azerbaijan. She recalled the much earlier visit by the two Americans, but had no clue that the young man she debated in 1955 was the assassinated brother of President Kennedy. Tears filled her eyes as she connected the dots.

In old St. Petersburg, the Soviets permitted the Kennedy group to view the Hermitage, which had previously been off-limits to all Westerners. My mother packed a tiny camera in a corsage provided by the CIA, and clandestinely snapped pictures of art masterpieces looted by the Nazis and later confiscated from Hitler's officers by Stalin's Red Army. For this minor intelligence coup she might have been jailed in a gulag, or even hanged, but she undertook the peril-ous mission nevertheless, probably motivated as much by her natu-ral inclination for mischief as she was by patriotism. "I took about a hundred pictures," my mother said. "When they debriefed me back in Washington we looked through all the photos. They had no idea about the extent of the Russian collection. They were all very excited, so that made me happy."

With Uncle Jack in the Senate, my father began his new job as chief counsel to the Senate Rackets Committee, a post he held from 1956 to 1959. He received national attention and used the Senate spotlight to grill crooked labor leaders like Teamster chief Dave Beck and his subordinate, Jimmy Hoffa, in the same Senate cham-ber where he had earlier confronted Roy Cohn. Grandpa Joe had warned my dad to avoid this crusade, concerned that an investi-gation of organized labor would kill Jack's presidential ambitions. "It was a screaming match," my aunt Jean Smith told me. "It was the worst disagreement they ever had." My father bucked his fa-ther and, worse, persuaded Jack to join him. My dad believed the "enemy within"—a dark force infiltrating American politics and business, unseen by the public, and out of reach of democracy and the justice system—posed a greater threat to our country than any foreign enemy. The union-Mafia nexus was a metastasizing dry rot undermining the hallowed halls of democracy. Swarms of reporters, television crews, and photographers jammed the hearing room. My mother attended every committee hearing for two years, often with us in tow. She would plant us in the front row and quietly explain the proceedings, while Uncle Jack sat abreast of committee chairman John McClellan, with my dad to his left. Sometimes Aunt Jackie and Uncle Teddy, who was still attending law school at UVA, would join us.

Jimmy Hoffa's cagey attorney, Edward Bennett Williams, my dad's principal nemesis, later enjoyed the distinction of being the only enemy of my father who became a friend of my mother. At the time, however, the relationship was stormy. When my father, as attorney general, eventually arrested Hoffa, my mother attended the arraignment. Williams asked her, "Why aren't you at home with your children?" My mother returned that slight by getting Williams fired from his job as Georgetown University's attorney. They later reconciled, but only after Williams became co-owner of her beloved Washington Redskins. Her tribalistic impulses would not allow her to both love the Redskins and loathe their owner.

As the Senate hearing expanded to examine the Teamsters' relationship with organized crime, the country was enthralled by its first exposure to the underbelly of the Mafia, whose existence FBI Director J. Edgar Hoover had long denied. His biographer offers many reasons to explain Hoover's reluctance to pursue the mob. Hoover was a gambler who enjoyed hobnobbing with hoodlums at their race tracks, resorts, and restaurants. Hoover gambled illegally in Washington, D.C., and timed his annual inspections of FBI field offices in Florida, New York, and California to coincide with horse-racing season. Furthermore, Hoover had no desire to concede the cancerous expansion of a ubiquitous, well-operated, national criminal syndicate during a time when he was carefully nurturing his image as the world's greatest crime fighter. Hoover preferred his FBI agents chasing bank robbers, car thieves, and white slavery rings, with their successes readily tabulated and touted in the FBI's annual crime data reports. Moreover, his clean-cut, fastidiously dressed, corn-fed, midwestern agents were ill suited for unknown work among mean-streak ethnic gangsters. Finally, Hoover wanted to avoid the inevitable corruption associated with the drug and gambling rackets and preferred that local police forces occupy the front lines. His solution was to claim that the Mafia was a myth. My father's hearings put a humiliating lie to that assertion.

Seated in the front row between Joe and Kathleen, I watched as my father summoned a parade of colorful villains to the witness stand: Anthony "Tony Ducks" Corallo; Joe "Little Caesar" DiVarco; Carlos "The Little Man" Marcello; John "Johnny Dio" Dioguardi, who punched a reporter during one session; and the murderous mob enforcer Crazy Joey Gallo, who, appearing in my father's office in a black shirt and black suit, knelt to feel the carpet and pronounced it "nice for a craps game." He then frisked a subsequent visitor, explaining, "No one is going to see Mr. Kennedy with a gun on him. If Kennedy gets killed now everybody will say I did it, and I'm not gonna take that rap."

Pointing at Teamster bosses Jimmy Hoffa and Dave Beck, and their shady mob associates, my mother declared, "Those are the bad guys." She didn't need to tell us that my father, Uncle Jack, and dour old Senator John McClellan were the good guys. When the gangsters pleaded the Fifth Amendment, my mother explained this was proof positive of their chicanery; Gallo "took the Fifth" 120 times, and from my front-row seat, I heard West Coast Teamster chief Dave Beck invoke the privilege when my dad asked him if he knew his own son. Beck went to prison based on the evidence uncovered during the committee hearings, and his tough, pugnacious five-foot-one underling, Jimmy Hoffa, took over Beck's vacated job. Hoffa's crime partner, Sam "Momo" Giancana, capo of Al Capone's Chicago outfit, took the Fifth thirty-three times. The following exchange was typical.

MR. KENNEDY: Would you tell us, if you have opposition from anybody, that you dispose of them by having them stuffed in a trunk? Is that what you do, Mr. Giancana?

MR. GIANCANA: I decline to answer because I honestly believe my answer might tend to incriminate me.

MR. KENNEDY: Would you tell us anything about any of your operations, or will you just giggle every time I ask you a question?

MR. GIANCANA: I decline to answer because I honestly believe my answer might tend to incriminate me.

MR. KENNEDY: I thought only little girls giggled, Mr. Giancana.

IN 1997, INVESTIGATIVE JOURNALIST SEYMOUR HERSH PUBLISHED AN outlandish story claiming Giancana had killed President Kennedy in revenge for JFK's failure to honor a "marker" owed by his father, Joseph Kennedy, to the Chicago gangster, for fixing the presidential election in Chicago. My grandfather, the yarn goes, sought the assistance of Giancana, whom he knew from his fictional bootlegging days with Frank Costello. According to historian David Talbot, the specious claim was Sam Halpern's contrivance. (As noted earlier, Halpern was a CIA official—Richard Helms's deputy—who took on the career-long task of tarring the Kennedy name.) Momo Giancana, a man known more for his braggadocio than for his honesty, told a less-detailed, self-serving version of that fable to his associates, prior to his 1975 murder. The absurdity of Halpern's fairy tale is evident to anyone reading my father's colloquy with Giancana, which occurred in 1957, when my uncle had already made his decision to run for president. Is it possible that my father—his brother's "control-every-detail" campaign manager—would authorize his father to conspire with a fanatically vindictive villain to commit a federal crime and fix a national election, having himself, only a few months earlier, not only picked a fight with this murderous gangster but also humiliated him on national TV? The scenario gets even more risible when one considers that my father, days after his appointment as attorney general, far from seeking a truce, made Giancana his top mob target. My father assigned a team of his top DOJ attorneys to put Giancana in jail, and placed a squad of the meanest G-men in the FBI to openly tail him 24/7. And is it believable that my grandfather, with his abiding sense of propriety, would have gone behind the back of his two sons to negotiate a reckless scheme with their sworn enemy—a caper that could have brought down the entire campaign if Giancana chose to expose it? The answer, of course, to all these questions, is no! In later chapters we shall see more of the CIA's role in spreading these fabrications.

AFTER TWO YEARS OF HEARINGS, THE COMMITTEE FOUND EVIDENCE of corruption in fifteen unions and fifty companies. Dave Beck went to jail, and the AFL-CIO expelled the Bakery and Confectionery Workers Union. My father received a steady stream of grateful letters from union members and others around the country who had been victimized by the Mafia; finally someone was standing up to the organized crime bosses. Although the committee also uncovered enough evidence to send Jimmy Hoffa to prison, the Eisenhower Justice Department refused to prosecute; Hoffa enjoyed a close friendship with Eisenhower's vice president, Richard Nixon, and had committed the Teamsters Union to endorse Nixon in the 1960 election. My father's Justice Department's "Get Hoffa" squad finally jailed Hoffa for jury tampering in 1967, three years after my father resigned as attorney general. President Nixon pardoned the imprisoned labor leader in 1971, after receiving large contributions from his handpicked successor, Frank Fitzsimmons, on Hoffa's instructions. Hoffa disappeared in 1975, soon after his release from prison.

In January 1960, after Uncle Jack announced his run for president, my dad left the Rackets Committee to run his brother's campaign.

Hickory Hill

When I was three, my family moved from Georgetown to Hickory Hill in McLean, Virginia, across the Potomac from the capital. The antebellum estate was formerly home to Robert E. Lee's father, "Light Horse Harry" Lee, a Revolutionary War hero, and later to General George McClellan, the Union Army's supreme commander during the early years of the Civil War. Supreme Court Justice Robert H. Jackson, who presided over the trials of the Nazis at Nuremberg, later acquired the house and sold it to Uncle Jack and Jackie. When they opted for a smaller Georgetown residence, my mother bought Hickory Hill from her brother-in-law in the spring of 1957. I grew up in that house, along with my elder brother Joe, my sister Kathleen, and our eight younger siblings.

Hickory Hill was a five-story white brick mansion, with a charcoal slate roof and a sprawling stone patio shaded by a grove of the oldest oaks in Virginia; a plaque erected by the Daughters of the American Revolution attested that the trees first sprouted in colonial times. My dad affixed a rope to one sweeping branch, allowing us to swing from the patio over a sloping expanse of clover and Kentucky bluegrass. At the base of the hill, an iron jungle gym, a pair of swimming pools, and a tennis court separated two large fields, ideally suited for football. A large barn, with attached stables and tack shed, abutted two half-acre pastures to the south.

At my father's request, the Green Berets built a ropes-and-obstacle course in the lower pasture for charity events. Later, they built a zip line, running from a high treehouse down to a conifer at the bottom of the hill, where daring riders slammed against football pads wrapped around the tall pine; it gave an exhilarating ride—and occasional trips to the Georgetown Hospital emergency room.

In 1961 there were seven of us children, all under ten. In the morning we would gather in our father's bathroom to help him shave. At the time his face was ruddy but not yet wrinkled or craggy, and his hair was still dark and neatly trimmed. We'd slather Noxzema on him and ourselves, then he would distribute empty razors so we could practice as he demonstrated. Having inherited Grandma's love of learning and self-improvement, my father played recordings of the plays of Shakespeare on his bathroom Victrola during these familial ablutions.

Most nights our dad came home from work at the Justice Department in time to eat with us, his sleeves rolled up past the elbows, his collar open, and his PT-109 tie clip dangling from his necktie. My mother kept bags of these gold-colored tie clips to replenish those he handed to well-wishers. He wore white sneakers or black loafers and, over our mom's objections, white woolen athletic sweat socks with both. When he worked late and missed dinner with us, he would sometimes wake us on his return and play freeze-tag, or cops-and-robbers, much to our nanny Ena's dismay.

My dad and mom drove a beat-up old station wagon, and later a black Chrysler convertible, roomy enough for seven kids and assorted large dogs. My dad loved milkshakes and chocolate cake, and he drank a Heineken on weekends, or a daiquiri or bloody Mary. Occasionally he would smoke a cigar. Each evening we knelt around our parents' bed to recite the rosary and our bedtime prayers. Then they would read from the Bible. I took note of the differences between the Old Testament—with its exciting tales of wars and heroes, and its arbitrary, vengeful, jealous, and occasionally genocidal God—and the New Testament, where ethics replaced tribalism as the centerpiece of religion, and "turn the other cheek" replaced "an eye for an eye, a tooth for a tooth." Afterward, we would all pile on their bed for a tickle-tumble.

My father's three brothers and seven sons all grew to be larger than him. At six-foot-one and 190 pounds, I'm about average for a Kennedy male. My dad stood five feet ten and weighed 160 pounds; Lem joked that he was the "runt of the litter." The shortness was in his legs; he had a long upper body, and powerful shoulders and forearms. My father was a graceful athlete, with lightning speed and coordination and endurance that lent him proficiency at all sports, from football, tennis, ice hockey, and skiing, and later in his life, to mountain climbing and kayaking.

My dad's relatively small stature relegated him, initially, to sixth string on the Harvard football team. But, with discipline, effort, and his ferocious competitive spirit, he fought his way onto the first string and earned a varsity letter. Harvard's legendary football coach, Dick Harlow, praised my father as "the toughest boy" on the largest, toughest, most talented football squad ever to play on Soldiers Field. His teammates were mostly seasoned World War II veterans from Notre Dame and Wisconsin, attending Harvard on the GI Bill, which allowed America's poorest kids to attend the best universities for the first time. They were blue-collar Micks, Italians, Poles, Greeks, and Armenians who didn't fit the old Harvard mold. Many would remain my father's lifelong friends. His laconic, sleepy-eyed pal Dean Markham was, my dad noted, the meanest, toughest lineman on the Harvard squad. Markham was a Marine Corps combat veteran who later worked for the DOJ Rackets Division, and played tennis at our house on weekends. We carpooled to school with his children as well as with the children of Jack's great friend and fellow PT boat commander Red Fay.

During dinner we played Botticelli, a game that tested our knowledge of history. Then my father would pour heavy cream over devil's food cake and tell us about his work at the Justice Department; or stories of battles that changed the course of history— the Colonial skirmishes at Lexington, Concord, and Bunker Hill, and the Civil War battles of nearby Manassas and Gettysburg. He told us about the revolutionary struggles of Simón Bolívar, whose daring military campaigns freed Latin America from Spanish rule in the early 1800s. He was a good military historian and he made the tales riveting by emphasizing details of individual heroics and self-sacrifice. To spotlight our Irish legacy, our dad told us about "the Wild Geese," who played pivotal roles in many of the military engagements that shaped the map of modern Europe. There were many Wexford Kennedys among the Wild Geese driven from their homes by the conquering armies of Oliver Cromwell who pledged to drive every Catholic in Ireland "to hell or Connaught." My father loved the poem that memorialized their grim exile:

War battered dogs are we gnawing a naked bone
Fighting in every land and clime
For every cause but our own.

Irish-Americans, my father pointedly told us, boasted more Medal of Honor winners than any other immigrant group. The American Revolutionary Army was so heavily weighted with Irish volunteers that Lord Montgomery reported to Parliament, "We have lost America through the Irish." Among the most prominent Irish fighters in the War of Independence was Commodore John Barry, a Wexford seaman whose exploits won him repute as "father of the American Navy." In addition to firing our interest in the past, our father discussed issues of the day: the arms race, *Sputnik*, Cuba, and the Civil Rights Movement.

We had to read an hour each day and record in our daybooks three current events from the newspaper. Sundays were poetry nights, and we spent the hour before dinner memorizing our stanzas. We had strict quotas for watching TV and an outright ban on cartoons, both of which we successfully evaded. In one treasured moment, Grandpa caught me watching cartoons. He came steaming into the den at Palm Beach to shut off the TV but got distracted by a topical plot involving Rocky and Bullwinkle foiling the villainy of some Cold War spies. I felt like the privileged member of a mischievous conspiracy when he sat me on his lap and we watched the whole episode together. The grown-ups did encourage us to watch the news, however, and we gathered each evening for Walter Cronkite and NBC's *Huntley-Brinkley Report*.

History was happening all around us. During the years that Uncle Jack was president, our home at Hickory Hill was a satellite White House. Many of the most momentous governmental decisions were made there, by men in swimming trunks on the pool-house patio, while children played around the jukebox. The house was the daily haunt of labor leaders, U.S. marshals, religious clerics, speechwriters, Cabinet members, congressmen, and government attachés strategizing about protecting Freedom Riders in Alabama, reining in treacherous steel titans, or freeing Bay of Pigs commandos from Castro's prisons. My father's staff and advisers came to Hickory Hill on weekends to swim and work. The clanging of the iron cattle guards signaled new arrivals, in a continuous parade that began before breakfast and continued till dark, and might on any day include Cuban refugees; Peace Corps volunteers; community organizers or handicap advocates; CIA spies, plotting Fidel's overthrow; or my father's top mob investigators, Carmine Bellino, La Verne Duffy, and Walter Sheridan; civil rights leaders, including John Lewis, Ralph Abernathy, and Harry Belafonte; and a steady stream of writers and journalists like Ted Sorensen, Arthur Schlesinger, Jr., Tony Lewis, Joe Alsop; foreign students from Iran, Africa, and Japan; and Latin American heads of state. Delegations of Navajo, Sioux, Hopi, and Cherokee came frequently to see my father, bringing gifts of headdresses, beaded moccasins, and stone peace pipes. My favorite days were when Major Francis Ruddy's Green Berets brought small units of the elite Army "special force" to Hickory Hill for demonstrations; they fired grappling hooks onto our roof, and rappelled down our home's five-story north face, wearing camouflage fatigues and black greasepaint.

During the Mississippi crisis, I sat behind a sofa and listened to my dad debate with his band of brothers—Nick Katzenbach, Lou Oberdorfer, John Seigenthaler, Ed Guthman, John Doar, Burke Marshall, and Byron White—tactics for getting James Meredith admitted to Ole Miss. White, a PT boat vet, had been a Rhodes scholar, an all-American football star, and a champion ski racer, credentials my father found irresistible. While listening to their passionate debates, I often heard references to our nation's moral obligations and the judgment of history. I never heard a cynical motive expressed, nor any plan based on personal advantage or private gain. In fact, despite his reputation as a tough political operator, my father despised partisanship, which he regarded as intellectually dishonest. He made a point of including many Republicans among his closest advisers, including Byron White, John Doar, Bob Mc-Namara, Douglas Dillon, and Jack Miller. He told me, "I never vote for the party; I always vote for the man." He surprised local New York supporters in 1967 by expressing satisfaction when a Republican won a crucial election. "Good," he said. "The Democrat was a crook."

During work breaks at Hickory Hill, Supreme Court justices, Cabinet secretaries, and Justice Department aides alike were shanghaied to play tennis or football. I remember Press Secretary Pierre Salinger and Assistant Deputy Attorney General Joe Dolan taking a break for a competitive game with Dave Hackett, my father, Joe, and me, from which they returned with grass stains on their dress shirts and suit pants. Uncle Jack occasionally came by for a swim or to toss a football. He threw with smooth, athletic grace to my father and Kenny O'Donnell, as they conferred on campaign matters or pending legislation. When they played touch football, my dad called Jack "Johnny." One Saturday I watched Jack impulsively swing up onto my mother's tall mare, Killarney, and canter bareback around the yard, holding her mane and steering the horse with his knees.

By age seven it was already my habit to escape the incessant activity at Hickory Hill to spend my weekends at the barn, playing with my brothers Michael and David. Even on school days I fled there at daybreak to check on my pigeons, relishing the smell of manure and cut grass, the distant call of bobwhite quail and the dawn coo of the mourning doves, abundant in nearby fields and forests before the construction of the Dolley Madison Highway. By seven a.m. I was saddled up on my pinto, Geronimo, behind Joe and Kathleen's ponies, pursuing my parents on our daily pre-breakfast ride. We rode most days if it wasn't pouring rain. Our mother raised us on horses and had us jumping fences, hedges, and even cars before our teens. Each morning before breakfast, she and my father shepherded us on breathtaking gallops through the Virginia woods and over fences near Pimmit Run, the Potomac

River tributary that ran near our house, or in the woodlands and hills surrounding the nearby CIA headquarters. Split-rail pastures bracketed the barn, but mostly the horses ran free. Touch football games commonly required us to dodge cantering ponies. Recalling the Hickory Hill heydays, my mother said, "It was so fun when you'd be walking up to the front door and a herd of horses would come galloping by."

My mother rode beside my father, mounted on his brown gelding, Attorney General. We followed, trotting down winding trails through the woodlands and meadows surrounding the newly constructed CIA complex in Langley, and cantering across the unbroken countryside, trailed by five or six panting dogs—setters, retrievers, a giant English sheepdog, and our immense Newfoundland, Brumus, their dangling tongues dragging through the timothy-grass. (Brumus gained rock-and-roll fame when his slobbering jowls highlighted the inner sleeve of Jefferson Airplane's *Crown of Creation* album.) Sometimes, as a concession to the later stages of her persistent pregnancies—she was pregnant ninety-nine months of her life—my mother followed us in a buggy, or, in winter, on a sled pulled by Joe's strawberry-roan quarter horse, Toby. On weekends after lunch my mother and father led us on long hikes through the woods up to Pimmit Run.

Our groom and gardener, Bill Shamwell, was a kind and dignified giant. He stood six-foot-four with the broad shoulders and braided musculature of a blacksmith. During World War II, Bill had served with the Seabees in the Pacific, because African-Americans were not allowed in armed combat until 1945. When we towed the horse trailer down to Middleburg for my sister's horse shows, he would give me money to purchase lunch at the roadside diners and we would eat together outside, because blacks then were not allowed to enter Virginia's segregated restaurants. Bill taught me how to make baling twine snares and box traps to capture starlings and English sparrows, and how to trap field mice by putting cracked corn in a Coke bottle, from which, once fattened, they could not escape. Sometimes I would hide in a dark feed bin until I heard mice emerging from their hole. Using a tin can top, I would quietly block their escape and leap up to catch them with my hands, ignoring the bites they gave me with their vicious little teeth.

Our neighbor, Mr. Nicholas Zemo, encouraged my interest in racing pigeons. He lived in a ramshackle clapboard cabin adjacent to our farm, where he kept homing pigeons, fighting cocks, and hunting dogs. He started me at age eight breeding Hungarian homers for competitions. On weekends he and I would load our best birds onto trains, having asked the conductor to release them in Delaware or Pennsylvania or southern Virginia. When they returned to our coops a few hours later, we would rush to get their bands time-stamped.

In the autumn of 1963, our mother chaperoned Courtney, David, Joe, Kathleen, and me on an afternoon gallop through the wooded campus surrounding the CIA's newly constructed Langley headquarters. On the way home we cut through Mr. Zemo's farm. Zemo was a wily old Confederate with a belly hanging over his britches, a straw planter's hat, a greasy smile, and a grizzled beard. My mother considered Zemo a flimflamming blackguard, and he always struck me as a shady operator, but I was nevertheless quite fond of him, and I was grateful for the time he had spent with me and felt that special bond that pigeon fanciers feel toward other pigeon fanciers, at least the ones they do not otherwise despise. However, I was about to find out he was a true villain. As we walked our horses through his property, I was silently admiring the snarling pit bull and roosters staked to the ground outside their hutches, when we heard a baleful whinnying that made my skin crawl. At our mother's direction, Joe and I broke into Mr. Zemo's decrepit barn to find seven pitiful yellow-eyed ponies in choke collars. All skin and bones, their fetlocks were bleeding from gaping holes in the floor. I could not believe that the torment was deliberate. Yet Mr. Zemo had hung buckets of oats just out of their reach.

At my mother's direction we returned in short order, under cover of darkness, towing our little horse trailer behind the station wagon. We spent the night loading Mr. Zemo's horses, two at a time, and driving them back to Hickory Hill. We put them in our stalls with abundant feed. One of them foundered and died before morning. My mother called a vet who arrived to treat the survivors as we left for school.

The two trials that followed our sortie on Zemo's barn provided us four years of circus comedy. With an unrelenting ASPCA riding him, Mr. Zemo got a fine and a six-month suspended jail term. Mr. Zemo, in turn, pressed charges for horse theft against my mother, asking $30,000, the supposed value of the sickly nag that expired. She finally went to trial in 1967 in the Fairfax County Courthouse, before a Virginia jury. Not surprisingly, she was in foal again—this time with my youngest brother, Douglas, who would soon be scared out of her by her run-in with my coatimundi. "Horse Lover, Ethel, Faces Thirty Thousand Dollar Suit" read the *Washington Post* headline. My father, then a senator, answered press queries about the equine abduction: "If you can figure out what Ethel is going to do next, be sure to let me know." The jury deliberated for a suspenseful two hours and twenty-five minutes before acquitting her. A few days later, the president of the National Press Club introduced Teddy, in a speech, as the "brother-in-law of an admitted horse thief."

In good weather my parents didn't permit children indoors during the day, so we played football, tennis, capture the flag, or went horseback riding. We built tree houses in the magnolia, and played for hours in the hayloft, making forts from hay bales. We invented our own games, mostly involving some element of risk, like "Tag on the Roof," where we leapt from atop the barn to the tack room, toolsheds, and horse trailers' roofs, or into a neighboring white pine. In a game we called "Combat," modeled after our favorite TV show, we battled Nazis with homemade bows and arrows and slingshots, or, using tin trash-can lids as shields, we pummeled each other with the hard green fruits that littered the ground beneath groves of black walnuts. We doused each other with squirt guns improvised from DDT pump canisters, a chemical then advertised as "safe as aspirin." When the DDT foggers came to our neighborhood, we would follow in their wake, ambushing each other, like commandos, from the clouds of toxic mist.

At Hickory Hill my life revolved around the seasons. In the springtime it was rare not to come across a box turtle with its blood-red eyes and brilliant patterned shell, or butterflies of a dozen species producing explosions of color as they danced among the wildflowers. Once teeming populations of bats, honeybees, amphibians, and flying insects have now dwindled to obscurity in northern Virginia. But in those days we could capture bats by lofting a bandanna-draped stone high into the twilight sky and netting the flying mammals as they chased the floating hanky to the ground. Salamanders and frog eggs crowded every roadside ditch and puddle, transforming them into bubbling cauldrons thick with tiny pollywogs. (Today, not just butterflies, but nearly one-third of the world's amphibian species have disappeared, impoverishing our planet, and our children's imaginations.) On weekends, I wandered nearby streams with David and Michael and my sister Kerry, searching for frogs and crayfish, snakes and mudpuppies; or spent time with my little brothers, digging in the yard for Civil War relics.

In summer, honeybees covered the clover in our yard, making barefoot play a hazard. David, Michael, and I would capture a dozen or so in a jar then let them go one at a time, triangulating to track them to their hive, and then smoking them into sedation to get the honey, without too many stings. On autumn weekends we would often visit Camp David. While my dad conferred with Uncle Jack, we explored the mountain woodland, sometimes with Secret Service guards, turning over logs and rocks, capturing red and dusky salamanders.

Winter, in those days, still brought snow to northern Virginia. David, Michael, and I spent long days building bobsled runs on Hickory Hill, or skiing at a neighboring farm that operated a small ski tow. We played pond hockey, or practiced barrel jumping on skates, a once popular sport that seems to have lost its mighty grip on the American imagination. After one unusually heavy blizzard, my father invited a group of Alaskan Inuits, who were visiting him at the Justice Department, to lunch at Hickory Hill, where they built a giant igloo in our backyard. One afternoon, as I was returning home from a long horseback ride in the snow, I watched my Uncle Jack ride a toboggan down Hickory Hill, standing with perfect balance, his hands in his overcoat pockets. He reminded me of George Washington on the Delaware.

While I still keep pigeons, my obsession with birds found its focus at age eleven. After reading T. H. White's Camelot saga, *The Once and Future King,* I bought every book available on falconry—the sport of training wild hawks for hunting. I followed the exploits of the Craigheads, a family of outdoorsmen and falconers, in *National Geographic,* and I wrote Jean Craighead George, when she published her best-seller *My Side of the Mountain,* asking her how best to find a kestrel nest. (When I finally met and befriended her, thirty years later, she still had my letter.) Whenever I visited the Justice Department or the White House, I'd look for eastern anatum peregrine falcons, who for decades nested on the cupola of the old post office in Washington, D.C. When I nearly severed my foot in a 1965 accident that kept me on crutches for months, my father bought me a pet-store red-tailed hawk, as consolation. I named her Morgan, after King Arthur's half sister.

Miraculously I found a local falconer, Alva Nye, to teach me the sport. Nye, who died in 1991, lived in Falls Church, five miles from Hickory Hill. An aviation engineer employed by the Pentagon, he was one of the pioneers of falconry in America. My father knew of Nye because the State Department occasionally asked him to entertain visiting Arab dignitaries, many of whom were crazy for hawks. Somehow I acquired a copy of Emperor Frederick II's manual on falconry. Emperor Frederick was one of history's great ornithologists, and perhaps the most humane, enlightened European monarch of the Middle Ages. Frederick filled his court with writers, artists, poets, scientists, philosophers, and theologians of every religion. He spoke six languages and revolutionized the legal system, the military, agriculture, currency, and the tax system. With all of his gifts and interests, Frederick considered falconry the high-est calling, and described, in detail, the qualities necessary to be-come a falconer. These became my blueprints for living. "He must be nimble and strong, able on horseback and know the habits and habitat of each hawk and its quarry. . . . He must possess a retentive memory, and be impervious to hunger, heat, and cold. He must have sagacity, book learning, and natural ingenuity. He must have acute hearing, and be able to identify the call notes of his hawk and other avian sounds. He must be a daring spirit, not fear to cross rough and broken ground, and be able to swim unfordable waters. He should be a light sleeper so as to hear his falcon bells, and have an even temper. He must avoid inebriation. Laziness in falconers is prohibited." Frederick's inventory became my credo. My brothers and I built obstacle courses, and climbed ropes, and rode horses, and learned birdcalls, and climbed cliffs and trees, and swam across Pimmit Run, and made the two-mile swim around the Hyannis jet-ties. My aim was to mold myself into a worthy falconer, according to Frederick's prescription.

On other fall weekends I would take my brothers and Kerry to trap vultures, crouching with them for hours beneath a piece of plywood draped with canvas suspended on two saw horses and covered with roadkill and scraps of meat. Sometimes we would lure in black vultures—Virginia was then at the northern rim of their range, and to me they seemed like exotic visitors from Latin America. On Friday nights we climbed with flashlights and burlap sacks in the rickety rafters of local barns, capturing pigeons to use for trapping hawks. Then, on Saturday mornings, Bill Shamwell would drive us south to trap redtail hawks in the rural farmland, now buried under Tysons Corner Center mall and the macadam runways of Dulles Airport.

In those days, whenever anybody asked me about my future plans, I replied without hesitation: I wanted to be a veterinarian or a scientist. Perhaps because of my love for the natural world, I was even then very conscious of pollution. We called K Street "Smelly Bridge Road" because of the ever-present stench from its cement factories. Their emissions sprinkled Georgetown daily with a fine sulfuric ash that stuck to our clothes and the car. I asked my father, "Why are they allowed to do that?" I launched a weekly periodical on the environment, *The Wildlife Report*, to which Jack Walsh, the boss of my cousin Caroline's Secret Service "kiddie crew," purchased a lifetime subscription. After a few issues, the unrelenting deadlines of the newspaper racket wore me down and I quietly ceased publication; but having spent my meager earnings on comic books and crickets, I was sadly unable to refund the subscriptions. Jack Walsh, who died in 2012, used to call me once a year to remind me that he was decades overdue for the next issue.

Boston-born political columnist Joe Alsop, a towering intellect, took a special interest in me because of his own love of nature. He patiently endured long tours of my pigeon coops and reptile collection, and invited me to his Georgetown townhouse for lunch to talk about animals and pollution. At his urging, I read Konrad Lorenz, and we talked about evolution and its impact on animal and human behavior. Alsop was one of the most trusted foreign-policy advisers to Jack and my father. Jack closed his Inauguration Day with a nightcap at Alsop's house. Among the many Washington journalists skillfully cultivated by the Dulles brothers, Alsop often used his column as a bullhorn to broadcast the CIA's Cold War propaganda. Once, during a trip to Moscow, he was entrapped by the Soviets, who filmed his tryst with a male KGB spy. To his credit, Alsop forestalled the Russian blackmail attempt by dutifully reporting the incident to the CIA and the FBI upon his return.

I also benefited greatly from the attention of two other friends of my parents who shared my zoological interest. One was Marie Harriman, the wife of Averell Harriman, the former New York governor, secretary of commerce, former ambassador to Britain and the Soviet Union, two-time presidential candidate, and the leader of FDR's cabal of "Wise Men." Averell's drooping jaws and watery red eyes always reminded me of a dignified basset hound. My friendship with Marie began one autumn when she was seated with Averell on the patio at Hickory Hill, watching me fly a kestrel in the yard.

The small falcon veered off to roost briefly on Marie's head before flying away with her wig, exposing a scalp that was very nearly bald. Instead of embarrassment, she responded with unrestrained laughter, and always took a keen interest in me after that.

The other enthusiast was Lem Billings, Uncle Jack's Choate roommate, who accompanied me on long collecting expeditions across the neighboring countryside, and who would become my surrogate father and bosom buddy after my father's death.

Ena

Lem was one of two adults in my life who lavished me with attention and unconditional love. The other was Ena Bernard, our nanny. For forty-four years, Ena was a serene island in a tempestuous sea. She cooked for and fed all eleven of us, wiped tears, bandaged wounds, changed diapers, read bedtime stories, tended us in illness, and if love is not boundless, then she loved us to its human limit. Despite our pleadings, ingenuity, and trickery, she always refused to tell us her birthday, and I did not discover it until she was in her early eighties. She was born on June 18, 1908, when Teddy Roosevelt was president. She died on July 23, 2013, at age 105, having met a majority of his successors.

Ena came to work for our family in 1951, following the birth of my parents' first child, Kathleen. Seven years later, Ena was watching over eight children. I cultivated a troubled relationship with my mother from the moment I could talk back, and the conflict mushroomed into an uninterrupted parade of skirmishes during my adolescence. I believe that the unconditional love I received from both Ena and Lem helped me develop my own capacity for loving relationships, despite my waxing jihad with my mother that eventually drove me, in my early teens, from my home.

Ena slept in a narrow twin bed next to the nursery on the third floor of Hickory Hill, often with a Kennedy infant. Crucifixes and a picket line of Catholic saints—punctured and bleeding men and luminescent and unsullied women—surveilled the room from wall frames. She spent summers on Cape Cod, Easter at Palm Beach, and she raised her daughter, Fina, in our house. I know that Fina dropped me on my head the day I came home from the hospital, because she always recalled the event fondly, grateful that my father hadn't scolded her.

On occasions when we children made a ruckus, Ena threatened a "powwow," Costa Rican for corporal punishment, but it was an empty threat. She motivated us instead with joy and humor, the harmless Spanish curse "¡Caramba!" and her pleas that our mischief was destroying her health. "You children are going to kill me," she warned. "And then what's going to happen to Fina?" That prospect was unbearable; I prayed daily to God that He not take Ena. She loved each of us fervently, and we all believed ourselves to be her favorite. My brother Douglas remarked at her funeral, "Her love was as close to God's love as I think I will ever experience."

Ena was descended from African slaves. Her father, Henry Schauschmidt, was a Jamaican crew chief for the United Fruit Company's railroad line in Costa Rica. At age nineteen, following the death of her four-month-old second child, Petronila, Ena fled poverty and a violent husband to the capital, San José. In 1945 she left Costa Rica, eventually finding a job at the Costa Rican embassy in Washington. In 1951 she arrived at our Georgetown home as a housekeeper and nanny, and remained with us for five decades.

I was enthralled with Ena's tales of her Limón Province village, one of United Fruit's company towns, where she lived in a tiny wooden house perched on stilts near the jungle's edge. She described volcanoes blackening the sky, chattering parrots, and playful "monos" (monkeys), toucans, macaws, tapirs, menacing tarantulas, the frightening midnight cough of the jaguar, and the lethal fer-de-lance viper that ambushed United Fruit workers from the banana bunch, killing those who didn't quickly apply a machete to lop off their poisoned hand. Her childhood adventures, including fleeing a hungry alligator, planted in me a determination to see the jungles of Latin America. Ena was generally frightened of jungle animals, so it was ironic that she ended up working in a home filled with Central American fauna: parrots, bush babies, a coatimundi, a kinkajou, spider monkeys, anacondas and boas, iguanas, horned toads, chuckwallas, and a spectacled owl.

At night Ena sat on my bed and read, in halting patois, *The Story of Babar*, Kipling's *Just So Stories*, *Brer Rabbit*, and *The Story of Little Black Sambo*. She loved flowers, and when I came home from the forests at dinnertime, I brought her magnolia and dogwood blossoms, lilacs, daffodils, buttercups, and black-eyed susans, wrapped in honeysuckle vines. She modeled for us a long portfolio of virtues that included humility, patience, tolerance, generosity, hard work, loyalty, honesty, humor, and courage. Ena's hands were perpetually busy, and she was nearly always on her feet. We made a game of trying to trick her into sitting down to watch TV with us, but we never succeeded.

Ena kept in her room dozens of small autograph books signed by the many entertainment, sports, and political celebrities who visited our home. She was particularly excited to gather these souvenirs from African American heroes like Arthur Ashe, Muhammad Ali, Andrew Young, John Lewis, Ralph Abernathy, Bill Cosby, Nat "King" Cole, Sidney Poitier, Sammy Davis, Jr., O. J. Simpson, Alex Haley, and Harry Belafonte. When Uncle Jack visited Costa Rica in March 1963, he brought Ena along as his special guest, introducing her to the nation's legendary president, José María Figueres—a visionary JFK greatly admired. Figueres imagined his country with no army, but with good health care and education, and a fair shake for the poor; his choices helped make tiny Costa Rica the most stable nation in Central America, its economy outstripping all of its larger neighbors.

At Hickory Hill, Ena later met and befriended another Costa Rican president, the peacemaker and poet Óscar Arias Sánchez, who was pals with my sister Kerry and a frequent guest between 1986 and 1990, when he was negotiating an end to Ronald Reagan's bloody Central American wars. When Ena visited San José with my brothers Joe, Chris, and Michael in 1982, while they were opening a solar-powered hospital built by their company, Citizens Energy, she upstaged them all. The Costa Rican press greeted her like a homecoming queen, and the Costa Rican government assigned her a limousine and driver, a police escort, and arranged for her to stay in the Presidential Suite of the San José Hilton. "At the press conference," Michael told me, "the press ignored us. They only wanted to talk to Ena."

Like my parents, Ena modeled physical courage. Apart from the dogs, Hickory Hill had no security, and Ena was often the first line of defense against the stalkers, kidnappers, vagabonds, ex-cons, mental patients, and floozies who regularly wandered onto the property—and into the compound in Hyannisport. One night I watched with admiration, from a chair in the TV room, as she broke a man's hand with a firm shoulder to the front door, preventing forced entry by a pair of menacing derelicts. When a would-be rapist ventured into five-year-old Rory's room from a third-story fire escape, Rory helpfully restrained her frantically barking Irish cocker and assured the man, "Spanky won't bite," as he stepped brazenly from the windowsill to her bed and demanded, "Where's your mother?" Rory directed him to my mother's room, but Ena, awakened by the racket, appeared just in time with a Masai spear she'd hastily stripped from my bedroom wall and a giant wooden mallet. She disoriented the intruder with a hammer blow, then cornered him in my mother's bedroom and kept him at bay with some keen spear prodding until the police arrived.

In the spring of 1998, I called Ena from Turkey—from the residence of the Patriarch Bartholomew, the Pope of the Greek Orthodox Catholic Church. We had finally learned that her birthday was June 18, and I was calling to bid her a happy ninetieth. She told me her spies had alerted her that I was in Turkey, and I asked her if she wanted me to bring her anything—I suggested she might like a eunuch from the sultan's harem. She made me explain what a eunuch was, and then she started laughing, saying, "That's really funny. You better be careful. They might do that to you." I told her, "That might solve a lot of problems for me, Ena!" She laughed some more. Then she said, "Bring me a church bulletin from one of those Greek churches. That's what I'd like. I always collects them from wherevers I go." She laughed, ending with one of her endearing grammatical flourishes. "That's why I have so much junks."

"Can you believe you're ninety, Ena?" I asked. She said, "I'm really lucky. God was very good to me. I got to see all you kids grow up and have your own kids. And I saw my grandchildren grow up, too. We only live once, so we must be happy. That's why God made us, so we could enjoy ourselves and let others have fun, too. That makes God love us."

When I asked Ena to serve as my son Bobby's godmother, together with my Aunt Jackie, she hesitated. I told Ena she was a saint, which she denied, but at the christening, she broke out with a beautiful poetic prayer, evoking my father's memory, and she cast her blessings on baby Bobby's future family and on mine. I was fortunate in my life to have a great many people I could look up to, but Ena gave me the gift of unconditional love. I I knew that no matter what mistakes I made, Ena would always love me with all her heart.

The Second Most Famous Address in Washington

The Broadway producer Leland Hayward once observed that Hickory Hill would make "the damnedest musical comedy in Broadway history." Despite my mother's efforts at discipline, the place was barely managed pandemonium. At that time, McLean, Virginia, was home to fewer than two thousand people and, like the Greenwich of my mother's childhood, it was rural farm and horse country. My mom brought to it the same boisterous and chaotic gestalt the Skakels had introduced to the Connecticut of her youth. Squadrons of paid and volunteer secretaries, supervised by her friends Gertrude Corbin and Leora Mora, opened correspondence, replied to mail, planned events, and raced to keep up with my mother while she barked orders. Small armies of decorators, designers, hairdress-ers, and tailors regularly penetrated the household perimeter. A loud clanging alerted us when guests arrived over the cattle guard.

By 1963 we had accumulated ten horses, eleven dogs, a donkey, two goats, pigs, my 4-H cow, chickens, pheasants, ducks, geese, forty closely related rabbits (I started with two), and a coop of Hun-garian homing pigeons, along with the hawks, owls, raccoons, snakes, lizards, salamanders, and fish in my personal menagerie. A pair of bush babies lived under the porte-cochère, a nocturnal honey bear slept away his days in the playroom crawlspace, while my coatimundi and my giant leopard tortoise roamed free in the house. A jill ferret fed her pups under the kitchen stove. The mail-man, retreating to his car, might be chased by goats, geese, or an imposing pack of barking dogs, where he might find a sea lion loll-ing on the vehicle's warm hood, playfully slathering the windshield with a fount of fishy saliva.

Aunt Jackie captured this bedlam in one of her watercolors, depicting frolicking children pursuing a football among a herd of galloping horses, a worn-out cook leaving the driveway with her overnight bags as her replacement arrives similarly encumbered, while canines pursue an accountant, tearing at his clothes.

There was a White House hotline in a green box accessible from the shallow end of the swimming pool, and another at the tennis court, installed by the Army Signal Corps in 1961 to facili-tate instantaneous communication between my father and Uncle Jack. More than a dozen other telephones, each with five lines, were widely distributed throughout the house. There were no air conditioners, so a profusion of ceiling and floor fans moved the muggy Virginia air through every room. Cruel coils of flypaper hung from our kitchen ceiling. Crucifixes and depictions of the Madonna crowded every bedroom, along with sterling silver holy water dispensers and small altars with prie-dieu kneeling benches, china lamps, clocks, and a television. In all eleven bedrooms, bookshelves groaned beneath stacks of histories, biographies, and science books, children's books, and encyclopedias. Embroidered pillowcases and linens adorned king-size beds. Hand-painted furniture separated clusters of Queen Anne tables bearing ubiquitous sterling silver frames with pictures of the family. Wall-size mirrors festooned every bathroom.

Hickory Hill was a giant, chaotic hotel. The house had twenty-eight rooms plus a pool house, which itself housed a movie theater. Weekend dinners seldom included fewer than twenty guests. My mother made Hickory Hill Washington's chic hot spot, rivaling the White House; columnists called her parties "the best in Washington history." She filled the house with interesting and, especially, funny people—comedians such as the Smothers Brothers, Alan King, Jonathan Winters, Bill Cosby, Buddy Hackett, and Phil Silvers; and performers like Judy Garland, Harry Belafonte, Sammy Davis, Jr., Richard Burton, Andy Williams, Tony Bennett, Carol Channing, and Motown artist Mary Wells. My mother's best friend through the years was the warm, round, and gentle humorist Art Buchwald, who shared our dinner table several times a week. Other houseguests included John Lennon, the Jefferson Airplane, George Plimpton, and Jamie Wyeth, great writers like John Steinbeck, and poets Robert Frost and Yevgeny Yevtushenko, and a host of sports stars—Stan Musial, Rosey Grier, Rafer Johnson, Sam Huff, Frank Gifford, and Mickey Mantle. Both my parents had a particular love for journalists, and the guest list invariably included David Brinkley and Sandy Vanocur from NBC, Roger Mudd from CBS, Warren Rogers from *Look* magazine, and columnists like Anthony Lewis from the *New York Times* (who had gone to school with my father), Ben Bradlee, Kay Graham, Rowland Evans, Joe Alsop, Joseph Kraft, and Mary McGrory.

Building on my grandmother's popular Greenwich lecture series, my parents launched the Hickory Hill seminars, attended by members of the Cabinet, the Supreme Court, key foreign ambassadors, White House aides, and occasionally preschool Kennedys. Distinguished speakers included economist John Kenneth Galbraith; economist and political theorist Walt Rostow; and National Security Adviser McGeorge Bundy; British philosopher A. J. Ayer;

Father John Cavanaugh, president of Notre Dame; and philosopher Mortimer Adler, founder of the *Great Books of the Western World* project, who should have been riveting, but whom my mother panned with her most devastating indictment: "He talked about himself." For me, the high point of that lecture series was an evening with my hero, marine biologist and best-selling author Rachel Carson.

On weekends, the lawn at Hickory Hill became a sea of candlelit tables. Joe, David, Michael and I crafted convincing dummies in our beds and snuck downstairs to clandestine nooks in the shrubbery to reconnoiter, as our mother shattered the protocols of stuffy Washington decorum. "People were uptight and too concerned about how they appeared," my mother remembers. To cure this contagion, she coaxed notables of different backgrounds into unfamiliar situations. A wizard at peer pressure, she compelled her guests to play charades, freeze tag, and capture the flag, and join in rope climbing and push-up competitions. She had Cabinet members fence with bamboo sticks on gangplanks spanning the pool. Something in her personality persuaded them to go along with it all. When Robert Frost visited Hickory Hill after Uncle Jack's inauguration, she made him judge a poetry-writing contest among government officials and celebrity guests.

At a party for Averell Harriman's birthday the guests came dressed as the Harrimans during some episode of their eventful lives. My mother borrowed life-size wax figures from Madame Tussauds of Harriman, FDR, Churchill, and Stalin at Yalta, and placed them unobtrusively around the living room to mingle with the crowd. In a game of sardines, six-foot-six mountaineer Jim Whittaker climbed into a tiny hat cabinet above my mother's wardrobe closet. Marie Harriman, who was on the evening side of sixty-five, hid under the piano and was soon joined there by most of the president's Cabinet. Governor Harriman chose a coal bin in an unlit basement storeroom, which, according to my mother, "smelled like a dead rat." At Russian poet Yevgeny Yevtushenko's party she made Jack's Cabinet play kick-the-can in the woods. When Yevtushenko launched into a long, overwrought toast at dinner, film director

George Stevens whispered to my mother, "Can't we play 'Kick the Commie'?"

She threw parties for André Malraux, the author and France's minister of culture; for calypso star Harry Belafonte—the first man I ever saw wear a pink shirt—who played tennis with her on weekends and helped fund the Civil Rights Movement; for Frank Sinatra, Richard Burton, Marlon Brando, Randolph Churchill, Liz Taylor, and Judy Garland. She hosted a celebration for Jim Whittaker after he summited Mount Everest in 1963. She staged a blowout adieu bash for the popular departing French ambassador Hervé Alphand, and his dazzling wife, Nicole, where the tables were so densely packed that the waiters had to come through the windows. While preparing for that evening, my mother's Guyanan cook, Ruby Reynolds, got in a running catfight with the French embassy's chef, who later that evening screamed to the assembled guests, "*Elle est folle!*" (She is mad!) Ruby had locked the kitchen doors to prevent his entry. Not even my mother could coax her from her lair. The Frenchman had cooked the meal in the garage, on hastily rented ovens. Those dramatics only added to the fun of a splendid evening, that ended with everyone singing the "Marseillaise" over profiteroles and sparklers.

My mother feted British prime minister Harold Macmillan shortly after he resigned. "Everyone loved him," she recalls. "He had a great sense of history." For General Maxwell Taylor's birthday, she hung papier-mâché paratroopers from the oak and elm trees on the front lawn to celebrate his heroic D-Day jump over Normandy with the 101st Airborne Division.

When Manuel Benítez Pérez, aka El Cordobés, the famous bullfighter, was gored during a practice match in Spain and was unable to attend a gala dinner in his honor at Hickory Hill, my mother, rather than cancel the party and disappoint her guests, dressed her swarthy friend Dean Markham in an electric light-blue matador's outfit. She had his thick black hair marbled and pulled back in a bun in the toreador style, by New York's most famous hairdresser, Kenneth, and passed Markham off as the wounded bullfighter. By maintaining a sulky silence and striking heroic poses, he proved a convincing substitute. My mother's beautiful six-foot-tall secretaries, Jinx Hack and Carol Gainor, distracted the guests as bikini-clad go-go girls dancing in gilded cages.

When she hosted astronaut Scott Carpenter after he orbited the earth, Alan Shepard and John Glenn attended, and hidden projectors bounced films of rockets exploding, from NASA's early days, against the walls to emphasize Carpenter's courage. His wife, Rene, became one of my mother's closest friends and part of the Hickory Hill "mafia." Rene and my mom, similarly athletic and toned, goaded one another into miniskirts and increasingly outlandish psychedelic plastic dresses. While Jackie was a clothes horse for haute couture, my mother and Rene went to town as All-American party girls in sixties-era electric pink and green skirts and blue plastic boots.

Celebrating John Glenn's return from space in 1962, my mother feted him and his wife, Annie, with a black-tie sit-down dinner on the swimming pool patio. Vice President Lyndon Johnson was there, as were Supreme Court justices Arthur Goldberg and Byron White, along with most of the Cabinet members. Angling for trouble, my mother perched the Glenns at a table for two on a narrow platform across the deep end of the pool. When the dancing started, much of the company, predictably, landed in the water, including my mother, presidential speechwriter Arthur Schlesinger, Senator Hubert Humphrey, Press Secretary Pierre Salinger, and New York's amiable, dignified Senator Kenneth Keating. The guest of honor, with the studied agility of an astronaut, narrowly escaped a dunking. Surveying the mayhem, Lem, still dry, commented disapprovingly, "That's just what people want to see, government officials dressing in tuxedos to go swimming." When the story hit the papers, Jack shared Lem's displeasure. He gently reproached my remorseful mother, "No more pool parties."

Even with aquatics out of bounds, the fun didn't stop at Hickory Hill. My memories are a montage of magical madnesses: swimming in the pool with a seal; watching my mom ride her tall mare Killarney through the house; Tina Turner teaching my sister to dance; John Lennon and my Uncle Teddy, sitting together on a piano bench, playing and singing till two a.m., with Rosey Grier accompanying on a giant zebra-skin drum; Marlon Brando at a formal sit-down dinner, eating his food, Indian style, on the floor; breakfast with the Smothers Brothers, Alan King, Andy Williams, and Buddy Hackett, all in pajamas; Rudolf Nureyev pirouetting on the patio; playing tag on the roof with Robert Vaughn, the *Man from U.N.C.L.E.*; seeing Jorma Kaukonen and Grace Slick, of Jefferson Airplane, take photos of Brumus, and building an obstacle course in the south pasture with Joe Theismann and the Washington Redskins, armed with picks, shovels, and posthole diggers.

There were many mishaps. In 1977, my mother borrowed an African elephant from the Barnum and Bailey Circus for her annual charity Pet Show. The pachyderm—named Suzy—threw her mahouts and stampeded through 1,500 panicking guests on our lawn. A contingent of Secret Service men drew their revolvers to protect President Carter's daughter, Amy. Suzy veered from the crowd. She crashed a fence and devoured our neighbor Mr. Ornstein's raspberry patch, and returned to our home for a long draught from the swimming pool. On another occasion I watched Vice President George Bush tumble off a "Slide for Life" zip line at a Special Olympics event. A few weeks later Muhammad Ali was swept away when a tether snapped on a hot-air balloon. He finally crashed on Ballantrae Mountain a mile away, dumping him from the basket— and so on, and so on. It seemed that every day brought new adventures. At my mother's thirty-fifth birthday bash in April 1963, my father and Steve Smith wrestled a heavy box bound in red ribbon through the crowded living room. When my mom unwrapped it, out popped Gene Kelly, who swung my mother around the dance floor with his twinkling smile while the band played "Singin' in the Rain."

When CIA director John McCone opened a tiny black box at his table setting during a St. Patrick's Day dinner, there was a small explosion, a puff of smoke, and a green hand reached out and closed the lid. In the early days after his appointment, McCone became very close to my mother. He was beyond grateful for her daily visits to his terminally ill wife at American University Hospital. During the Cuban Missile Crisis, his relationship with my parents soured. Thereafter, McCone suffered a kind of nervous breakdown. He vacillated in his positions and ultimately joined the military mutiny against Jack. Neither Jack nor my dad ever trusted him thereafter. McCone became more embittered after Jack's confrontation with the steel titans. He publicly excoriated the Kennedys for being tough on Wall Street and the oil and steel industries. In a gesture of extraordinary insubordination, he openly fought Jack on the Nuclear Test Ban Treaty, lobbying Congress against the White House position. McCone failed disastrously in his charge to change the culture at the CIA, which continued to overthrow leaders of other nations quietly while Jack was president, and openly after his death.

An Agency Out of Control

One of the defining features of the Camelot years was the Kennedys' running battle with the CIA over control of American foreign policy. My most prized possession at Hickory Hill was a stuffed Sumatran tiger, mouth wide in a mute snarl that unnerved visitors to my bedroom. The tiger, which I still have today, is a reminder of the persistent enmity between my family and the CIA. It was a gift to my father from Achmad Sukarno, the fire-breathing nationalist who led Indonesia's fight for independence from the Netherlands. Sukarno was one of the world's most outspoken leaders for Third World independence. He coined the term "Third World" at the first conference of Non-Aligned Nations, which he hosted in Indonesia in 1955. Because of his gift of the tiger, and because I knew that Indonesia was home to the world's largest reptile, the Komodo dragon, I became fascinated by Sukarno.

Sukarno made clear at the outset that he would manage his country's resources for the benefit of his people. He nationalized

Dutch companies and expropriated their resources. As the Dutch withdrew, U.S. corporations, drooling over Indonesia's oil, timber, copper, and lavish mineral treasures, mobilized to divvy up the spoils. During the Eisenhower administration the CIA endeavored to dispose of Sukarno in the same manner they helped rid the world of Lumumba—by assassination. Deputy Director Richard Bissell, whom Jack later fired for his role in planning the Bay of Pigs, explained the Agency's rationale. "Lumumba and Sukarno were two of the worst people in public life I've ever heard of. They were mad dogs." Bissell argued that assassinating such leaders was "not bad morality" unless it was unsuccessful.

The CIA arranged a coup against Sukarno in 1958, supplying arms to rebellious army officers and using a squadron of disguised CIA planes to bomb Sukarno loyalists. One CIA bomber dropped its load onto a church and marketplace, killing many civilians. Indonesian gunners shot down the plane and captured its CIA pilot, forcing Eisenhower to admit to U.S. involvement in the attempted coup, which until then he had forcefully denied.

Jack wanted to see if it was possible to work with Sukarno rather than murdering him. Hoping to keep Sukarno non-aligned, Jack hosted him at a White House dinner in 1961—delighting Sukarno, who afterward muted his anti-American diatribes. Seeing the world through Sukarno's eyes, Jack remarked to a friend, "When you consider things like the CIA's support to the 1958 coup, Sukarno's frequently anti-American attitude is understandable." Sukarno somehow learned of that remark and he gratefully invited Jack to Indonesia, promising "the grandest reception anyone ever received here."

The improved relations allowed Jack to open a dialogue with Sukarno. In February 1962, he sent my dad to Indonesia, with the assignment to avert a war between the Netherlands and its former colony over the disputed West Irian region. My father brokered this settlement by persuading both sides to allow West Irian residents to vote on their future alignment with Indonesia. Sukarno was so delighted by the outcome that he commuted the downed CIA pilot's death sentence and released him to my father. He also gave my father the stuffed tiger that he himself had killed. My mother, who accompanied my father to Indonesia, was not so keen on Sukarno's personal traits. His earthy crudity repelled her puritanical streak, and the vulgar celebration of sexual promiscuity in the murals, statues, and wood carvings adorning his palace alarmed her. But Sukarno was genuinely fond of my dad and gave him his trophy Sumatran tiger, which he kept for a while in his Department of Justice office, before giving it to me.

Just before he left for that trip, I asked my dad to try to bring back a Komodo dragon. My father told Sukarno of my interest in the Indonesian reptiles. A few weeks after my parents returned home, Sukarno sent over a pair of Komodo dragons. They were full-grown adults, ten and twelve feet in length, and probably capable of eating a child my size, and certainly any of my younger siblings. Still, I was disappointed that we were not able to keep them at Hickory Hill. My dad diverted them to the Washington Zoo, where I worked after school and on weekends. I cleaned their cages and fed them frozen piglets and rats.

My father's peaceful resolution of the Irian crisis with Sukarno alarmed the CIA. Richard Bissell argued that the peace might "inadvertently help to consolidate a new regime which is innately antagonistic toward the United States." As a good-faith gesture to Sukarno, JFK pledged to increase foreign aid to Indonesia, further aggravating the spooks at Langley. He ordered his State Department and his rebellious CIA to look for more amicable ways of dealing with Sukarno. On November 19, 1963, three days before his own assassination, Jack accepted Sukarno's invitation to visit Indonesia for the following spring. He intended his junket to be the most dramatic demonstration yet of U.S. support for Third World independence. A few months after Jack's death, my father made a second visit to Sukarno, this time as an emissary of President Lyndon Johnson. On that trip he persuaded Sukarno to settle a potentially bloody border war with Malaysia. A short time later, in the thrall of an intelligence apparatus rid of Jack Kennedy and once again flexing its muscles, LBJ cast aside Jack's pledge to increase foreign aid and instead cut off all economic support to Indonesia. The CIA then proceded with a successful coup against Sukarno.

Ralph McGehee, a twenty-five-year CIA veteran, would later write about the coup in his book *Deadly Deceits*. "The Agency seized this opportunity to overthrow Sukarno and to destroy the Communist Party of Indonesia (PKI), which had three million members. . . ." The CIA's Indonesia station chief later testified that his people had assisted U.S. embassy officials in Jakarta to compile death lists of thousands of PKI members, for use by the Indonesia Army command, which systematically massacred the people, as CIA and U.S. embassy personnel checked off the names of those killed and captured. Estimates of the number of deaths that occurred as a result of this CIA operation ran from one-half million to more than one million people." A quarter century later, Robert Martens, one of the U.S. government officials who took part in the massacre, recounted the events: "It really was a big help to the army. They probably killed a lot of people and I probably have a lot of blood on my hands, but that's not all bad. There's a time you have to strike hard at a decisive moment."

The CIA replaced Sukarno with Mohamed Suharto, a rightwing, reliably pro-American dictator. Suharto established a brutal military control over every facet of political and private life in the country. He forced students to swear loyalty oaths to the government and attend mandatory indoctrination classes. The government muzzled the press and banned long lists of books. Dissenters endured death, exile, or torment on the prison island of Burk. Suharto's free-market economic policies helped run inflation rates up to 600 percent by 1966, and made this resource-rich country one of the poorest in Asia.

———

DAYS BEFORE HIS MURDER, AS MY FATHER PULLED AHEAD IN THE CALifornia polls, he began considering how he would govern the country. According to his aide Fred Dutton, his concerns often revolved around the very question that his brother asked at the outset of his presidency, "What are we going to do about the CIA?" Days before the California primary, seated next to journalist Pete Hamill on his campaign plane, my father mused aloud about his options. "I have to decide whether to eliminate the operations arm of the Agency or what the hell to do with it," he told Hamill. "We can't have those cowboys wandering around and shooting people and doing all those unauthorized things."

CIA's covert operations function—the so-called clandestine services—had worried civil and constitutional rights advocates from the moment the public became aware that the intelligence agency had mysteriously acquired powers to conduct covert operations never authorized by its charter. Critics warned that the "tail" of the covert operations branch would inevitably wag the dog of intelligence gathering (espionage). And, indeed, the clandestine services quickly subsumed the CIA's espionage function as the Agency's intelligence analysts increasingly provided justification for the CIA's endless interventions. If the clandestine operations branch wanted to overthrow a government or rig an election; if an oil company wanted to punish a non-compliant head of state or derail a young nation's democratic impulses; or if a U.S. president needed to gin up a provocation for a "splendid little war" in Vietnam, the Dominican Republic, Nicaragua, Grenada, or Iraq; the Agency could be counted on to "fix the intelligence" appropriately. Rather than evaluating the costs and benefits of its interventions, the Agency also became complicit in concealing from the American public the ruinous costs of a blowback from its covert operations. Both my father and uncle recognized the inevitable conflict of interests that arose when the intelligence-gathering function—which aims to keep the globe stable and America out of wars—becomes subservient to a paramilitary branch that actually increases its power, prestige, and resource flow from destabilization and war.

A permanent state of war abroad and a national security surveillance state at home are in the institutional self-interest of the CIA's clandestine services. My father and uncle realized that the CIA's

covert-action branch should therefore not have the ability to manipulate our leaders and public opinion by gathering, dispensing,
and organizing the intelligence that informs our foreign policy.
They understood the critical importance of separating those functions so that our leaders receive a flow of disinterested information
upon which to base rational decision making.

MY MOM AND DAD'S VISIT TO SUKARNO, IN FEBRUARY 1962, WAS PART
of a round-the-world tour, with stops in Japan, Thailand, India, and
Berlin. Setting the pattern for his future foreign travels, my father
defied embassy schedulers and ducked the official black-tie functions with the oligarchs and the power brokers, visiting, instead,
grade schools, hospitals, union halls, and universities. He walked
the streets and markets and mingled with factory workers, university students, and small merchants, and, as always, the children.
He enjoyed fiery debates with Communist students and left-wing
labor leaders following impromptu speeches at universities and
labor forums. My father readily admitted America's faults, while
reaffirming her idealism and our desire to be a force for good in
the world. His friend John Seigenthaler told me, "He loved debating the most anti-American students, particularly the most radical
Marxists. He admired them for the idealism that had radicalized
them." My father sought out the marginalized people, whom every
other politician wanted to avoid, ignore, arrest, or kill. Said Seigenthaler: "Bobby thought he could engage them, and even win them
over! He made it his practice to acknowledge America's flawed history. But he still believed we had the best, most idealistic and most
pragmatic system, and that, even with all our warts, he could win
debates about the superiority of a free, open society, on the merits.
He thought our cause was lost if we were frightened to challenge
the Communists to open discussions."

This candid hands-on and idealistic foreign policy offered a
fresh impression of the United States. The Indonesian embassy
telegraphed a summary of the trip to Johnson's State Department:

"Young audiences which were initially cold or even hostile rapidly warmed to his frank, man-to-man approach, [his] admission that all was not perfect in [the] U.S., [his] demand for [a] grown-up approach to [our] inevitable differences of opinion, and [his] assurances that [the] U.S. did not seek to impose its system on [the] world." Protesters had stoned the U.S. embassy in anticipation of my father's visit. A few days after his arrival, however, tens of thousands of cheering Indonesians followed him as he walked from the marketplace to the university. People mobbed him in the streets, reaching out to shake his hand, touch, or greet him, shouting, "Hello, Kennedy!" The embassy communiqué grudgingly concluded, "No short term visitor to Indonesia has had impact on the public that Attorney General Kennedy did during [his] visit."

My dad made the same fresh impression in Japan. While affirming America's idealism, in a series of speeches, he acknowledged our imperfections and the mistakes that left us falling short of our own ideals. His refreshing candor blunted that anti-American hostility that had forced Eisenhower to cancel a planned trip to Japan three years before. But now Japanese lined the streets by the thousands to cheer him. When fierce heckling by Communists threatened to disrupt his talk to 3,000 students jammed in a tiny auditorium at Waseda University, he invited the Communists up to the dais to debate him. Their leaders cut electric power to the jam-packed auditorium, but a quick thinking pro-democracy student, Yoshiro Mori, averted a potential calamity by climbing the dais through the darkness and leading the audience in a rousing chorus of the Waseda University fight song. When the lights returned, my father stood his ground, renewing his challenge that the Communists stand and debate him. He eventually won over the hostile elements in the audience. Even the Communists shouted friendly goodbyes as he left the hall, and enthusiastic crowds engulfed him.

In 2000, I had the opportunity to meet with the enterprising student leader, Yoshiro Mori, whose singing had saved the day, and who was, by then, Japan's prime minister. When I visited him that year to discuss environmental problems in Japan's rivers, he recounted that "Your father's willingness to stand and debate strengthened Japan's pro-democracy movements and those of us who wanted to believe in American idealism and your system of values." He sang me a few bars of the Waseda fight song and sent his regards to my mother. "I met President Clinton and President Bush and many other heads of state," he told me. "But my favorite person has always been your mother." Following my father's visit, an embassy deputy chief wrote, "The overwhelming consensus of opinion is that the Attorney General's 'whirlwind' visit was the most successful accomplishment by the United States in its postwar relations with Japan."

On the airplane home, my father reflected to Arthur Schlesinger his central revelation from his journey. I believe it remains relevant to American foreign policy today: He told Schlesinger that America could make contact with the youth and intellectuals *only* as a progressive nation. "I kept asking myself," he said, "what a conservative could possibly say to these people. I could talk all the time about social welfare and trade unions and reform; but what could someone say who didn't believe in these things? Can you imagine Barry Goldwater in Indonesia?"

When our parents came home to Hickory Hill, we put on a skit for them, on a makeshift stage in the basement, with costumes from each country they visited. Afterward, they had us sit in the chairs while they climbed onto the stage, and my mother sang "Twinkle, Twinkle, Little Star" in Japanese. Then my dad joined her in a stirring rendition of the Waseda University fight song. My mother knows the words to this day (in Japanese, of course) and will, with minimal prompting, sing the song.

CHAPTER 6

● ● ● ● ●

Attorney General

As long as there is plenty, poverty is evil. Government belongs wherever evil needs an adversary and there are people in distress.

—ROBERT F. KENNEDY

ON THE DAY MY FATHER TOOK THE OATH AS THE NATION'S CHIEF LAW enforcement officer, an Arlington County, Virginia, Circuit Court judge fined my mother forty dollars, in absentia, for driving 70 mph in a 40 mph section of the Georgetown Turnpike, and for failing to have a Virginia license, which she had forfeited following a previous encounter with the police. Despite her legal difficulties, my mother's adoration of my dad had already transmuted into a reverence for his causes that would eventually become a complete devotion to public service. But in those days politics was still something of a sideline for my mother. Her main racket was cranking out babies.

By the time I was born, in 1954, I had two siblings, Kathleen, who was older by two and a half years, and Joe, who had me beat by a year. Surprisingly, I have gauzy memories of my younger brother David's christening in Hyannisport when I was barely two years old myself, mostly because it marked the arrival of a tall, stern, pear-shaped French nanny whom we addressed as "Nu-Nu," or "Mademoiselle." Nu-Nu became a fixture in the third-floor nursery during my mother's fecund years. She wore a white uniform and capped her severe ginger bun with a traditional nurse's tiara adorned with a red cross. We soon ascertained that corporal punishment was fashionable in France, as she beat us mercilessly with wooden spoons and spatulas. In her strong French accent she menacingly swore to "red your bottom when you get home" each morning as she handed us our school lunch boxes, a warning that seemed also to serve as a formal adieu.

I don't recall much about Courtney's September 1956 birth, which took place in Washington. I do remember some excitement when my parents brought Michael home in February 1958. Joe and I marveled at his webbed feet. We were, of course, filled with envy, and surprised that our mother seemed alarmed by this genetic advance; she forbade us to discuss it. Kerry popped out in September 1959. My mom was in a Boston hotel campaigning for Uncle Jack when her water broke, and two Boston beat policemen rushed to the rescue. She remembers them as giant Irishmen, "wonderful and kind," who strapped her to a stretcher, flipped her upside down, and loaded her in a Black Maria for the race to the hospital. "It was a paddy wagon for transporting hoodlums and ne'er-do-wells to prison," my mother recalls.

Christopher emerged on Independence Day 1963 (Kathleen was an earlier Yankee Doodle Dandy baby). My mother was playing tennis at Hyannisport when the contractions began. Dr. Janet Travell, the White House physician, told her she needed to get to the hospital immediately. Grandpa ordered her to Boston to deliver. The Cape, in those days, carried the stigma of the refuge of Brahmin swanks and Swamp Yankees. "No grandson of mine is going to be born on Cape Cod," he proclaimed. Jack's Marine helicopter was on the front lawn, but the pilots weren't familiar with the route to Boston, so Teddy came off the tennis court, sweating in his white shorts. A pilot himself, Teddy directed the Marines up the coast and along the Charles River to land on the practice field beside Harvard

Stadium. From there a police car transported my mother and Teddy to Mass General, where reporters and cameramen already jammed the lobby. "Unfortunately, the police had to use their radios, so the press found out about it," my mom remembers. "That's embarrassing—when you're having a baby and everyone is there."

Eventually my mother would have eleven children, including five by C-section, and had memorized all of Dr. Seuss's standard works. "By the time Rory came around," she recalls, "everyone was pretty sick of *Green Eggs and Ham*." My grandmother Rose Kennedy, who foreswore childbearing after nine children, observed wryly, "I didn't know it was a competition." But everything was a contest for my mom, and she was a born broodmare; all that parturition would cost her neither her health nor her athletic figure, and if my father hadn't died so young, who knows how many siblings I'd have? We would all be wearing name tags at holiday gatherings.

Our father often asked us to join him at the Justice Department for lunch and, when he worked very late, for dinner. Sometimes we visited on a lark. My mother loaded seven kids into the convertible, with nearly as many dogs, for the ten-mile ride to his office across the Potomac, which took only fifteen minutes, thanks to sparse D.C. traffic in those days, and my mother's heavy foot. Our finger paintings covered his walnut-paneled walls and hung beneath the sailfish mounted over the mantel. We capered in the marble corridors, and horsed around in the secret dining room concealed behind fake panels in the rear of his office. We slid down brass bannisters and played in the elevators. We checked the teeth on the stuffed Sumatran tiger. The rivulets of saliva cascading from beneath my father's desk announced that Brumus, our 200-pound Newfoundland, was in attendance. The giant canine suffered from abandonment anxiety, and my dad couldn't bear to hear all the whining and crying when he tried to take leave of Brumus most mornings. Furthermore, Brumus often went next door to bite the neighbor's dog in revenge for being left behind. We watched our father talk on the phone and toss the football with one of his "band of brothers"—Ed Guthman, Nick Katzenbach, John Seigenthaler,

John Doar, or Byron "Whizzer" White. Our mom showed us the button my father had installed on his desk to summon FBI chief J. Edgar Hoover.

Hoover was his most troublesome employee. "The Director," as he preferred to be called, was a poster child for abuse of power. Calvin Coolidge had appointed him to head up the Bureau of Investigation, the FBI's predecessor, in 1924, the year before my father was born, and Hoover organized the Agency's growth into a massive secret police apparatus that operated in the shadows of our democratic system and routinely used dubiously obtained information to blackmail political targets, suppress dissent, and enhance his own power. The Bureau accounted for over half the DOJ's 30,000 employees and nearly half its budget. Hoover kept files and spied on nearly everyone who interested him, including his bosses and their families. One member of Congress commented that Hoover's files contained "the greatest depository of dirt in the world." William Sullivan, the FBI's number two man, called Hoover "the greatest blackmailer in history." Hoover ordered his agents to suppress political heterodoxy and liberal thought whenever it emerged. The FBI opened dossiers on citizens who dared to pen mild complaint letters to their congressmen. The Bureau infiltrated antiwar, civil rights, women's, and labor groups, opening their mail, listening to phone calls, and subjecting them to violence. Hoover spied for a decade on Albert Einstein and instructed agents that any black man was a subject of investigation. By 1961, he was already spying on Uncle Teddy, suspicious that during a trip to Latin America he had talked to "intellectuals," a term Hoover used interchangeably with "Communists."

Hoover had been the dark puppeteer behind Joe McCarthy, providing the Wisconsin senator with most of the sketchy information "Tail Gunner Joe" deployed to ruin the lives of hundreds of thousands of American citizens branded as "Reds" or Communist sympathizers. Hoover's carefully nurtured image as a national icon, his public popularity, and his cozy relationships with powerful journalists and congressmen made my father and Jack reluctant to fire him in their first term. Nevertheless, my dad regarded Hoover as a dangerous right-wing megalomaniac who saw Communists behind every progressive cause. A fervent racist, Hoover considered the American Civil Rights Movement a Stalinist plot. He refused to read the *Washington Post* or the *New York Times*, which he considered Bolshevik rags, or even the *Wall Street Journal*, which he deemed pink.

My father tried to rein in this rogue agency by severing Hoover's direct access to the White House—a sacred prerogative—instructing the FBI director that all communications with the president go through the attorney general. My dad installed the summons button, and furnished his desk with a direct phone to Hoover's office, another indignity. He also jumped into Hoover's sandbox by becoming the first attorney general to visit the FBI's local field offices and talk directly to agents, even accompanying them on raids—sometimes with local city vice cops. (One night he brought us home some trophies from a midnight raid on a Mafia gambling den—loaded dice and red-tinted glasses that magically revealed invisible symbols on the backs of marked cards. This paraphernalia proved a big hit on the Sidwell Friends School playground.) My father made it clear that the Justice Department was not a division of the FBI and that the Bureau's director worked for the attorney general. He caused a storm by demanding Hoover provide him a list of all his African American agents. There were none. My father then ordered J. Edgar Hoover to immediately hire black G-men.

Worst of all, my father ridiculed Hoover's obsessive preoccupation with the Red Menace. Despite the decline of the American Communist Party, Hoover—like the Wizard of Oz—used smoke, mirrors, and hot air to inflate the party into a malevolent bugaboo, poised to deal a death blow to American democracy; and of course only Hoover, wielding unlimited police power, possessed the genius and character needed to avert this calamity. Edgar—as he asked my father to call him—told the skeptical attorney general that the Communist Party posed "a greater menace to the internal security of our nation today than it ever has since it was first founded in this country in 1919." Maintaining this fiction required Hoover to keep secret the party's plummeting membership rolls, which only amplified my dad's skepticism. He stung Hoover publicly by commenting to the *Times* of London that the U.S. Communist Party "couldn't be more feeble and less of a threat," adding that "its membership consists largely of FBI agents."

All these insolent invasions of the director's sacred province earned my father Hoover's loathing. Hoover, who regarded himself an American hero, was indignant at his new underling status and at the disappearance of the deference he had enjoyed from previous AGs and presidents. He told his confidant, Clyde Tolson, that the three people he most despised were Robert Kennedy, Martin Luther King, and Quinn Tamm (a former dissident assistant director of the FBI). The slobbering dogs, the marauding children, and my father's casual appearance in shirtsleeves further infuriated the fastidious Hoover, who enforced a strict grooming and apparel code among his agents. We kids certainly did our part to torment the director. Left alone to play in my dad's office, we mercilessly pushed the Hoover button on my father's desk until Dad's secretary, Angie Novello, ordered us to never do it again.

There was little doubt in my mind that Hoover was aggravated when one day he walked in unexpectedly and startled me in his anteroom, where I stood perched on his coffee table, in my soaked shirt and blazer, both arms submerged in his fish tank. I felt like a bootlegger caught in the act. Nearby, my brothers Joe and David were standing on Hoover's swivel chair, attempting to crack his office safe, which they had discovered on the wall behind his desk. He ordered us out, and we departed tout de suite!

Sometimes my mother would take us through the underground tunnel linking the Justice Department to the FBI shooting range to watch G-men blast away at black silhouettes of villains. On one occasion she mischievously stopped near the director's office for a piece of hijinks; she placed an anonymous note in Hoover's suggestion box recommending Los Angeles's Chief Parker—a man despised by Hoover—to replace the intensely paranoid Hoover as FBI director. After some hasty sleuthing, Hoover sent the document up to my dad, who found the offending note—in his wife's readily recognizable Sacred Heart scrawl—on his desk when he returned to his office.

Hoover was also distressed by my father's insistence he investigate the Mafia, which Hoover had long claimed didn't exist. Hoover wasn't the only party angered by my father's relentless efforts to unravel the tentacles of organized crime. The creation by my father of an elite "Get Hoffa" squad had Jimmy Hoffa in a murderous rage. Ironically, my father's successful 1957 investigation of Hoffa's former boss, Dave Beck (I was front row at the Rackets Committee hearings to watch Beck take the Fifth 140 times in answer to my father's questions), was the catalyst that precipitated Hoffa's ascension to the presidency of the International Brotherhood of Teamsters, America's most powerful union. Jimmy Hoffa, the tough bulldog, had taken control by deploying his army of gorillas to break heads, break rules, and fix votes. Using mob muscle, rigged elections, stolen ballots, and crooked votes from "paper locals" run by Mafia capos such as New York's Johnny Dio (DioGuardi), James Riddle Hoffa, the tough, brassy high school dropout, won control of the Teamsters. Because of his well-documented corruption and mob affiliations, the AFL-CIO immediately voted to expel the Teamsters unless the union removed Hoffa from its presidency. Hoffa refused to step down. Whenever my dad drove us through Washington, he would point to the Teamsters' opulent gold-and-marble headquarters at the foot of Capitol Hill, the so-called Marble Palace. The structure was Jimmy Hoffa's lavish monument to his own meteoric rise. Once, as we drove home late following dinner at the Justice Department, my dad pointed to a lit window on the top floor. "That's Jimmy Hoffa's office," he told us. "He's still at work!" Not to be outdone, my Dad dropped us at Hickory Hill and returned to work at the Justice Department.

Hoffa's Teamsters were America's largest and most muscular union. Hoffa had judges, politicians, and business leaders on his payroll, and enjoyed the awesome power to shut America down by calling a national strike. Hoffa had corrupt deals with businesses that he squeezed for kickbacks and protection payoffs to avoid union strikes. An enforcement team of head-breaking thugs beat, and sometimes killed, union members who complained about Hoffa selling out his members. For my father, who experienced the world as a struggle between moral absolutes, the existence of a thoroughly corrupt and immensely powerful figure like Hoffa was not merely a threat to democracy. It was confirmation of the existence of Satan. Hoffa, in turn, saw my father as the pampered agent of an aristocracy of wealth with whom he had engaged in a lifelong slugfest. The imbalance of power, he felt, justified his use of dirty street-fighting tactics. "We were like flint and steel," Hoffa wrote in his biography. "Every time we came to grips, the sparks flew."

By 1962, the "Get Hoffa" squad had filed half a dozen criminal complaints against Hoffa, on charges ranging from jury tampering to taking kickbacks on pension fund investments. Hoffa had turned the Teamsters' billion-dollar pension fund into a piggy bank for the hoodlums. He regularly invested in shady deals with his Mafia partners, including Sam Giancana, Moe Dalitz, Carlos Marcello, and Santo Trafficante, to develop hotels, casinos, resorts, and golf courses from which he would collect personal kickbacks. Hoffa and his mob cronies had more to fear than jail from federal prosecutors. A conviction of the Teamsters leader would remove Hoffa from his union pension fund leadership position and cost the mob its unfettered access to the Teamsters lucre.

So Hoffa and his associates fought back. Convoys of Teamster-driven rigs would detour down the rural two-lane road in front of Hickory Hill—backfiring, gunning their engines, and grinding their gears—as we children played football on the front lawn. We knew the big semis were trying to intimidate us, and we pointedly ignored them. We often saw slow-moving black cars driven by burly goons in fedoras and flat caps casing our house. We assumed they were Hoffa's goons or spies.

A Teamsters local chief from Baton Rouge, Ed Partin, Hoffa's trusted confidant, told Justice Department investigators that during a Marble Palace meeting in August 1962, Hoffa had asked him to use plastic explosives to kill my father. "I've got to do something about that son of a bitch Bobby Kennedy. He's got to go." My father, Hoffa informed Partin, drove around in a convertible with no bodyguards and swam alone in the Hickory Hill pool. Our house was unguarded. Hoffa proposed "the possible use of a lone gunman with a rifle with a telescope sight . . . on orders without any identifiable characteristics to the Teamster organization or Hoffa himself." Hoffa suggested that the assassin could shoot my dad in the pool or toss a bomb into his open convertible—but Hoffa preferred the option of bombing our home with an incendiary explosive, so that anyone not killed by the blast would be incinerated. Partin testified against Hoffa in the DOJ's Nashville jury tampering case that ultimately got Hoffa convicted and jailed.

We got plenty of hate mail at Hickory Hill from southern racists, but we assumed that at least some of the more earnest ultimatums were probably emanating from the Marble Palace. "Do you know what hydrochloric acid does to your eyeballs?" asked one caller to Hickory Hill. We learned to ignore those phone calls and the periodic threats among the bushel bags of correspondence arriving daily from the post office. That said, one of those letters made such an impression on me that I shared it with David and Michael. It was an anonymous memo promising "slaughter time" for my parents and "the ten little piggies." Reading that line aloud to my younger brothers, I told them menacingly, "That means *you!*" They looked at each other and then back at me. "Oink, oink, oink," they squealed. In response to the deluge of threats, Kathleen, Joe, and I sat in the principal's office until our mom arrived to fetch us after school, instead of waiting in the sidewalk carpool line with our schoolmates. That was pretty much the only precaution I ever saw.

We never had bodyguards. "The Kennedys can take care of themselves," my father said. In retrospect, that was hubris. The Kennedys were making plenty of homicidal enemies beside Hoffa. We now know that Hoffa, exhausted by his persecution at my father's hands, sent a message via his lawyer, Frank Ragano, in July 1963, to gangsters Carlos Marcello and Santo Trafficante, asking them to kill my uncle. Ragano said in his autobiography, *Mob Lawyer*, that the two capos took his message seriously. Both hoodlums had their own reasons for hating the Kennedys.

Besides the prospect of losing access to Hoffa's rich piggy bank, gangsters such as Sam Giancana, Johnny Roselli, Carlos Marcello, and Santo Trafficante had other reasons to kill my dad. In addition to his twenty-six-lawyer "Get Hoffa" squad, my father had appointed one hundred ambitious, energetic, and experienced attorneys to put an end to organized crime. The crime bosses suddenly found themselves under an unrelenting assault that cut into their profits and, in some cases, disrupted their daily lives. During his first three years, RFK's Justice Department filed 673 indictments against organized crime, up from 22 during the Eisenhower years. Author John Davis wrote, "Given another five years in office, the Kennedys could conceivably have exterminated the Cosa Nostra entirely or at least crippled it beyond repair." Florida mob boss Santo Trafficante, who had lost his $20 million Havana casino empire when Fidel took Cuba, blamed the Kennedys not just for failing to evict Castro but also for the heightened scrutiny and diminished profits from his Florida numbers rackets.

My father's Organized Crime Division had arrested two of Trafficante's cousins and indicted both of his brothers for tax evasion and was now threatening his golden goose—the Hoffa-controlled Teamsters pension fund. In September 1962, Trafficante discussed murdering President Kennedy on my father's account with Cuban exile leader José Aleman. "Have you seen how his brother is hitting Hoffa, a man who is not a millionaire, a friend of the blue collar? Mark my words, this man Kennedy is in trouble and he will get what's coming to him." When Aleman predicted JFK's reelection, Trafficante interrupted, "No, José, he is going to be hit."

New Orleans and Dallas crime boss Carlos Marcello had repeatedly invoked the Fifth Amendment to stonewall my father's questioning during his Rackets Committee grilling in 1957. After 1960,

Robert Kennedy's Justice Department forcibly deported Marcello, an illiterate Tunisian-born Sicilian. Marcello, who had a Guatemala passport, had entered the United States illegally. After his capture, the FBI dropped him, barefoot and penniless, in a Guatemalan jungle. He made it back to the States, allegedly with the help of a CIA-affiliated New Orleans pilot named David Ferrie. Marcello, by that time, was in a state of volcanic apoplexy. My father immediately ordered a new deportation proceeding. Prior to his suspicious 1967 death, Ferrie was the principal target for New Orleans Prosecutor Jim Garrison's Grand Jury investigation of the JFK assassination.

In September 1962, while discussing with his associates a hit on my father, Marcello uttered the traditional Sicilian death warrant: *"Livarsi na petra di la scarpa"* (Take the stone out of my shoe!) Marcello then said, "Don't worry about that little Bobby sonofabitch. He's gonna be taken care of."

SAM GIANCANA FELT THE MOST HEAT. MY FATHER'S TEAM HAD Giancana's charismatic Las Vegas and Hollywood front man Johnny Roselli under surveillance and Giancana himself under a suffocating "blanket stake-out." Only when inside his residence was he ever out of sight of his FBI surveillance team. And they were never out of *his* sight. Agents waited in cars and vans outside his home. They trailed him whenever he ventured out. When Giancana dined at a restaurant, the FBI was at a neighboring table. They pursued his car, bugged his clubs, his home, and his phone; and when he went to a movie, six of the FBI's orneriest agents filed into the seats directly behind him. They ate popcorn and loudly muttered epithets, curses, and insults at him throughout the show.

Everyone in the family got excited when Budd Schulberg, the author of *On the Waterfront*, Marlon Brando's classic film about Mafia infiltration of the New York Longshoremen's Union, tried to make a movie based on *The Enemy Within*, my father's best-selling book on his battle with the mob. Now my dad would have a movie about *his* exploits, just as Uncle Jack did with *PT 109!* However, the Mafia killed the project. Mobsters and Teamsters threatened violence and strikes against Twentieth Century Fox and any other studio that tried to produce the film, and Schulberg eventually relented. We knew the mob was linked to the Teamsters. No one suspected at the time that the Mafia and the CIA were equally intertwined and that the CIA had for years worked specifically with Trafficante, Roselli, Giancana, and Marcello on various assassination projects.

In addition to forcing the FBI to investigate Hoffa and the mob, the Kennedy brothers seemed to be perturbing Hoover in every fashion they could devise. Jack deployed the famously independent journalist Edward R. Murrow to run the United States Information Agency (USIA), charging him with the task of transforming that agency from a CIA propaganda organ into a forum for honest debate about world issues, as well as a scaffolding for elevating American culture. That appointment especially angered Hoover. In 1954 Murrow had helped end the reign of terror of Hoover's friend, and red-baiting ally, Joe McCarthy. Murrow's program *See It Now* had aired a groundbreaking exposé of McCarthy's witch hunts against supposed Communist sympathizers. Now my father wanted him to wrest control of the USIA from the CIA and put it to work promoting democratic ideals.

Murrow was an outspoken critic of communism but a supporter of civil liberties—which Hoover equated with communism. Hoover's obsessive investigation and surveillance of Murrow filled a 200-page file. Hoover tried to block Murrow's appointment to the USIA, tarring his name with the Senate confirmation panel prior to his confirmation. The senators grilled Murrow for some of his work at CBS, in particular a farmworker documentary that exposed the dark underbelly of crony capitalism and corporate agriculture. Murrow, however, prevailed in the public forum with his calm and reasoned defense; people who live in a free society should tell the Americans' entire story, including its flaws. When Murrow came to the USIA, he fired all the Cold War hard-liners and filled their posts with McCarthy's blacklist victims. My father liked Murrow, and we sometimes went to the USIA to watch films with him and director George Stevens.

I had my own run-in with Murrow, so I can testify that he was indeed one tough cookie. Somehow he'd been permitted to film a segment of his popular TV show *Person-to-Person* at Hickory Hill. In 1957, three years after he dispatched McCarthy, Murrow turned his camera on me in a pioneering piece of "gotcha" journalism. Murrow's cameras caught me immediately after I'd been rousted from my Sunday-night ritual of watching *Walt Disney World* with my siblings. The newsman stripped off my Mickey Mouse cap and endeavored to interrogate me under glaring lights on the front hall staircase. David, then two, who'd just suffered a similar indignation, was already in tears. I lay facedown on the stairs in an act of civil disobedience. They forced me to sit up, but I refused to cooperate, erupting in tears and holding my stuffed elephant, "Bursey," between my face and the camera. Finally, my father scooped me up in his arms and took me back upstairs. Fortunately that episode is preserved forever on celluloid. It was my first nationally broadcast TV interview. I can't say I embraced the camera, but the performance was genuine.

Appointing independent men like Murrow reflected Jack's confidence in American values. With thousands of heated dinner-table debates behind them, Jack and my father believed America ought to be able to win the argument for democracy on the merits, without having to silence opponents or use propaganda like the Nazis and fascists. They believed we could engage in debate with Communists without being contaminated by their ideology, and that we ought to show we could compete with them and beat them—not in an arms race, but in the realm of ideas.

In the same vein, my dad irritated the State Department by urging Dean Rusk to end travel restrictions, which he thought silly and an affront to American traditions. (My own first passport included a page listing the places I was legally prohibited from visiting, among them China, Cuba, and North Korea.) At the time of Jack's death, my father was seeking to lift the travel restrictions to Cuba, which he thought asinine and counterproductive. Having been encouraged by his father to visit the Soviet Union as a boy, my dad said, "If I were a young person, Cuba is the first place I'd want to visit!"

The DOJ in the South

My father's next—and greatest—crusade was even more offensive to Hoover than the fight against organized crime: civil rights. He directed Hoover to investigate the Ku Klux Klan and prosecute civil rights violations in the former Confederate states, which the agency, under Hoover's direction, had hitherto ignored. Hoover was openly racist. He believed all civil rights groups were Communist fronts and refused, as a policy, to investigate violence against blacks, explaining at one point that he would not "send the FBI in every time some nigger woman says she's been raped." On his first day at work at the Justice Department, my father had noticed that there were very few blacks in the sea of white faces and ordered the department integrated. My father's Justice Department hired more African American lawyers than at any other time in its history. Among his first calls was a suit forcing the Washington Redskins to integrate. But initially, at least, civil rights was not a central preoccupation for the incoming Kennedy administration. Before long, however, a six-year-old Supreme Court decision that ended Rosa Parks's Montgomery bus boycott by outlawing segregation in public transportation and bus terminals would thrust the Justice Department into the center of the civil rights fray.

To test the court ruling, young black activists calling themselves Freedom Riders boarded Greyhound buses in northern cities to tour the South. On Mother's Day, May 14, 1961, Alabama bigots burned a Greyhound bus on the highway outside Anniston and beat its occupants. Birmingham's commissioner of public safety, Bull Connor, promised the Ku Klux Klan fifteen free minutes with the Freedom Riders when they disembarked in his city, and Klansmen took the opportunity to savage the activists with baseball bats, ax handles, and chains. Warned in advance of the Klan's planned melee and ordered by their superiors not to intervene, Hoover's FBI stood on the sidelines jotting notes and snapping pictures. They deliberately kept the Justice Department out of the loop.

Following the donnybrook in Birmingham, Alabama governor John Patterson, who had been a Kennedy political ally during the 1960 campaign, promised my father that he would protect the Freedom Riders as they proceeded south to Montgomery. Later Patterson would explain his failure to do so by claiming he lacked the authority to intervene within the capital city limits. Floyd Mann, the courageous chief of the Alabama State Troopers, who became my close friend years later, gave the activists safe passage on the highways, but once they reached Montgomery the troopers no longer had jurisdiction. Montgomery's bus station became the site of a bloodbath. While FBI agents scribbled in their logbooks, my father's friend John Seigenthaler waded into the melee to rescue a black woman from the deadly mob. As he pulled the woman out of harm's way, a rioter asked him, "Who the hell are you?" Seigenthaler later told me he gave the dumbest reply possible: "I'm a federal man!" The Klansman's response was to club John with a pipe. When he awoke in the hospital, my father was on the phone. "Let me give you some advice," Seigenthaler told him groggily. "Never run for governor of Alabama." When he returned home to Washington, John showed us his U.S. marshal helmet with the big dent in it.

By late 1961 my father's consciousness had evolved. Burke Marshall, who ran the DOJ's civil rights division for my father, told me, "When he began to finally realize how blacks were being systematically abused in the South, he got angrier and angrier. He just couldn't stand bullying. It grated on him. By the end of that year he was so furious that it displaced organized crime as his principal concern. It was just inconsistent with everything he believed about America." By 1962, my dad had decided that the most important domestic issue facing the nation was civil rights. At dinnertime he talked with us about pitched battles in Alabama, Mississippi, and Georgia.

In 1954, the year I was born, another Supreme Court decision, *Brown v. Board of Education*, declared that the Southern system of "separate but equal" schools was unconstitutional and ordered Southern schools desegregated. In January 1961, a young black high school senior, inspired by Uncle Jack's inaugural address, applied to attend the Confederate redoubt Ole Miss. Despite James Meredith's obvious academic qualifications, the University of Mississippi, at Oxford, denied him admission. With help from NAACP lawyers, Meredith obtained a federal court order requiring the school to admit him, but Governor Ross Barnett, a close political ally of the Kennedys during the 1960 campaign, refused to comply. In a recorded colloquy on September 25, 1962, my father reminded Governor Barnett of his duty to obey the national law by admitting Meredith to Ole Miss.

BARNETT: That's what it's going to boil down to—whether Mississippi can run its institutions or the federal government is going to run things.

RFK: I don't understand, Governor. Where do you think this is going to take your state?

BARNETT: A lot of states haven't had the guts to take a stand. We are going to fight this thing. . . . This is like a dictatorship. Forcing him [Meredith] physically into Ole Miss, General, that might bring a lot of trouble. You don't want to do that. You don't want to physically force him in.

RFK: You don't want to physically keep him out. Governor, you are a part of the United States. . . .

BARNETT: We have been a part of the United States but I don't know whether we are or not.

RFK: Are you getting out of the Union?

BARNETT: It looks like we're being kicked around—like we don't belong to it. General, this thing is serious.

RFK: It's serious here.

BARNETT: Must it be over one little boy, backed by Communist front, backed by the NAACP, which is a Communist front? . . . I won't agree to let that boy get to Ole Miss. I will never agree to that. I would rather spend the rest of my life in a penitentiary than do that.

The Kennedy administration's fight to get Meredith admitted to the University of Mississippi would culminate in the last great battle of the Civil War. Truckloads of rednecks were arriving in Oxford by the thousands, armed with rifles, shotguns, ax handles, and clubs. They came from all over the South and as far away as Texas and California—summoned by former general Edwin Walker, whom Jack had evicted from the military for force-feeding American soldiers his right-wing diatribes and John Birch Society literature. Walker's army of hate-mongers joined segregationist forces from across the South for what became the largest confrontation between state and federal power since Appomattox.

My father had promised Meredith and his courageous mentor, Medgar Evers, that he would protect them. Finding Nick Katzenbach, Ed Guthman, John Doar, and Dean Markham in their Justice Department offices on Saturday afternoon, he asked if they had Sunday plans. When they answered no, he sent them to Oxford, Mississippi. "If things get rough," he joked, "don't worry about yourself; the president needs a moral issue." By Sunday evening my dad's four friends were holed up with Meredith in the university's Lyceum under fierce assault by a savage mob numbering in the thousands, wielding bricks and bottles and firing at them with pistols and rifles. One hundred and sixty federal marshals, led by my father's trusted deputy, U.S. Marshal Jim McShane, were wounded, and two people lay dead as the besieged survivors awaited rescue by 16,000 federal troops, whose prompt deployment had been bungled by inept or insubordinate commanders. Jack and my dad were furious at what they assumed was deliberate foot-dragging.

In Jackson, the U.S. marshals on the scene, their numbers supplemented by hastily deputized prison guards and immigration officers, included some of the best marksmen in America. All night long, McShane's besieged men were begging him for permission to open fire. Jack and my dad would only give the marshals permission to use tear gas, but under no circumstances to shoot, except to preserve Meredith's life. The following morning, when federal troops finally arrived to disperse the mob, Nick Katzenbach escorted James Meredith to the registrar, making him the first African-American ever admitted to the University of Mississippi.

Whenever he was away from us for prolonged periods, holed up in his office or traveling, my father called home every day; we could also reach him through the White House switchboard, where the famously competent operators recognized our voices. My dad also sent each of us regular letters, which included excerpts from poems and updates on his political battles. That day, he wrote to me simply: "Yesterday we sent soldiers to Mississippi to bring James Meredith to school. Governor Barnett opposed us, but this morning we resisted him."

Medgar Evers, the self-possessed, outspoken World War II veteran who had counseled James Meredith and organized peaceful resistance to a brutal campaign waged by Jackson's segregationist mayor, Allen C. Thompson, continued to lead a series of demonstrations and parades, despite daily death threats and attempts on his life. He matter-of-factly predicted his own death in a nationally aired CBS interview a few days before he was shot and killed in his driveway on June 12, 1963. His wife, Myrlie, kept her promise to continue marching, and the violence escalated. Only an act of insanely reckless heroism by John Doar, one of my father's aides, who stepped between marching blacks and white police at Medgar Evers's memorial service in Jackson, prevented a full-blown riot.

My father helped Medgar's family cut through bureaucratic red tape to get him buried at Arlington National Cemetery. My dad also gave Medgar's brother, Charles, his private phone number, telling him to call whenever he needed help. Charles Evers, who was more radical and less patient than his brother, called my father day and night, and remained a lifelong ally and family friend. During the U.S. Senate election in 1964, when LBJ was quietly pressuring the NAACP to support my father's rival, the liberal incumbent

Republican senator Kenneth Keating, Charles Evers elbowed his way into a critical NAACP meeting and gave an impassioned speech that swung the wavering organization behind my father's senatorial bid. In 1965 he toured the Mississippi Delta with my dad; he was with him the day my father died, and rode with us in the funeral train. I often visit Charles and James Meredith in Mississippi, where Charles is a revered hero and a prickly but popular radio host.

My father and Uncle Jack had been shocked by the hatred Southern whites unleashed during the battle over Ole Miss. In its aftermath, the cohorts of bigotry and hatred directed their poison at the Kennedys. In Birmingham, my dad orchestrated a fragile peace by bringing together business leaders with black preachers, stilling violent rioting that was beginning to envelop the city. Still, Birmingham mayor Arthur Hanes cursed my dad for empowering blacks, saying, "I hope that every drop of blood that's spilled he tastes in his throat, and I hope that he chokes to death!"

Alabama's new governor, George Wallace, made oft-repeated promises to block with his body the door of any schoolhouse that the federal government tried to integrate. Some thirteen years later, in 1976, I would take my junior year off from college to work on a biography of George Wallace's law school classmate and nemesis during the Civil Rights Movement, Federal Judge Frank Johnson, who decided the Montgomery Bus Boycott case and the Selma March and made many other milestone rulings. I spent a lot of time with Governor Wallace, visiting him at his office in the Alabama State House, over which the Confederate flag still flew, in defiance of federal laws. Wallace welcomed me in my first meeting with words that surprised me, "Your daddy was a good man," he said. "I loved the Kennedy boys. They was good for America," he added thoughtfully. "I could never figure out how to get to them." That was not at all how he felt in 1963. Standing beneath the Confederate flag to deliver his inaugural address in the exact spot where Jefferson Davis had sworn his oath of office in 1861, Wallace had directed a solemn vow at the Kennedy administration. "From this cradle of the Confederacy, this very heart of the great Anglo-Saxon

Southland . . . I draw the line in the dust and toss the gauntlet before the feet of tyranny. And I say: Segregation now! Segregation tomorrow! and Segregation forever!"

The inevitable clash came when black students applied to the University of Alabama, in Tuscaloosa. In an effort to defuse the confrontation and avoid another Oxford, my father reached out to Wallace, who refused his phone calls, denouncing him with profanities that Wallace's aides deemed too obscene to repeat. Finally the Alabama governor agreed to meet my dad in Montgomery. Upon arriving at the capitol, my father threaded a gauntlet of state troopers in steel helmets emblazoned with Confederate flags. One giant trooper in Wallace's praetorian guard hit him in the belly with a nightstick. "The point," my father later said, "was to try to show that my life was in danger in coming to Alabama because the people hated me so much."

Inside the executive office, Wallace placed a tape recorder on the table between them and endeavored to goad my dad into threatening to send federal troops into the state, a provocation Wallace could use to political advantage. My father danced around the trap while Wallace vowed that never in his lifetime would he allow a Negro student into any white Alabama school. He later confided in a phone conversation to Uncle Jack that all the problems were caused by Martin Luther King, Jr., who, he informed the president, was competing with civil rights leader Reverend Fred Shuttlesworth to see "who could go to bed with the most nigger women, and white and red women too!" The confrontation ended with Wallace's choreographed retreat on June 11, when Nick Katzenbach confronted the governor before the national press on the university steps. With the cameras rolling, Wallace theatrically declared, "I can't fight bayonets with my bare hands," and backed down, allowing the first African American students through the door to register.

There were no bayonets. With typical restraint, my father and Jack had delayed federalizing the National Guard until the final hours. My dad wrote me this letter: "At 1:30 today, Jack nationalized the Alabama National Guard and at 5:30 we were able to register the two Negro students. Love Daddy. P.S. When you are in college I hope this will all be over."

That evening, at my father's urging, Jack decided to give a televised speech on civil rights to the nation. My father had been working on a draft with the then-unknown novelist Richard Yates. The final speech was hastily composed by Ted Sorensen and Uncle Jack, using some of the concepts outlined by Yates, it was one of the great speeches of American history, according to historian Arthur Schlesinger. Historian Andrew Cohen notes that Sorensen completed his draft less than one hour before air time. Jack had only a few minutes to look it over before going on camera. He delivered it without a teleprompter. Jack declared, "We are confronted primarily with a moral issue . . . as old as the scriptures and as clear as the American Constitution." He asked his fellow Americans to put themselves in the shoes of their fellow black citizens. "Who among us would be content to have the color of his skin changed and stand in his place? Who among us would then be content with the counsels of patience and delay?" During the last ten minutes Jack left his notes, speaking extemporaneously, directly to the camera. It was way past time, he told Americans, "for this nation to fulfill its promise [of freedom]." His initial attempts at comprehensive civil rights reform had died in a hostile Congress, but he promised now, "I shall ask the Congress . . . to make a commitment it has not fully made in this century to the proposition that race has no place in American life or law."

With the exception of my father, the entire Cabinet and all of Jack's political advisers opposed the legislation, believing it would destroy the Democratic Party and derail the administration's entire legislative program. Among them was Commerce Secretary Luther Hodges; Jack told him, "There comes a time when a man has to take a stand, and history will record that he has to meet these tough situations and ultimately make a decision." Both my father and Jack believed the legislation might sink them politically, but both of them committed to it with all their energies. Civil rights leader Joseph L. Rauh, Jr., declared it "the most comprehensive

Civil Rights Bill ever to receive serious consideration from the Congress of the United States."

The obstacles were enormous, and Dr. King proposed a march on Washington to bolster support. After initial jitters within the administration, my dad and Jack became convinced that a large public demonstration would help passage of the legislative package. They encouraged the civil rights community to mobilize and worked with organizers to make the march a formidable success. At the same time, they tried to avert incendiary rhetoric that might torpedo the bill among moderate senators, stationing Harris Wofford, a lifelong civil rights advocate who was then special assistant to the president and later a U.S. Senator from Pennsylvania, by the giant sound system at the Lincoln Memorial, ready to pull the plug if the polemics overheated. On August 28, 1963, a quarter million people flooded the Mall to hear Dr. Martin Luther King, Jr., describe his dream: "Free at last, free at last, thank God almighty, we are free at last." And in the wake of Jack's death, in February 1964, the House of Representatives passed the Civil Rights Bill. The Senate followed in March, after enduring a fifty-seven-day filibuster led by Senator Robert Byrd (Democrat of West Virginia) and eighteen Southern Democrats. Uncle Teddy gave his maiden speech in the Senate in support of the bill as a memorial to his slain brother. President Lyndon Johnson signed it into law in July, reportedly assessing to an aide after he did so the devastating rift that his action would impose on his beloved Democratic Party: "We have lost the South for a generation."

Driven by his idealistic belief in American democracy and his inherent sympathy for the underdog, my father turned the Justice Department toward addressing what he called the disparities in access to justice that made our legal system favor "the rich man over the poor." He introduced legislation to reform the bail system, and instructed U.S. Attorneys to refrain from seeking bail in certain classes of crime. He braved Hoover's disdain for "coddling prisoners" by creating halfway houses and training opportunities focused on rehabilitation. He closed Alcatraz, which was America's most notorious prison, prior to Guantánamo. For me, one of the most touching tributes after his death was the thousands of letters our family received from incarcerated inmates, thanking my dad for improving their conditions in prison. At Sing Sing, hardened prisoners of every race wore black armbands after his death.

At the Justice Department my father worked feverishly on policies affecting juvenile delinquency, particularly among Hispanic and Native American children, a crusade that put him, once again, at odds with Hoover, who complained that my father was coddling "beastly punks." My father saw juvenile delinquency, according to Schlesinger, as a symptom of poverty—the "unconsolidated rebellion of an underclass" promised the American dream and then denied any genuine opportunity to achieve it.

Washington, D.C., particularly drew his attention, and he took us for drives in his convertible through the grimmest slums in Northwest and Southeast D.C. to identify abandoned lots he could transform into parks, swimming pools, skating rinks, and recreation facilities. Then he would solicit contributions from wealthy friends, twist the reluctant arms of local and federal bureaucrats, and get those projects built. He called Bob McNamara to have the Pentagon deliver an old bomber, a tank, or a retired fighter jet for the kids to play on. He made Hickory Hill and the Justice Department regular party venues. Inner-city kids flooded our home on weekends and holidays, splashing in the pool and playing on the obstacle course in the yard.

My mother's energy turned the DOJ Christmas party into an annual children's extravaganza that included sleigh rides on fake snow in the cobblestoned Justice Department courtyard. Dressed as clowns, my father's college football teammates circulated among the children. We led kids around the courtyard on pony rides or in a mule-drawn buggy, and helped distribute popcorn and ice cream, presents, and candy. Jack's PT boat shipmate Barney Ross played Santa. Entertainment included mimes, musicians, magicians, and appearances by the Washington Redskins. A variety show MC'd by Jim Symington—one of my dad's friends, his executive assistant, and later, U.S. senator from Missouri—featured Carol Channing, Harry Belafonte, and the Smothers Brothers in the DOJ's Great Hall, with its Art Deco lighting fixtures and terra-cotta tile floor.

The Case Against Big Steel

Among the battles that make me most proud of the Kennedy administration was my uncle's fight against Big Steel. As an advocate who has spent three decades watching powerful polluters dominate American government and political leaders from the president on down kowtow to corporate fat cats, the story of JFK's confrontation with the steel barons makes me most nostalgic for Camelot. I can't think of any other president in the last half century willing to battle—and then to win—a confrontation with one of America's biggest industries.

My father absorbed from Grandpa a healthy skepticism about the willingness of Wall Street to manage the economy for the public good, and in 1962 my uncle and dad had their own historic confrontation with the tycoons. On April 10 of that year I wrote in my daily journal: "The steel companies made hire prises [sic]." Four days earlier, JFK had personally intervened to settle a national steel strike; he'd persuaded company chiefs and the United Steelworkers Union to cap both prices and wages in order to hold down galloping steel costs that threatened to stampede inflation. On the day of my journal entry, U.S. Steel chairman Roger Blough asked to visit Uncle Jack in the Oval Office. Instead of explaining the reason for his visit, Blough handed the president a press release announcing a six-dollar-a-ton hike in steel prices, "effective 12:01 a.m. tomorrow." It was a brazen double cross. Jack eyed him coldly. "You've made a terrible mistake." Dismissing Blough, he remarked to his advisers, "My father always told me that all businessmen were sons of bitches, but I never believed it until now." Then he phoned United Steelworkers Union president David J. McDonald to break the news. "Dave, you've been screwed, and I've been screwed."

Six other companies joined U.S. Steel's insurrection, while six held the line. To Wall Street's surprise, Jack threw the full power of the Oval Office at the corporate mutineers. Mounting an attack that would shock the robber barons with its raw power, Jack ordered the Pentagon to cancel its shipbuilding contracts with the seven rogue companies and shift those contracts to the six smaller companies that had not joined the hike. The steel industry's biggest customer was the Pentagon. If necessary, he told the Defense Department to fill orders from foreign sources to avoid dealings with Blough and his treacherous frondeurs. Even as they reeled from these billion-dollar contract hits, my father struck them again by convening a federal grand jury to prosecute price-fixing and antitrust violations by the steel manufacturers, cases that would eventually net maximum fines. He ordered the FBI to raid corporate offices, confiscate steel company documents, subpoena their personal and company records, and round up executives for questioning. The following morning, brigades of G-men ransacked companies across the country, searching desks and hauling off account books and filing cabinets on dollies.

Jack had his lawyer, Clark Clifford, call Roger Blough to promise a battery of new assaults, "including tax audits, anti-trust investigations, and a thorough probe of market practices." My uncle took to the airwaves, making his case that the steel barons were enemies of the republic. Referring to his inaugural address, he said, "Some time ago I asked each American to consider what he would do for his country, and I asked the steel companies. In the last twenty-four hours we had their answer." In an April 11 press conference he scourged the executives, adding that "The American people will find it hard, as I do, to accept a situation in which a tiny handful of steel executives, whose pursuit of private power and profit exceeds their sense of public responsibility, can show utter contempt for the interest of 185 million Americans." Wall Street was stunned by the unsheathed force of the attack. The nation's most powerful financial behemoths found themselves castigated as traitors in the biggest assault on U.S. businessmen since Teddy Roosevelt broke up

Standard Oil, and probably the fiercest ever unleashed. On April 13, the rogue steel barons surrendered, with all six companies rescinding their price hikes. And the White House's tough tactics worked to break the larger inflationary spiral: U.S. inflation held at 1.2 percent in 1962, and 1.6 percent in 1963.

Characteristically, with the battle won, Jack "permitted no gloating by my administration spokesmen, and no talk of retribution," according to Sorensen. He went out of his way to be gracious to Roger Blough, inviting him often to the White House for consultations. Nevertheless, Jack and my dad were thereafter despised on Wall Street, and by its allies within the arms industry and military-industrial complex, for their willingness to use government power to rein in corporate greed. Jack had only been in office fourteen months, but Wall Street and industry loathed him now, as they had FDR, considering the Kennedys to be traitors to their class. Publisher Henry Luce praised Blough as a "business statesman" and condemned Jack for exercising "jawbone despotism" over steel prices. The *Christian Science Monitor* excoriated Robert Kennedy for deploying "agents of the security police" in a tyrannical demonstration of "naked power." The *Wall Street Journal* editorial opined that "It was dangerously wrong for a President to loose arsenals of power for the purpose of intimidation and coercing private companies." Jack would remark to Sorensen, Schlesinger, and O'Donnell, "I understand better every day why Roosevelt, who started out such a mild fellow, ended up so ferociously anti-business. It's hard as hell to be friendly with people who keep trying to cut your legs off." He finally gave up on the moguls, determined to represent the interests of the rest of America. On May 8, 1962, he told the United Automobile Workers, "Harry Truman once said, 'There are 14 or 15 million Americans who have the resources to have representatives in Washington to protect their interests. And that the interest of the great mass of the other people—150 or 160 million—is the responsibility of the President of the United States, and I propose to fulfill it.'"

For his part, my father transformed the Justice Department.

He shared his brother's driving conviction that government was owned by the people, and so it should work for them. My dad's hard work, enthusiasm, and willingness to take on tough issues attracted some of the most talented young lawyers in the country to the Department of Justice, and inspired them to work overtime and on holidays and weekends. The *New York Times* and *Washington Post* had opposed my father's appointment as attorney general, but when he left that office three years later, the *Washington Post* summed up his record as follows: "He has guided more important legislation through Congress than did any of his predecessors in the past thirty years. He has made the Federal Government, for the first time, a vigorous enemy of organized crime. He has pushed equal rights for all Americans."

● ● ● ● ●

CHAPTER 7

JFK: In Pursuit of Peace

Thomas Paine said in the Revolution of 1776 that the cause of
America is the cause of all mankind. I think in 1960 the cause of
all mankind is the cause of America. If we fail, I think the case
of freedom fails, not only in the United States, but every place.
　　　　—JFK, AT THE STEELWORKERS' CONVENTION,
　　　　　　　　SEPTEMBER 19, 1960

If we cannot now end our differences, at least we can help
make the world safe for diversity.
　　　　　　—JOHN F. KENNEDY

MY GRANDPARENTS MADE THE PURPOSEFUL EFFORT TO ENCOURAGE
their children and grandchildren to see the world from the perspective of others. Jack's foreign travels as a youth, his wartime experiences, his intense love of reading, and his natural curiosity all
inclined him to build bridges with both friends and enemies in an
effort to incline us away from war. Jack's initial interest in politics
came neither from a warrant to "carry the torch" after his brother
Joe's death nor from his father's guiding hand and ambition, as popular mythology speculates. It arose during his experience as a journalist observing the inaugural United Nations Conference in San

Francisco and his attendance at the Potsdam Conference, which lit his intense interest in America's capacity to steer the course toward an idealized postwar world. In the words of his fellow veteran and closest political adviser, Kenny O'Donnell, "Jack realized that, like it or not, politics was the place where you personally could do the most to prevent another war." As early as 1953, as a U.S. senator, Uncle Jack was considering paths to find common cause with the Soviets based on the blasphemous notion of coexistence with Communist nations. In 1959 he endured scathing criticism from Republicans, including "moderate" Nelson Rockefeller, for calling for a Nuclear Test Ban Treaty. In 1960, after the Soviet Union shot down an American U-2 spy plane over its territory, derailing Eisenhower's U.S.–Soviet Arms Summit, Jack criticized fellow Democrats for faulting Eisenhower for trying to revive peace talks with Soviet premier Nikita Khrushchev. He believed that a president *should* take political risks for the goal of peace. In this spirit, and with high hopes, he arranged for his own summit conference with Khrushchev in Vienna in early June 1961.

Despite Khrushchev's shoe-pounding belligerence, Jack was optimistic he could find some mutuality with the former Bolshevik. In truth, Khrushchev was very different from his predecessor Joseph Stalin, whose extreme policies had resulted in the deaths of almost three million Russians and the imprisonment of tens of thousands of dissidents in gulags. Khrushchev freed thousands of Stalin's political prisoners, and astonished delegates to the 1956 Communist Party Congress by denouncing Stalin. Aleksandr Solzhenitsyn later described Khrushchev's risky decision to indict the Soviet icon as "a movement of the heart, a genuine impulse to do good." A veteran of the Nazi siege of Stalingrad, Khrushchev had no appetite for another war and little imperial ambition, beyond insulating the Soviet frontiers with a buffer of friendly satellite states. Khrushchev's history hinted at a predisposition toward peace, but Jack's first encounter with him was a crushing disappointment.

JFK opened the Vienna summit with an olive branch. He presented Khrushchev with a scale model of "Old Ironsides," the USS

Following our morning ride with Grandpa at his Osterville, Massachusetts, farm. *From left:* Joe Kennedy II, Kathleen Kennedy, Joseph P. Kennedy, RFK Jr., Maria Shriver, and Bobby Shriver. *(Kennedy family collection)*

Watching Teamsters leaders Dave Beck and Jimmy Hoffa and a parade of mobsters at the 1957 Senate Select Committee hearings. *From left:* Kathleen Kennedy, RFK Jr., Ethel Kennedy, and Joe Kennedy II. *(Bettmann/Getty Images)*

Opposite, top: Grandma Rose dancing with RFK at Jack's wedding to Jackie. *(Jacques Lowe)*

Opposite, bottom: Crossing JFK's "Big House" football field with Kathleen and our Saint Bernard, Menolin, with RFK Jr. on RFK's shoulders. *(Jacques Lowe)*

Following pages: RFK Jr. during the "dead salamander" visit with Uncle Jack in the Oval Office. *(Abbie Rowe)*

RFK Jr. with his sister Kerry and his red-tailed hawk at Hickory Hill, 1967. *(Kennedy family collection)*

Opposite: Ethel Kennedy taking a fence at Hickory Hill on Kathleen's white hunter, Atlas. *(Jacques Lowe)*

Following pages: In Kraków's market square, after mass, 1964. Despite efforts by Poland's Communist regime to censor news of our visit, vast crowds appeared at every stop to cheer us. *From left:* Kathleen Kennedy, Ethel Kennedy, Joe Kennedy II, RFK Jr., and RFK. *(Kennedy family collection)*

Opposite, top: RFK confers with FBI director J. Edgar Hoover in his office. My father's most troublesome employee, Hoover bridled against enforcing a corrupt civil rights violation on the Mafia. *(Jacques Lowe)*

Opposite, bottom: Snake River, 1969. *From left:* Chris Lawford, RFK Jr., David Kennedy, Carl and Scott Whittaker, and RFK. *(Harry Benson)*

Above: JFK and RFK at the Biltmore Hotel in Los Angeles after Jack's convention victory, discussing Jack's choice of LBJ as his vice-presidential running mate. *(Jacques Lowe)*

Following pages, left: Ethel Skakel Kennedy's Valentine's Day card, on the Lincoln Memorial Reflecting Pool. *(Cecil Stoughton/Kennedy family collection) Top right:* RFK campaigning in Philadelphia, 1968. *(Constantine Manos/Magnum Photos). Bottom right:* RFK in Watts, 1968, with his bodyguards, prizefighter Tony Zale and the Oakland Raiders' Fearsome Foursome: Rosey Grier, Deacon Jones, Merlin Olsen, and Lamar Lundy. *(Hulton Archive/ Getty Images)*

RFK Jr. and Ethel in Forest Hills, New York, 1972. *(Ron Galella)*

Opposite, top: Jack and Jackie's wedding at the Bouviers' Hammersmith Farm in Newport, Rhode Island. *(Toni Frissell/Library of Congress) From left:* Jackie Kennedy, Robert F. Kennedy, Eunice Shriver, Patricia Kennedy Lawford, Ted Kennedy, Jean Kennedy Smith, and John F. Kennedy.

Opposite, bottom: Lem Billings and RFK on Salmon River, Idaho, 1967. *(Kennedy family collection)*

Following page: Views from the funeral train, 1968. *(Paul Fusco)*

Constitution berthed in Boston Harbor, which his mother had taken generations of Kennedy children to see. Pointing to the ship's tiny cannons, Jack observed that modern weaponry could now exterminate "seventy million people in ten minutes" and suggested to Khrushchev that they shared a joint responsibility to avert such a catastrophe. After the meeting, he told Lem Billings, who accompanied him to Vienna, "Khrushchev just stared at me like he didn't give a damn." Years later, Lem told me that Jack had had high hopes of leaving the meeting with an agreement that both countries would refrain from military efforts to expand their spheres of influence, encourage emerging nations to remain non-aligned, and discard the ruinous arms race. Jack also hoped Khrushchev would agree to a nuclear test ban.

Instead of agreeing to retreat from the nuclear brink, Khrushchev met Jack's proposals with bombast and a patronizing recital of America's history of imperialism and hypocrisy. The pugnacious Soviet premier rejected Jack's proposal for Third World neutrality. He threatened to cut off and suffocate West Berlin, which at the time was a free, democratic enclave surrounded by Soviet-controlled East Germany. Khrushchev refused to budge on the nuclear test ban and practically dared Jack to make war. Over lunch that day, Jack asked Khrushchev the significance of a medal he was wearing. Khrushchev told him it was the Lenin Peace Medal. Jack commented wryly, "I hope you get to keep it."

Right-wing mythology holds that Khrushchev manhandled Jack in Vienna, but despite Khrushchev's posturing, the Soviet leader came away with a healthy respect for Jack that would bear fruit in later days. Khrushchev would write that Jack "had obviously studied all the questions we were likely to exchange views on, and he had absolutely full command of the material. This was not by any means what I had observed in the case of Eisenhower. This spoke in Kennedy's favor, of course, and in my eyes he grew in stature. Here was a counterpart to whom I could relate with enormous respect, even though we held completely different positions and in fact were adversaries." And Khrushchev did make a critical concession on a point of immediate importance. He agreed to allow Laos to choose its own government, which kept the great powers out of immediate war in Southeast Asia. Nevertheless, Lem told me that Jack was dismayed by the Soviet leader's seeming indifference to the threat of a worldwide thermonuclear apocalypse. On the long flight home from Europe a despondent President Kennedy jotted down a paraphrased quote from Abraham Lincoln: "I know there is a God. I can see the storm coming. If He has a plan for me, and I think He has, I believe I am ready."

Beyond Dr. Strangelove

After the failure of Vienna, Jack launched a military buildup and inaugurated a fallout-shelter program that had us all digging fortified bunkers in our backyards. My father was one of the lonely voices opposing the program. He didn't think we should encourage the crackpot delusion that nuclear war was winnable. Jack subsequently came to agree with him, but the military intelligence apparatus over which Jack now presided considered nuclear war both survivable and desirable. Knowing that we had a decisive nuclear weapon advantage that would only diminish with time, the Joint Chiefs were spoiling to fight the war to end all wars. And the sooner the better.

The Civil Defense program helped usher us all into the desired mind-set. Emergency broadcast tests regularly interrupted our favorite TV shows, reminding us that nuclear annihilation was only minutes away and we should prepare ourselves to survive the blasts. At school we practiced "duck and cover" routines. I learned to remove sharp objects from my desktop at the first flash of a nuclear detonation, and to tuck my head between my legs until the percussion took out the school windows. If this happened during gym period, I knew to stand with my head against my locker. Then, as mushroom clouds formed over Our Lady of Victory, my fellow students and I were to march single file into the basement, where we would feast by candlelight on the dehydrated food and canned fruit cocktail then stored in elephantine canisters against the cellar walls. Fruit cocktail being my favorite dessert, there was part of me looking forward to cracking those glass containers. I sealed deals with my close friends for reciprocal access to our imaginary bomb shelters when we heard each other's frantic knocks. All bets were off in cases of visible mutations.

Both sides in the Cold War seemed pumped for a showdown. On July 20, 1961, at a meeting of the National Security Council six weeks after Vienna, CIA director Allen Dulles and the Joint Chiefs of Staff unveiled their proposal for the U.S. version of the "Final Solution," a full-scale preemptive thermonuclear attack to obliterate the Soviet Union, to be launched sometime late in 1963. This was three months after the Bay of Pigs, and, fortunately, Jack had by now lost all faith in the brass. Nevertheless, he went through the motions of convening them occasionally. On hearing their proposal, he stood and walked out of the meeting in disgust, remarking scathingly to his secretary of state, Dean Rusk, "And we call ourselves the human race."

That summer, Joint Chiefs chair General Lemnitzer and General Curtis "Bombs Away" LeMay demanded Jack's authorization to use nuclear weapons in Berlin and in South Asia. Again Jack departed abruptly, telling Defense attaché Ros Gilpatric, "These people are crazy." Sterling Hayden's portrayal of LeMay as the paranoid, warmongering, stogie-chomping, psychopathic general Jack D. Ripper in Stanley Kubrick's 1964 spoof *Dr. Strangelove* actually fell short of the mark. In World War II, LeMay had devised the tactic of targeting civilians in the Pacific theater, where his massive firebombing of more than sixty-five cities incinerated at least 220,000 Japanese citizens with newly invented napalm. Now he advocated using the latest nuclear weapons in Cuba, Germany, Laos, and Vietnam, not to mention Russia and China, in order to end the Communist threat and establish America's unchallenged global hegemony once and for all.

LeMay and Lemnitzer were Cold War fanatics with strong ties

to ultra-right-wing leaders, including powerful senators like John Stennis, Strom Thurmond, and Barry Goldwater, whom they accompanied on hunting and fishing outings, and to whom they provided military aircraft for junkets and home visits. Those connections and their status as World War II heroes made them troublesome to fire. They had won the war with Germany and Japan and prepared the world for an era of U.S. domination fortified by a nuclear advantage they had no intention of yielding. During the Vienna Conference the two generals fulminated against Jack's "treasonous" negotiations with the Soviet premier. Jack responded by ordering his new chair of the Joint Chiefs, General Maxwell Taylor, and Robert McNamara, his secretary of defense, to depoliticize the military and bring the insubordinate generals under civilian control. (My father had General Edwin Walker—the John Birch Society lunatic—locked in a mental institution.) During their three-year battle to achieve this goal, McNamara and Max Taylor became better friends to Jack and my parents than any other high-level appointees in the administration. McNamara, his wife, Margie, and their children visited us on Cape Cod each summer, and often joined us on our winter ski vacation. Max Taylor played tennis with my parents on weekends. My little brother is his namesake.

Jack and my father held office at a time when there was an existential threat that civilization itself might evaporate in a nuclear Armageddon. But unlike their generals, they were convinced that America could just as surely lose the Cold War by compromising our ideals. "There is little value," Jack observed, "in insuring the survival of our nation if our traditions do not survive with it." In the early 1960s, conservative and right-wing broadcasters, often supported by conservative foundations, occupied a dominating presence on the airwaves, drumming up fear, attacking dissenters, and applauding military aggression. In October 1963, just before his assassination, Jack was encouraging Americans to file Fairness Doctrine petitions with the Federal Communications Commission to counter the uninterrupted flow of right-wing commentary by the three television networks on the publicly owned airwaves.

Then as now, conservatives gravitated toward strong authoritarian leaders who put security ahead of civil or constitutional rights. Ted Dealey, the conservative publisher of the *Dallas Morning News*, led the chorus of voices on the far right, calling for an American Napoleon. "America," he said, "needs a man on horseback" with the courage to lead our country in a hot war against the Soviets. Jack characterized those calls for a supreme leader as a betrayal of democracy: "They call for a 'man on horseback' because they do not trust the people. They find treason in our churches, in our highest court, and even in the treatment of our water. They equate the Democratic Party with the welfare state, the welfare state with socialism, and socialism with communism. They object quite rightly to politics intruding on the military—but they are anxious for the military to engage in politics."

In the summer of 1961, all the grown-ups at Hyannisport were reading and talking about Barbara Tuchman's *The Guns of August*, which chronicled how inept European political leaders and incompetent generals had blundered, through sanctimony, venality, and self-righteous pride, into the charnel house of World War I. They passed the book around the compound, read from it aloud, and talked about it during our daily picnics on the *Marlin*. "All wars start from stupidity," Jack told Kenny O'Donnell. They also read *The Manchurian Candidate*, Richard Condon's novel about an assassination attempt orchestrated by intelligence agencies using a hypno-programmed assassin, and the best-selling novel *Seven Days in May*, which portrayed a military coup against the U.S. government by top brass. Speculating about whether that could occur during his administration, Jack told Ted Sorensen, "I know a couple of generals who might wish they could." On another occasion he predicted that a few more failures like the Bay of Pigs could embolden his generals to embark on such an adventure. Kathleen, Joe, and I watched a screening of director John Frankenheimer's film adaptation of *The Manchurian Candidate* in the White House theater, along with Jack, Jackie, and our parents, in August 1962. Afterward, Jack encouraged Frankenheimer to make *Seven Days in May* into a movie "as a warning to the republic," then forced a reluctant Pentagon to cooperate with the film, and even volunteered to leave the White House for Cape Cod in order to facilitate filming there. Even as he considered how to control his bellicose brass, Jack found himself overtaken by new events that threatened to annihilate the human race.

The Berlin Wall

In August 1961 the Iron Curtain went from a Churchillian rhetorical device to stone, mortar, and barbed wire as Khrushchev erected the Berlin Wall overnight to stanch the hemorrhage of Germans fleeing Communist-controlled East Germany to the West. Each week brought fresh reports of people killed while trying to cross the newly created barrier that turned East Germany into a grim prison camp. During Jack's three years in office, Berlin would be the fulcrum of the Cold War; the Soviets considered the U.S. occupation of West Berlin a gaping breach in their border security. If the Soviets invaded West Berlin, our treaties and other obligations would force us to respond militarily, leading inevitably to nuclear war. In every foreign policy conflict from Cuba to Vietnam, Berlin was the tripwire most likely to trigger global thermonuclear Armageddon.

On August 30, 1961, the Soviets exploded a hydrogen bomb over Siberia, the first of thirty-one atmospheric detonations including the largest explosion in history—58 megatons, or 4,000 times more powerful than the bomb dropped on Hiroshima. Dismayed by the Soviet tests, Uncle Jack first responded with diplomacy, challenging the Soviets at the UN on September 25, 1961, not to "an arms race, but a peace race." He proposed to compete, instead, in the marketplace of commerce and ideas, and in science—contests that would benefit all humanity. When the Soviets spurned his offer and continued their tests, JFK reluctantly resumed America's own atmospheric testing in April, while continuing to advocate peace and disarmament. Beginning that month, on the evening news we watched mushroom clouds rising over idyllic tropical atolls as the United States detonated twenty-four bombs in the South Pacific. I wondered what became of the fishes, birds, the crabs and lizards and coral reefs, and all the tiny animals and palm trees in that tropical paradise.

Then, on October 27, the Allied governor general of West Berlin, American general Lucius Clay, made an unauthorized attempt to knock down the Berlin Wall using tanks equipped with bulldozer plows. Clay was trying to provoke the Soviets into a reaction that would justify a nuclear first strike. The Kremlin responded by sending its own tanks to foil Clay's demolition project. Clay's armored division met the Russians at the border crossing known as Checkpoint Charlie, in a perilous sixteen-hour face-off. The world trembled. The nuns had us practicing our "duck and cover" drills in double time. Using back-channel communications, Jack secretly asked Khrushchev to break the lethal deadlock by withdrawing his tanks first, promising that the United States would then follow suit within twenty minutes. Against the excoriations of the Kremlin's own hard-line militarists, Khrushchev courageously agreed.

General Clay bitterly complained to Secretary of State Dean Rusk that America had missed a crucial opportunity. "Today, we have the nuclear strength to assure victory at awful cost," Clay explained, voicing the Joint Chiefs' fear that the clock was ticking against America's first-strike advantage. The United States, he told Rusk, must be "prepared instantly to follow the next provocation with a nuclear strike. It is certain that within two or more years, retaliatory power will be useless as whoever strikes first will strike last." But Jack balked at a strategy that involved killing millions of innocent people. "What right do I have to touch that wall?" Jack asked his friends. "It's in East German territory!" He told Kenny O'Donnell that it seemed a "particularly stupid risk killing Americans over access rights to the Autobahn. . . . Before I back Khrushchev against a wall . . . the freedom of all Western Europe will have to be at stake." That summer after a family dinner at the Hyannisport compound he told Red Fay, "We're not going to plunge into an irreversible action just because a fanatical fringe puts so-called national pride above national reason. Do you think I'm going to carry on my conscience the responsibility for the wanton maiming and killing of children like our children we saw here this evening?" During Thanksgiving that year, he told my parents and Lem Billings, "I'd rather have a wall than a war."

After the cliff-edge confrontation at Checkpoint Charlie, Moscow relaxed its pressure on Central Europe. Khrushchev's good-faith maneuver, and the trust it engendered between Jack and Khrushchev, foreshadowed the resolution of the Cuban Missile Crisis showdown a year later. While the two leaders continued to denounce each other publicly, they discreetly developed a personal friendship that helped avoid the hot war craved by hard-liners in both Washington and the Kremlin. The principal vehicle for peace was a remarkable exchange of twenty-six secret back-channel letters between Jack and Khrushchev that began during the Berlin Crisis and continued until Jack's assassination.

Even before initiating his secret correspondence with Uncle Jack, Khrushchev had reached out in personal gestures clearly intended to mute the impacts of his sharp attack in Vienna. In June, Khrushchev sent Jack an upbeat and friendly note accompanied by two gifts. The first, a response to Jack's gift to Khrushchev of a scale model of the U.S.S *Constitution*, was a precise walrus-bone model of a New England whaling ship crafted from memory by a Siberian Inuit. The second was a totally unexpected gift for Caroline, a puppy of indeterminate pedigree whose mother was the famous cosmo-canine, Laika, whose ride on *Sputnik 2* made her the first dog in space. Pushinka joined the White House menagerie, which included Caroline's Shetland pony Macaroni, John's pony Leprechaun, a kitten, parakeets, rabbits, and hamsters, and a half dozen dogs, among them my favorite, Shannon, a blue roan Irish cocker spaniel that was a gift from the Irish descendants of Commodore John Barry—the American Revolution naval hero who hailed from the Kennedys' seacoast home village in Wexford. Pushinka never really embodied the warm personal congeniality that Khrushchev hoped to evoke with his gift. Apparently, the Soviet space-shot laboratory that whelped Pushinka did not give careful attention to her socialization. She was an ornery bitch who growled and bit all of us. President Kennedy had to finally impose a quarantine. Caroline's Welsh terrier, Charlie, somehow ran the blockade and Pushinka had her own pups in the summer of 1963. The president called the dogs "pupniks" and ordered them distributed to some of the five thousand children who wrote the White House requesting a Pushinka pup. Khrushchev followed Pushinka with the first of his secret letters three months later.

Khrushchev sent his first letter on September 29, 1961. Four days after Jack delivered a speech at the UN suggesting that each side take tandem baby steps toward universal disarmament. He used that speech to propose a strategy that seemed maniacally reckless to his spooks and fruit-salad brass. "It is therefore our intention to advance together step-by-step, stage-by-stage, until general and complete disarmament has been achieved."

The Soviet spy Georgi Bolshakov smuggled the first of Khrushchev's letters to Jack hidden in a newspaper delivered to Pierre Salinger. Bolshakov, the top Soviet spy in Washington, was a stocky bon vivant brimming with humor and athletic grace. To the U.S. State Department's horror, he became friendly with my parents after they met him at a Soviet embassy function. He was a frequent guest at Hickory Hill, and a favorite among us children for his loud laugh and Cossack dancing, but most of all because we knew he was a Russian spy and, in the heyday of James Bond, it was a thrill to have him in our home. My mother once egged Bolshakov into an arm-wrestling contest with my dad. "It was one of those times Daddy got really mad at me," she would later say. (Two others were when she wore a mink coat on a visit to the troops in the frozen trenches of South Korea—"I was freezing to death!" she explained—and when she used a special home-brewed treatment to cure LBJ's laryngitis just before an Al Smith Dinner where he was to have spoken opposite my father, shortly before the 1968 election.)

Ted Sorensen told me thirty years later that by using Bolshakov to skirt official channels, Khrushchev was "taking an enormous risk." The Kremlin brass would have been furious had his end run been discovered. But my father's friendship with Georgi—and Jack's correspondence with Khrushchev—kept the world from vaporizing in mushroom clouds.

The letters reveal a Soviet premier under dire pressure by Kremlin hard-liners to remain uncompromising toward his American counterpart. Khrushchev began his first missive with a description of his Black Sea dacha—his personal refuge. He continued with a fairly direct apology for his belligerence at Vienna. "My thoughts have more than once returned to our meeting in Vienna. I remember you emphasized that you did not want to proceed towards war and favored living in peace with our country while competing in the peaceful domain. . . . The whole world hopefully expected that our meeting . . . would turn relations in our countries into the correct channel and give the people confidence that at last peace on earth will be secured. To my regret, and, I believe, to yours, this did not happen." He told Jack he had been impressed by the accounts of visiting Soviet journalists who, after meeting Jack, had described him to Khrushchev as a man of "inquisitiveness, modesty and fearlessness," qualities, Khrushchev said, "which are not often found in men who occupy such a high position." Finally, he appealed to Jack for a path to peace—oddly, for an avowed atheist, using the biblical metaphor of Noah's ark. "We have no other alternative," Khrushchev wrote. "Either we should live in peace and cooperation so that the Ark maintains its buoyancy or else it sinks."

Three weeks later, on October 16, 1961, Jack replied from his own seaside retreat at Hyannisport. He drafted the letter as he looked out over the field where we all played baseball and football. "My family," he wrote Khrushchev, "has had a home overlooking the Atlantic for many years. My father and brothers own homes near my own; and my children always have a large group of cousins for company. So this is an ideal place for me to spend my weekends during the summer and fall, to relax, to think, to devote my time to major tasks instead of constant appointments, telegrams, calls and details. Thus, I know how you might feel about the spot on the Black Sea." He embraced Khrushchev's image of the ark, observing that all of us must be "determined that it stay afloat." And he reminded Khrushchev that our political systems had not prevented cooperation between our nations in the past. "Whatever our differences, our collaboration to keep the peace is as urgent—if not more urgent—than our collaboration to win the last world war."

During the Berlin Crisis, Khrushchev wrote a second secret letter to Jack, hinting that he was at odds with his military and the hard-liners within his own government. "I have no ground to retreat further, there is a precipice behind." Khrushchev requested in that communiqué that Jack de-escalate his Cold War rhetoric, which was complicating Khrushchev's efforts with his own hard-liners. Beneath the obligatory Cold War saber rattling, the two men developed a bond fortified by their knowledge that they both were isolated, surrounded by hard-liners and fighting a flood tide sweeping their nations toward war.

Laos

On January 19, 1961, during JFK's transitional briefing with Eisenhower, the outgoing president urged Jack to intervene to save Laos from Communist Pathet Lao guerrillas. Under Eisenhower, the Allen Dulles CIA and the Pentagon had deposed Laos's democratically elected government of Souvanna Phouma and replaced him with a brutal military dictator, General Phoumi Nosavan, who was now battling for survival against the Communist guerrillas. After listening to Ike's pitch, Jack told Kenny O'Donnell and Dave Powers, "There he sat telling me to get ready to put ground troops into Asia, the thing he himself had been carefully avoiding for the last eight years." Despite his eloquent warning against the military-industrial complex, Eisenhower had done little to rein in the power of that syndicate. The outgoing president now appeared to be baiting a trap for Uncle Jack that threatened to transform America into an imperium.

Jack was predisposed against sending U.S. troops to Asia. He had traveled extensively in the region in the 1930s and early 1950s and greatly admired the anticolonial nationalists fighting for independence from European power. He equated them with the American Revolutionaries. "If one thing was born in me as a result of my excursions in the Middle as well as the Far East," he said after a trip to Vietnam with my father in 1951, "it is that communism cannot be met effectively merely by force of arms." General Douglas MacArthur, still America's most respected military figure and the undisputed authority then on fighting ground wars in Asia, fortified Jack's natural instincts during an Oval Office meeting. Referring to Allen Dulles's mischief throughout the Third World during the Eisenhower years, MacArthur told Jack, "The chickens are coming home to roost and [you] live in the chicken coop." MacArthur added a warning that ought to resonate still today: "Anyone wanting to commit American ground forces to the mainland of Asia should have his head examined."

General Maxwell Taylor recalled that MacArthur's advice "made a hell of an impression" on Jack, who routinely threw those words back in the faces of his generals when they asked him for ground troops in Laos or Vietnam. "He'd get this military advice from the Joint Chiefs or from me or anyone else, he'd say 'Well now, you gentlemen, you go back and convince General MacArthur, then I'll be convinced.'"

Instead of heeding Eisenhower's counsel that he plunge into Laos, JFK withdrew backing for the CIA's puppet dictator, and announced support for a neutral independent Laotian government to be chosen by the Laotian people. He dispatched his ambassador-at-large Averell Harriman to negotiate a treaty with thirteen nations, including the Soviet Union, to support a new Laotian government that would be unaligned with East or West. By June 23, 1961, all parties had signed.

Openly mutinous CIA and Pentagon cold warriors, for whom

Laos was the front line in the war against communism, condemned Jack's agreement as appeasement. But the Bay of Pigs fiasco had confirmed Jack's skepticism toward his military and intelligence bureaucracies, and their scorn only fortified his determination. He told Arthur Schlesinger, "If it hadn't been for Cuba, we might be about to intervene in Laos. I might have taken this advice seriously."

Vietnam

Jack was wary of American involvement in Vietnam from the outset, and was moving to extract us at the time of his death. Jack's struggle to avoid that quagmire provides a valuable handbook for modern presidents under similar pressure to send American troops to die in Asia and the Middle East. The arguments of warmongers are hauntingly familiar, yet the case against America committing ground troops to Asia remains overwhelming. History's judgment will be harsh against those leaders who ignore its lessons.

Vietnam was a CIA caper. The South Vietnamese nation and its corrupt government were "essentially the creation of the United States," according to the Pentagon Papers, the Defense Department's own internal analysis of the war. Vietnam's independence leader Ho Chi Minh had fought alongside and supported American troops to evict the Japanese from South Asia. He saw America as a former colony herself, as the reliable friend of little nations in their idealistic struggles against imperialism. His admiration for the United States was so genuine that he named his guerrilla movement "the Viet American Army" and recited long tracts from the Declaration of Independence in his inaugural address to the Vietnamese people. After the French fled their former colony, the Communists would have taken control of the entire country had not President Eisenhower and Allen Dulles installed their corrupt proxy, Ngo Dinh Diem, as ruler of the government, derailing the democratic elections provided for in a four-nation covenant signed in Geneva in 1954. The CIA then infiltrated the country with U.S.

advisers, running shadow governments in twenty of forty-one provinces and installing spies at every level of public, private, and military life.

In 1959, Eisenhower had sent the first of 685 military advisers into Vietnam to counsel the Diem government in its battle against a few thousand South Vietnamese guerrillas known as the Vietcong and a division of North Vietnamese troops sent by Ho Chi Minh, who had evicted the French from the North and was intent on reunifying his country. Eisenhower justified American intervention by advancing what would become known as the "domino theory." As Ike explained it, "The loss of South Vietnam would set in motion a crumbling process that could, as it progressed, have grave consequences for us." But Eisenhower's 685 advisers were not sufficient; with little support from its population, South Vietnam's crooked regime was only tenuously clinging to power.

During Jack's first months in office, the Pentagon requested 3,600 ground troops. Jack reluctantly sent 100 additional advisers, and only with the understanding that Vietnam had an army able to defend itself against Communist aggression. Roswell Gilpatric, who headed the White House's Vietnam task force, said that Jack "showed at the very outset an aversion to sending more people out there." Jack told *New York Times* columnist and family friend Arthur Krock in October 1961 that American troops had no business in the Asian mainland. He would eventually allow the number of advisers to grow to 16,000 (fewer soldiers than he sent to Mississippi to quell the Ole Miss rioters), but he steadfastly refused to put combat troops in Vietnam, earning him the antipathy of both liberals and conservatives, who rebuked him for "throwing in the towel" against international communism. His critics included his most trusted advisers and friends—General Maxwell Taylor, Defense Secretary Bob McNamara, Secretary of State Dean Rusk, National Security Adviser McGeorge Bundy, and Jack's own vice president, Lyndon Johnson.

When Johnson visited Vietnam in May 1961 at Jack's request, he returned adamant that it was insufficient to send military advisers and equipment: victory required U.S. combat troops capable of independent action against guerrilla fighters. Virtually every one of Jack's advisers concurred, yet the president steadfastly resisted, saying that we could support the South Vietnamese but we could not do their fighting for them. Thinking about it later, Taylor would observe, "I don't recall anyone who was strongly against [sending combat troops to Vietnam] except one man, and that was the President. The President just didn't want to be convinced that this was the right thing to do. It was really the President's personal conviction that U.S. ground troops shouldn't go in."

Following Johnson's disappointing assessment, JFK stalled his hawkish generals by sending Maxwell Taylor to Vietnam on another fact-finding mission. Taylor knew exactly what Jack was looking for. My father and his brother personally drafted for Taylor the findings that they wanted Taylor to give them when he returned from his trip to Vietnam. As CIA historian and former intelligence officer John Newman shows in his book, *JFK in Vietnam*, those edits, drafted into Taylor's report by Victor Krulak and Fletcher Prouty, would lay the groundwork for Jack's subsequent National Security order calling for the withdrawal of all American troops from Vietnam, and Jack and Daddy had their edits couriered to Honolulu to intercept Taylor on his way home from Saigon. Nevertheless, Taylor too was persuaded by military and intelligence experts across the Pacific, and came back urgently recommending intervention. To prevent the fall of Vietnam, he said, we needed a "U.S. force of the magnitude of an initial 8,000 men," expanding to as many as six divisions of combat troops "or about 205,000 men." Frustrated by Taylor's report, Jack next sent a reliable dove, Harvard economist John Kenneth Galbraith, to Vietnam to make the case for nonintervention. Jack confided in his old friend that he felt politically vulnerable and isolated. "There are limits," Jack told Galbraith, "to the number of defeats I can defend in one twelve-month period. I have had the Bay of Pigs and pulling out of Laos, and I can't accept a third."

"They want a force of American troops," Jack told Arthur

Schlesinger, referring to Vietnam. "They say it's necessary in order to restore confidence and maintain morale. But it will be just like Berlin. The troops will march in; the bands will play; the crowds will cheer; and in four days everyone will have forgotten. Then we will be told we have to send in more troops. It's like taking a drink. The effect wears off and you have to take another." By the summer of 1963, as the corrupt South Vietnamese regime continued to crumble, Jack privately told his trusted friends and advisers that he intended to get out as soon as the 1964 election was over. He told the Vietnam War's most outspoken Senate critic, Mike Mansfield: "I can't do it until 1965. After I'm re-elected." Later that day, he explained to Kenny O'Donnell, "If I tried to pull out completely now, we would have another Joe McCarthy red scare on our hands. But I can do it after I'm re-elected." His political rivals in the upcoming election, Rockefeller and Goldwater, were uncompromising cold warriors who would have loved to tar Jack with the brush that he had lost not just Laos, but now Vietnam. Goldwater was campaigning along the lines of General LeMay's stated desire for "bombing Vietnam back into the Stone Age," a lyrical and satisfying construct to the Joint Chiefs and the CIA. "So we better make sure I am re-elected," Jack told Kenny.

JFK was fighting a public clamor, which in the summer of 1963 ran two-to-one for escalation. The *Washington Post* and the *New York Times* were bawling to expand the U.S. military commitment. The Joint Chiefs, already in open revolt against Jack for failing to unleash the dogs of war in Berlin, Cuba, and Laos, were howling for a massive influx of ground troops. The brass regarded talk of withdrawal as treasonous. Langley's mood was even uglier. Journalist Richard Starnes quoted a high-level U.S. diplomat who described the insubordinate agency as a "malignancy." Starnes cited another U.S. official, who warned, "If the United States ever experiences a *Seven Days in May* [coup], it will come from the CIA and not the Pentagon. . . . [The CIA] represents tremendous power and total unaccountability to anyone."

By now, Jack and my dad realized how truly isolated they were;

their own government was in open revolt against them, and they were beginning to lose control. Already Jack had ordered McNamara, in the spring of 1962, to order the Joint Chiefs to plan a phased withdrawal, to disengage the United States from Vietnam altogether by December 1965. On May 8, 1962, McNamara instructed General Paul Harkins "to devise a plan for bringing full responsibility [for the war] over to South Vietnam." Defiant, the general ignored the order until July 23, 1962, when a furious McNamara again commanded him to produce a withdrawal plan. The foot-dragging brass returned ten months later with a half-baked scheme that fell short of complete withdrawal by the end of 1965. McNamara ordered them back again to the drawing boards. McNamara told his assistant secretary of defense, John McNaughton, that the president intended to "close out Vietnam by '65 whether it was in good shape or bad."

On September 2, 1963, Jack laid the groundwork for his unpopular decision by telling CBS News' Walter Cronkite in a televised interview, "In the final analysis, it is their war. They are the ones who have to win or lose it. We can help them, we can give them equipment. We can send our men out there as advisers, but they have to win it, the people of Vietnam." On October 11, 1963, five weeks before his death, JFK bypassed his own National Security Council and issued National Security Action Memorandum 263, making official the withdrawal from Vietnam of "1,000 U.S. military personnel by the end of 1963" and "the bulk of U.S. personnel by the end of 1965."

On November 20, 1963, two days before his trip to Dallas, Jack announced at a press conference a plan to assess "how we can bring Americans out of there. Now, that is our objective, to bring Americans home." The following morning he reviewed a casualty list for Vietnam indicating that seventy-three Americans had died there to date. Shaken and angry, Jack told his assistant press secretary, Malcolm Kilduff, "After I come back from Texas, that's going to change. There's no reason for us to lose another man over there. Vietnam is not worth another American life."

On November 24, 1963, two days after Jack died, Lyndon

Johnson met with the American ambassador to Vietnam, Henry Cabot Lodge, whom Jack had been on the verge of firing for insubordination. LBJ told Lodge, "I am not going to be the president who saw Southeast Asia go the way China went." Eventually 500,000 Americans, including many of my friends, entered the paddies of Vietnam, and 58,000 never returned. The war endured for a decade, and ended a year after I reached draft age.

Over dinner at my Hudson Valley home in 2015, Daniel Ellsberg told me of a meeting with my father in 1967. At the time, Ellsberg, a former Marine first lieutenant and wavering war hawk, had just returned from two years in Vietnam and was working for the Rand Corporation as a Pentagon analyst writing a top-secret account of the Vietnam War that would later be known as the Pentagon Papers. Ellsberg had read the bundles of warmongering memos from 1962 and '63 bellowing for JFK to escalate. He asked my dad how he and his brother had managed to stand against those overwhelming currents. My dad launched into a rote explanation that JFK did not want to fight France's leftover colonial battles and wanted no part of "a white man's war against Asians, a war of rich against poor, a war for imperialism against nationalism and self-determination." Pressing my father, Ellsberg asked whether JFK would have accepted a South Vietnamese defeat. "We would have handled it like Laos," my dad told him. My father explained that it would have been a simple thing to put in a South Vietnamese government that would decline further American support. Jack Kennedy gave a similar answer to the same question from Kenny O'Donnell and Dave Powers six years earlier. "Easy," he told them. "Put a government in there that will ask us to leave." My father told Ellsberg, "My brother was determined to never send American soldiers to Asia." Ellsberg was intrigued. "I asked your father, somewhat impertinently, 'What made you and your brother so clever?'" Ellsberg told me that he would never forget my father's reaction. "He slammed his hand on the desk with such force that I leapt in my chair. Whap! His voice raised in volume so that he was practically shouting, 'Because we were there!' He pounded the desk again. Whap! 'WE were there in

'51! We saw what was happening to the French. We saw it. We were determined never to let that happen to us.'"

As a young congressman, Jack had visited Vietnam with my father. The two brothers had marveled at the fearlessness and competence of the French Legionnaires, and the absolute hopelessness of their cause. On the same trip, Charles de Gaulle had warned Jack that America should avoid the trap. Soon after he returned, Jack isolated himself in the Senate with his outspoken opposition to squandering American blood and treasure in this "hopeless internecine struggle."

During this period, Jack read voraciously everything he could about Vietnam's colonial history. Sadly, there was very little available in English. Jack got his hands on four French histories of Vietnam and began plodding his way through them using his own high school French and the help of a beautiful young reporter he had recently met who was fluent in the language. It was during this project that Jaqueline Bouvier Kennedy's romance with my uncle first blossomed, a fact reported by none of her biographers. Among those histories were two by French sociologist and historian Paul Mus (who would be a key source for Frances Fitzgerald's influential book *Fire in the Lake*) and France's proconsul in Vietnam, Jean Santenay. Jack was particularly taken by Jackie's translation of a conversation Santenay reported having with Viet Minh leader Ho Chi Minh. Ho told him that if the French chose to fight, "you will kill ten of us for every one we kill of you, but in the end, you will be the one exhausted." In 1976, long before that book was ever translated to English, Daniel Ellsberg had a conversation about Santenay with Jackie. When Ellsberg began reciting Ho's quotes, Jackie completed it for him. "How could you possibly know that quote?" Ellsberg asked. Jackie told him that it was over Santenay that she and Jack had fallen in love.

Three years later, in April of 1954, he made himself a pariah even in his own party by condemning the Eisenhower administration for entertaining French requests for U.S. assistance in Indochina, predicting that fighting Ho Chi Minh would mire our country in France's doomed colonial legacy. "No amount of American military assistance in Indochina can conquer 'an enemy of the people' which has the sympathy and covert support of the people."

Like Grandpa, Uncle Jack recognized that hubris corrodes great powers. He warned that the illusory notion that we could pick and choose governments for foreign nations would "drain our treasury and erode our moral leadership." Worse, such interventions abroad would strangle our democracy at home, and ruin us as an exemplary nation.

The Cuban Missile Crisis

Following the abortive U.S.-backed invasion at the Bay of Pigs, Fidel Castro requested sophisticated Soviet weaponry to deter further incursions from the United States. From Khrushchev's perspective, the tactic of placing missiles in Cuba was rational. Russia, after all, was surrounded by nuclear missile bases in immediate proximity to its borders. In contrast to the United States, which had never been invaded and had only the tiny impoverished island of Cuba to fret about, a tight gauntlet of hostile nations encircled the Soviet Union. One of these, Nazi Germany, had invaded Russia, killing millions of her citizens and occupying her cities. "Our [Cuba-based] missiles would have equalized what the West likes to call the 'balance of power,'" Khrushchev later explained. But to Americans, Khrushchev's decision to position nuclear missiles and 60,000 support troops in Cuba was an outrageous act of aggression intended to upset the equilibrium that Jack had advocated for at Vienna. On October 16, 1962, Jack received aerial photographs proving that the Soviets had installed nuclear missiles in Cuba. For thirteen days— the most perilous in human history—the world trembled at the brink of nuclear obliteration.

JFK immediately assembled an executive committee of national security advisers, which became known as Ex Com. Predictably, the Joint Chiefs were urging a full-out preemptive nuclear strike. Even Jack's closest, most trusted counselors concurred with the brass. During the heated Ex Com debates, only McNamara and my father strongly opposed a preemptive strike. In that quaint age, preemptive wars were still considered unconscionable acts of barbarism; Americans of that post–World War II era took it as gospel that Japan's surprise attack on Pearl Harbor was the apogee of underhanded barbarism. My father passed a note to Jack, as their colleagues plotted a first strike: "Now I know how Tojo felt." My father argued that preemptive war was both morally reprehensible and contrary to American values. He later wrote about those deliberations: "I could not accept the idea that the United States would rain bombs on Cuba, killing thousands and thousands of civilians in a surprise attack." Early into the debate, my father engaged in a heated screaming match with former secretary of state Dean Acheson, who insisted that an air strike was the only prudent course. My father ended the donnybrook by shouting, "We're not going to make my brother the Tojo of the 1960s." Upset that the president found his brother's arguments persuasive, Acheson abandoned the Ex Com, retreating to his Maryland farm to sulk. According to Kenny O'Donnell, "Bobby continued to hammer on the moral question of a sneak attack by a big nation on a small independent country, and his persistent crusading finally turned several members of the group." Douglas Dillon similarly recalled, "What changed my mind was Bobby Kennedy's argument that we ought to be true to ourselves as Americans, that a sneak attack was not in our traditions."

An air attack also risked triggering retaliatory nuclear attacks from Cuba against the United States, and from the Soviet Union against Berlin, developments that could rapidly cascade into all-out nuclear war. My father favored action that would make clear the seriousness of the United States' determination to get the missiles out of Cuba, while still allowing the Soviets to save face. JFK adopted my father's and McNamara's recommendation for a maritime quarantine, or blockade, of Cuba.

On October 22, 1962, Jack took to the airwaves with a televised speech announcing the discovery of Soviet missiles in Cuba and the U.S. imposition of "a strict quarantine of all offensive military equipment under shipment to Cuba." He demanded that Russia dismantle and withdraw all nuclear weapons from the island. His Joint Chiefs were apoplectic. Even JFK's own handpicked ally at the CIA, John McCone, and his vice president, Lyndon Johnson, saw the blockade as a sign of weakness. When Jack queried his military brass as to whether a U.S. nuclear strike on Cuba would prompt a retaliatory nuclear strike or a Soviet occupation of West Berlin, Admiral George Anderson and General Curtis LeMay assured him the Russians wouldn't dare to retaliate. But Jack pushed back skeptically: "They can't let us just take out, after all their statements, take out their missiles, kill a lot of Russians and not do anything."

Thirty-five years later, at a 1997 conference in Havana that included Russian, U.S., and Cuban participants, the Russians revealed that one hundred of their Cuban missiles had already been armed with nuclear warheads. More disturbing, the Soviet Presidium had authorized individual missile battery commanders to fire the missiles at will, if they came under attack. Daniel Ellsberg told me, "That virtually guaranteed an escalation to full-bore nuclear exchange between the Soviet Union and Western forces. Russia and Europe would have been annihilated in minutes. America would survive the initial damage, but with sixty million Americans dead. But then nuclear winter, which we didn't know about, would have wiped out all but about one percent of the world population." The only survivors, according to Ellsberg, would have been a handful of South Pacific islanders who could forage mollusks.

With 3,000 nuclear missiles and bombs under his command, and itching for a fight, LeMay exploded at Jack's restraint. "He was just beside himself," Jack told Kenny O'Donnell. Demanding an air strike, LeMay scornfully told Jack that the blockade was "almost as bad as the appeasement at Munich." That was a potent insult, referencing Grandpa's support of Chamberlain's discredited peace efforts before World War II. "I just don't see any other solution except direct military intervention right now," added LeMay. "I think that a blockade and political talk would be considered by a lot of our friends and neutrals as being a pretty weak response to this, and for sure a lot of our citizens would feel that way too." Afterward, while describing the colloquy to Special Assistant Kenny O'Donnell, Jack commented, "Those brass hats have one great advantage in their favor. If we . . . do what they want us to do, none of us will be alive later to tell them they're wrong."

We didn't see my dad during most of the Cuban Missile Crisis. For thirteen days and twelve nights he stayed at the White House. At the height of the crisis, the government made plans to evacuate public officials and their families. We all knew that the capital would be vaporized in the first minutes of a nuclear exchange. Jack asked Jackie to go with John and Caroline to the bunkers in West Virginia. She refused, saying she would not want to live without him. When U.S. marshals arrived at our home to transport our family to the bomb-proof subterranean metropolis, my father called Hickory Hill to forbid the move. Joe and I complained to him on the phone. We pleaded to be allowed to at least go see the giant caverns. My dad told us that we needed to be "brave soldiers" and show up at Our Lady of Victory School, that our absence would be noticed and might cause panic in the capital. This was a great disappointment to me; I was anxious to eyeball the setup firsthand.

During that phone call from the White House, my father told us that if there was a nuclear war, none of us would want to be alive anyhow. While I idolized my dad, I just couldn't go along with him on that one. What about all that planning and practicing for the apocalypse at Our Lady of Victory? What was the point of all those drills? Like most other kids my age, I'd watched lots of post-apocalyptic science fiction movies and TV shows. I considered my father's view overly pessimistic.

Those Hollywood pieces inclined me toward the Joint Chiefs' position that even all-out nuclear war wouldn't be so bad. The idea of living in the adventure-filled postapocalyptic Armageddon was nearly as heady to me as it was to the Joint Chiefs, who of course, would be running the show. According to their assessment, we could declare victory if we lost only sixty million people, so long as the Soviets lost 130 million. It was giddy up there on the steep wave of what McNamara later called the "Cold War mob psychosis."

I, for one, intended to be among the survivors. Nuclear holocaust, I reasoned, could hardly be worse than school. I had never really caught on to academics, and my grades were disappointing at best. I was pretty confident that nuclear war would end school and obliterate my "permanent record." I felt like I did my best work amid calamity; I tended to prosper in chaos. I knew I could thrive in the woods. I would eat birds' eggs, snakes, frogs' legs, freshwater mussels, crayfish, mudpuppies—exactly the circumstance in which I seemed to flourish. I was ready to fight off mutants and do my part in recreating civilization after the apocalypse. Were we truly going to waste all that canned fruit cocktail?

In hindsight it all seems surreal. Here we were, looking at the end of the world, engaging in debates in Catholic school about the morality of locking your friend out of your fallout shelter. Hours of civic drills had us all participating in the greatest act of madness in human history. During those weeks, we were all suspended in a kind of science fiction insanity, looking out at our fragile world, knowing that false pride or a single misstep by political leaders might make it all disappear. On October 24, at the height of the Cuban Missile Crisis, two Soviet ships and their submarine escort cruised toward the quarantine line, where U.S. helicopters and aircraft waited with depth charges and cannon. The globe held its collective breath. Sitting together in the Oval Office, my dad and Uncle Jack feared that they had lost control of events. The world was spiraling toward apocalypse. As my father described the scene in his diary, "We stared at each other across the table. For a few fleeting seconds it was almost as though no one else was there and he was no longer the President. Inexplicably, I thought of when he was ill and almost died; when he lost a child; when we learned that our oldest brother had been killed; of personal times of strain and hurt." Suddenly, the Soviet ships stopped dead in the water just short of the demarcation line.

Khrushchev's brave decision to end the standoff was rooted in the relationship he had cultivated with Jack; two days later, on October 26, he sent Jack a secret letter offering to remove the missiles if the United States would pledge not to invade Cuba. The crisis seemed resolved, but worse moments lay ahead, for Khrushchev was still under tremendous pressure at home.

On October 27—Black Saturday—a Cuban surface-to-air missile (SAM) crew commander, acting without orders, shot down an American U-2 reconnaissance aircraft, killing its pilot, Major Rudolf Anderson, Jr. American military and intelligence experts saw the act as a deliberate provocation by the Kremlin. The Joint Chiefs crowed that the blockade strategy had failed. They had deemed it a sissified response from the outset, and now they were barking at Jack for a forceful retaliation that would destroy the Soviet missile sites. Meanwhile, the Kremlin military machine, egged on by a frantic letter from Castro, was pushing Khrushchev to hold a muscular Soviet position. Khrushchev sent Jack a second, more bellicose letter, demanding that, in addition to the pledge not to invade Cuba, the United States must remove its Jupiter missiles from the Russian border in Turkey. In fact, Jack had ordered the obsolete Jupiters removed months before, but a defiant Pentagon had delayed the decommissioning. Now America could hardly be seen by our NATO allies as removing them under threat.

At the White House my father debated the Ex Com team and the Joint Chiefs, who argued that Khrushchev's letter meant the blockade had failed. My father and Jack seemed to be fighting a losing battle to avert nuclear conflagration. My dad later wrote of Jack: "The thought that disturbed him the most and that made the prospect of war much more fearful than it would otherwise have been, was the specter of the death of the children of this country and all the world—the young people who had no role, who had no say, who knew nothing even of the confrontation, but whose lives would be snuffed out like everyone else's. They would never have a chance to make a decision, to vote in an election, to run for office, to lead a revolution, to determine their own destinies."

Painted into their respective corners, Khrushchev and Jack were searching for paths out. It was my father's idea that Jack send a letter accepting Khrushchev's milder earlier proposal and to ignore Khrushchev's later letter demanding the removal of the U.S. missiles from Turkey. JFK sent a letter pledging that the United States would not invade Cuba. He asked my father to meet with Soviet ambassador Anatoly Dobrynin to secretly assure him that the missiles in Turkey would be withdrawn after a cooling-off period. Dobrynin met my dad in his Justice Department office, where, surrounded by his children's finger paintings, my father told the ambassador that he and his brother had their backs against a wall. The Soviets, he said, had drawn first blood. The United States would be forced to bomb the Cuban launch pads and follow up with an invasion if Russia didn't remove the missiles within forty-eight hours. It sounded like an ultimatum, but my father told Dobrynin it was not. Dobrynin interpreted this to mean that my father and JFK feared a coup if they did not act. Dobrynin's follow-up memo to the Soviet Foreign Ministry described my father as haggard, looking like he hadn't slept for days. "Taking time to find a way out is very risky," Dobrynin recorded my father as saying. Because many unreasonable heads among the generals, and others, were "itching for a fight . . . the situation might get out of control, with irreversible consequences." In a carefully calibrated approach, my father conveyed Jack's written pledge to not invade Cuba and secretly promised to remove the Jupiter missiles from Turkey, providing the withdrawal was not publicly linked. If the secret leaked, the deal was off.

That night many people in our government went to sleep wondering if there would be a tomorrow. But my father's meeting and Jack's artfully drafted letter, cowritten by Ted Sorensen, did their work. On Sunday, October 29, the world stepped back from the brink. As soon as he received the message from Dobrynin, Nikita Khrushchev announced he would withdraw the Soviet missiles in Cuba in exchange for JFK's pledge not to invade. True to his word, Jack quietly removed the U.S. missiles from Turkey six months later.

My dad was not exaggerating to Dobrynin the fragility of the president's hold on power. At the boiling point Jack instructed the Navy to stay away from Cuba to avoid risking an accidental provocation, but in yet another act of blatant disobedience, Admiral George Anderson sent battleships and cruisers to probe the Cuban shore, clearly hoping to provoke a confrontation. Acting without presidential authority, General LeMay's deputy, General Tommy Power, elevated alert levels to DEFCON 2—one step below full nuclear war—and broadcast his provocation to the Russians. In direct disobedience of a presidential directive, Bill Harvey, the CIA's bug-eyed, gun-toting, Cuba project director dropped sixty elite commandos into Cuba to prepare the ground for the invasion they deeply hoped would follow. My father went through the roof when he learned of this act of insubordination in the midst of the hair-trigger missile crisis. "You were dealing with people's lives, and then you're going to go off with a half-assed operation such as this?" he exclaimed. A defiant Harvey replied, "We wouldn't be in this jam in the first place if your brother had had the balls to invade." A Harvey protector, CIA Deputy Director Richard Helms, realized he had to beat my father to the punch and transfer Harvey out of Washington. Harvey ended up in Italy, where the insubordinate spy continued to mingle with mobsters and plot against the Kennedys. Even when the Cuban Missile Crisis subsided, General LeMay, spoiling for a war to end all wars, hoped all was not lost. According to my father, he ventured: "Why don't we go in and make a strike on Monday anyway?"

With the crisis abated, Jack, referencing Lincoln's fateful visit to Ford's Theatre to celebrate the successful end to the Civil War, joked to my dad, "This is the night I should go to the theater." My father replied, "I want to come with you." My father would later say of the thirteen Ex Com people in the room that they were "perhaps the most able, bright and energetic in the country, and if any one of a half dozen of them were President the world would have been plunged into catastrophe." Jack forbade his staff to gloat or to speak of the Cuban Missile Crisis resolution as a victory. He wanted to allow Khrushchev to honorably retreat and avoid rhetoric that would humiliate the Soviet leader.

Jack graciously hosted a White House reception for the Joint Chiefs, giving them some undeserved credit for the crisis's peaceful resolution. The Joint Chiefs grudgingly attended the party but spurned the olive branch. Disgusted, LeMay taunted, "But the administration was scared to death [the Russians] might shoot a missile at us." LeMay raged at Jack to his face, "It's the greatest defeat in our history." Both the Pentagon and the CIA smoldered in a state of revolt. Bill Harvey, the CIA officer who had green-lighted the commandos, railed to his colleagues that Jack's failure to launch was "treasonous." Incensed military brass were in a state of "disbelief." Pentagon defense analyst Daniel Ellsberg described the atmosphere in Arlington: "There was virtually a coup atmosphere in Pentagon circles." Nixon publicly condemned Jack for pulling "defeat out of the jaws of victory." However, Dean Acheson, Truman's former secretary of state, who had been the most hawkish member of the Ex Com group, was impressed, praising Jack for his "leadership, firmness and judgment."

When I attended Harvard, thirteen years later, my professor, the revisionist historian Steven Therston, argued that Jack should have left the Soviet missiles in Cuba. As I sat in his class, Therston complained that JFK's "schoolboy machismo" had brought the world to the brink of destruction. But liberal historians like Therston take no account of the politics of that day. The presence of nuclear missiles in Cuba put America's southern states a few seconds away from nuclear annihilation and eighty million Americans within reach of Russia's warheads, a morally and politically untenable position for any American president. Southern conservatives would have been in bloody revolt. If JFK had simply allowed the missiles to stay in Cuba, "You would have been impeached," my dad told Jack. "That's what I think," Jack replied. "I would have been impeached."

Jack Kennedy was the right man in the right place at the right time. As Arthur Schlesinger told an interviewer, "JFK had a great capacity to resist pressure from the military. He simply thought he was right. Lack of self-confidence was never one of Jack Kennedy's problems. We would've had a nuclear war if Nixon had been President during the missile crisis. But Kennedy's war hero status allowed him to defy the Joint Chiefs." It's equally likely that if Lyndon Johnson had been president, the world would have ended. Johnson consistently opposed the blockade and supported a preemptive strike.

Khrushchev said to American political journalist Norman Cousins that, following the Missile Crisis, the Kremlin's military hardliners regarded him as a traitor. But American citizens and people around the world breathed relief. That happy ending was a resounding victory for humanity. Furthermore, the near miss firmed the two leaders' resolve to mitigate the risks of armed conflict in the future. In his memoirs Khrushchev would write that Kennedy had won his "deep respect" during the crisis. "He didn't let himself become frightened, nor did he become reckless. . . . He showed real wisdom and statesmanship when he turned his back on right-wing forces in the United States who were trying to goad him into taking military action against Cuba."

Toward Nuclear Disarmament

Sobered by their glimpse at Armageddon, both Jack Kennedy and Nikita Khrushchev now embarked on a path to end the Cold War. After the Missile Crisis, Uncle Jack installed a special hotline to the Kremlin on his Oval Office desk, and another red telephone at the compound in Cape Cod. Today the old phone wires still protrude from basement walls in the former summer White House, now owned by my brother Chris. Both Jack and my father recognized that the greatest obstacle to easing global tensions was the ability of the Soviet premier and the American president to communicate candidly, and privately. While Jack continued to publicly challenge the Soviet leader to give freedom to his people, he was also working to pave the road to détente. Against the advice of Vice President

Johnson and his own State Department, Jack authorized sales of wheat to the Soviet Union in 1963, a transaction that Barry Goldwater and the Republican right reproved as more treason.

That spring, at Khrushchev's request, Jack ordered a crackdown on Cuban refugee gunboats then conducting raids on Cuban and Russian ships from bases in Miami. He wrote secretly to Khrushchev that he was "aware of the tensions unduly created by recent private attacks on your ships in Caribbean waters, and we are taking action to halt those attacks which are in violation of our laws." The authors of these attacks were mainly members of two CIA-sponsored commando groups, the Directorio Revolucionario Estudiantil (DRE) and Alpha 66. Alpha 66 had been conceived by the CIA Cuba Operations chief David Atlee Phillips—allegedly an Operation 40 associate—and was led by the Cuban accountant and violently anti-Castro activist Antonio Veciana. When I met him in Miami in 2017, Veciana told me that the CIA's purpose for the raids was "to publicly embarrass President Kennedy and force him to move against Castro." Phillips was the CIA's propaganda czar for both the Bay of Pigs invasion and the CIA's 1954 coup against the democratically elected Guatemalan government. When the House Select Committee on Assassinations (HSCA) investigated the JFK assassination in 1977, investigators considered Phillips and his then boss, William Harvey, as central suspects in Jack's murder. Veciana, a leader of the Cuban refugee efforts to assassinate Castro, testified to the HSCA that he had met with his CIA handler, Maurice Bishop, and Lee Harvey Oswald in Dallas in early September 1963. Veciana subsequently told me that "Maurice Bishop" was the nom de guerre of David Atlee Phillips.

Jack ordered the CIA's Alpha 66 commandos confined to Miami and had the Coast Guard impound the group's vessels and arrest their crews. My dad sent the FBI to Cuban exile headquarters around Florida, seizing dynamite and bombs. Jack and my father worked with the British and Bahamian governments to shutter Alpha 66's bases in Anguilla, forty miles off the Cuban coast. Veciana told me that Phillips, who had hated Jack since the Bay of Pigs, was in a murderous fury toward him after JFK shut down Alpha 66.

Ted Sorensen said that after the Cuban Missile Crisis the president had a kind of epiphany on the subject of disarmament. Jack told his childhood friend British ambassador David Ormsby-Gore, "You know, it really is an intolerable state of affairs when nations can threaten each other with nuclear weapons. This is just so totally irrational. A world in which there are large quantities of nuclear weapons is an impossible world to handle." Ormsby-Gore, who was a habitué at Hickory Hill and Cape Cod, shared British prime minister Harold Macmillan's horror at the prospect of nuclear war. Their intimate discussions greatly fortified Jack's own convictions on the issue. My father also matured. The crisis forced him to confront the rectitude of any use of nuclear weapons. In his journal he posed the question: "What, if any circumstance or justification, gives this government or any government the moral right to bring its people and possibly all people under the shadow of nuclear destruction?"

In December 1962, a month after the Missile Crisis, Jack's chief science adviser, Jerome Wiesner, whom I would later befriend when I lived in Cambridge, gave Jack a memo assessing the arms race as "an unmitigated disaster for the national security of the United States," as recalled by another administration aide, Marcus Raskin. Wiesner had been among the atomic weapons gurus who had originally decried the so-called missile gap during the 1960 election campaign, an alleged deficit originally ballyhooed by an Eisenhower commission of graybeard defense experts. After Jack's inauguration, Wiesner informed Jack that their appraisal had been badly mistaken: the Soviets would never achieve anything close to first-strike capability. Moreover, he told Jack that U.S. force levels already far exceeded levels required to maintain a secure deterrent. Worse, the arms race was not only bankrupting our economy, it was actually *elevating* the risk of all-out thermonuclear war.

Jack asked Wiesner about the effects of radioactive fallout, thinking, no doubt, of the many nuclear tests conducted that same year, including the two dozen bombs the United States detonated in the South Pacific in response to Russia's Siberian tests. When Wiesner told him it returns to the earth in raindrops, Jack turned to the window, where a light rain was falling on Washington. "You mean there might be radioactive contamination in that rain out there right now?" he asked. "Possibly," Wiesner replied. Jack was silent. After Wiesner left, Jack sat quietly, staring out the window. Kenny O'Donnell entered the room and left, writing later that he'd never seen Jack so depressed.

All of us were thinking about radioactive fallout back then. It was the subject of our dinner-table discussions and a persistent theme in the sort of science fiction films that captured my interest. I first became aware of Bob Dylan when I heard "A Hard Rain's a-Gonna Fall." The album sleeve explained that Dylan had written the song at the time of the Cuban Missile Crisis. A few months later Jack spoke to the American people about the danger of radioactive fallout, lamenting "the number of children and grandchildren with cancer in their bones, with leukemia in their blood, or with poison in their lungs. . . . This is not a natural health hazard—and it is not a statistical issue. The loss of even one human life, or the malformation of even one baby—who may be born long after we are all gone—should be of concern to us all. Our children and grandchildren are not merely statistics toward which we can be indifferent."

During their first meeting, Aunt Jackie, then a journalist, asked my Uncle Jack, a first-term senator, what he considered his best virtue. He surprised her by answering: "Curiosity." And what is curiosity but empathy, the quality that allowed him to put himself in Khrushchev's shoes? From his travels as a naval officer and with his family and with Lem Billings, he knew that foreign leaders labored under the same forces as Boston politicians: the demands of powerful special interests, the narrow perceptions of political constituents, and the ambition of political rivals all constrained their options. Similar imperatives defined the narrow space in which Khrushchev could operate, and limited the realistic opportunities of American foreign policy. But once these universal political limitations were recognized, a conversation could become possible, and negotiations might begin.

Jack had come to read Khrushchev's strategic choices as rational. While Jack understood that evil existed in the world, he did not accept Cold War theology, which demonized every Socialist or Communist. While he loved our country and considered us an exemplary nation, Jack understood the importance of leading with humility and compassion. "All virtue does not reside on our side," he told a small meeting of Quaker peace activists in the Oval Office in 1962. Finally, Jack recognized the insanity of his own national security apparatus and the truth of Eisenhower's warning of the antidemocratic thrust of the military-industrial complex.

The American University Speech

Pope John XXIII, who played an important but unheralded role in the settlement of the Cuban Missile Crisis, was likewise alarmed at the planet's brush with Armageddon. Afterward, the Pope sent his informal emissary, *Saturday Review* editor and disarmament advocate Norman Cousins, to serve as a peace diplomat between Jack and Khrushchev. Cousins drank vodka and played badminton with Khrushchev at his Black Sea dacha, then flew to Washington to brief Jack. Jack observed to Cousins, "One of the ironic things about this entire situation is that Mr. Khrushchev and I occupy approximately the same political positions inside our governments. He would like to prevent a nuclear war but is under severe pressure from his hardline crowd, which interprets every move in that direction as appeasement. I've got similar problems. Meanwhile . . . the hardliners in the Soviet Union and the United States feed on one another, each using the actions of the other to justify its own position." Bearing all this in mind, Jack asked Norman Cousins to draft a speech that would shatter the Cold War boilerplate and build a bridge with the Soviet leader. Cousins wrote for Jack the first draft.

One week after Pope John's death (and twenty-four hours after his momentous civil rights speech), Jack gave his greatest speech ever at American University, a ten-minute drive (with my mother at the wheel) from Hickory Hill. Jack asked Arthur Schlesinger and Ted Sorensen to prepare the address from the original draft by Cousins. The commencement speech, which Robert McNamara later described as "one of the greatest documents of the twentieth century," called for an end to the Cold War. In it, Jack painted the heretical vision of America living and competing peacefully with Russian Communists.

JFK challenged Cold War fundamentalists, who cast the world as a clash of civilizations, in which one side must win and the other be annihilated. He suggested instead that peaceful coexistence with the Soviets was the most expedient path to ending totalitarianism. "Some say that it is useless to speak of world peace . . . until the leaders of the Soviet Union adopt a more enlightened attitude. I hope they do. I believe we can help them do it." He challenged American prejudice against the Soviets with what Schlesinger termed "a sentence capable of revolutionizing the whole American view of the Cold War." He asked for a national self-examination, requesting Americans to explore their own presumptions and to take responsibility for their own part in the stand-off. "Let us reexamine our attitude toward the Cold War remembering that we are not engaged in a debate, seeking to pile up debating points. . . . Our attitude is as essential as theirs."

Two days earlier, in his spontaneous televised civil rights address, he asked white Americans to imagine what it feels like to be black. Now he asked all Americans to imagine they were Russian. Jack urged an empathy toward the Soviets that could only come from understanding their history. Instead of vilifying and demonizing the Russians, he sought to humanize them. He asked Americans to consider that Russians had borne the brunt of the Allied effort to defeat Hitler, including a massive military invasion and devastation on a scale that most Americans could not comprehend. Our schoolbooks, movies, and television shows—including my beloved

Combat, with Vic Morrow—left no doubt in the American mind that it was the United States that had won Europe's war against the Nazis. Now Jack contested that narrative, suggesting that perhaps the Russians had done even more to defeat Hitler. The Russians, he proposed, had shown unimaginable bravery and endured unthinkable suffering to end the Nazi threat. He asked Americans to "question our own worldview" and to put ourselves in Russians' shoes. "No nation in the history of battle ever suffered more than the Soviet Union suffered during the course of the Second World War. . . . At least 20 million lost their lives. Countless millions of homes and farms were burned or ransacked. A third of the nation's territory, including nearly two-thirds of its industrial base, was turned into wasteland—a loss equivalent to the devastation of this country east of Chicago."

Jack asked Americans to consider how, in light of their World War II experience, the Russians might reasonably regard as aggression the very conduct we viewed as self-defense. And he acknowledged that now. "Above all, while defending our own vital interests, nuclear powers must avert those confrontations which bring an adversary to a choice of either humiliating retreat or a nuclear war. To adopt that kind of course in the nuclear age would be evidence only of the bankruptcy of our policy—or of a collective death-wish for the world." Focusing, then, on our common concerns, rather than our differences, he argued that both the United States and the Soviet Union "have a mutually deep interest in a just and genuine peace and in halting the arms race." Then he made an audacious pitch for coexistence. "If we cannot end now our differences, at least we can help make the world safe for diversity. For, in the final analysis, our most basic common link is that we all inhabit this small planet. We all breathe the same air. We all cherish our children's future. And we are all mortal."

JFK shocked the military-industrial complex by announcing his intention to pursue America's "primary long-range interest," which he said was a "general and complete disarmament" designed to take place in stages, permitting parallel political development to build the new institutions of peace that would take the place of arms. He also announced his decision to unilaterally end atmospheric testing. "I now declare that the United States does not propose to conduct nuclear tests in the atmosphere so long as other states do not do so. We will not be the first to resume." This was the first step in the process he went on to outline for bringing the Cold War to an end.

Republicans lambasted the American University speech. Appalled generals considered it the final proof of JFK's treason. But it electrified a world terrified by the prospect of nuclear exchange. Jack's recognition of Soviet sacrifice and the legitimacy of the Soviet point of view had an immediate salving effect on U.S.-Soviet relations. Sorensen described how Jack's compassionate description of the Soviet's worldview instantly deescalated the Cold War: "The full text of the speech was published in the Soviet press. Still more striking was the fact that it was heard as well as read throughout the U.S.S.R. After fifteen years of almost uninterrupted jamming of Western broadcasts, by means of a network of over three thousand transmitters and at an annual cost of several hundred million dollars, the Soviets jammed only one paragraph of the speech when relayed by the Voice of America in Russian . . . Then did not jam any of it upon rebroadcast—and then suddenly stopped jamming all Western broadcasts, including even Russian-language newscasts on foreign affairs. Equally suddenly they agreed in Vienna to the principle of inspection by the International Atomic Energy Agency to make certain that Agency's reactors were used for peaceful purposes. And equally suddenly the outlook for some kind of test-ban agreement turned from hopeless to hopeful."

Khrushchev, deeply moved, later told treaty negotiator Averell Harriman that the American University address was "the greatest speech by an American president since Roosevelt." Khrushchev met Jack's proposal by agreeing in principle to end nuclear testing, and by proposing a nonaggression pact between NATO and the Soviet satellite countries of the Warsaw Pact. And he agreed to welcome test ban negotiators to Moscow.

The Test Ban Treaty

Knowing that America's military-industrial complex would oppose him, Jack kept the text of his American University speech secret. He did not submit the speech or his proposal for a unilateral test ban to the State Department, his National Security team, the Pentagon, the Defense Department, or the CIA. Even worse, by their lights, prior to the speech he surreptitiously laid the diplomatic groundwork for détente. Working secretly with British prime minister Harold Macmillan in the month ahead of his American University speech, Jack arranged top-level discussions of a permanent test ban. As a gesture of good faith he had suggested the treaty talks take place in Moscow—and Khrushchev had accepted. Then Jack ordered the unilateral termination of atmospheric testing by the United States.

JFK sent Averell Harriman (for whom my brother Douglas Harriman Kennedy would be named), the former U.S. ambassador to the Soviet Union, to negotiate the treaty, though Jack personally supervised every detail of the negotiation. He worked at unprecedented speed to end-run his furious adversaries at the Pentagon. On July 25, 1963, Jack approved, and Harriman initialed, the treaty, and less than one month later, he signed it jointly with Khrushchev. It was the first arms control agreement of the nuclear age. Historian Richard Reeves has observed: "By moving so swiftly on the Moscow negotiations, Kennedy politically outflanked his own military on the most important military question of the time."

The military-industrial complex responded predictably. A nuclear test ban was heresy to an arms industry and intelligence apparatus. Right-wing hard-liners were heavily invested in the continuing arms race as a source of their political and financial power. Caught off guard, these forces quickly mobilized to derail the treaty. Announcing that they were "opposed to a comprehensive ban under almost any terms," mutinous Joint Chiefs and Jack's own CIA director, John McCone, defied their boss, openly lobbying

against the agreement in the Senate. The Pentagon tried to sabotage its passage by hiding information about the ease of detect-
ing underground tests. National Republicans, including liberals
like Nelson Rockefeller, savaged the agreement as treachery. The
right-wing propaganda machine found plenty of arable ground in
the American national consciousness to fertilize with fear. Initially
congressional mail ran 15-to-1 against the treaty. Jack believed the
chance for passage by a two-thirds Senate majority was "about in
the nature of a miracle." On August 7, he told his White House
confidants that only divine intervention could achieve Senate ratification.

Against the counsel of his political advisers and his secretary
of state, Dean Rusk, Jack turned to grassroots politics for support. "That's the only thing that makes any impression to these
god-damned Senators. . . . They'll move as the country moves," he
told Rusk. Jack ordered White House staff to pull out every stop
to mobilize the population, saying he would "gladly lose the 1964
re-election if it meant getting the treaty passed." The following day
he appeared before the country on TV to launch a national whistle-
stop tour. "This treaty," he told Americans, "can symbolize the end
of one era and the beginning of another—if both sides can, by this
treaty, gain confidence and experience in peaceful collaboration."
He explained that the treaty "is particularly for our children and
our grandchildren, and they have no lobby in Washington." He organized church and women's groups and magazine and newspaper
publishers. Jack reached out to the Catholic Church, drafting Cardinal Cushing, who helped mobilize an ecumenical movement to
support the treaty, including the National Council of Churches and
the Union of American Hebrew Congregations. Jack called in chits
with labor unions and business leaders, and drafted scientists and
academics to the cause.

Against all predictions, JFK's whistle-stop tour promoting the
test ban proved an astonishing success. Jack found a hunger for
peace in the American countryside. Even in Republican strongholds, large crowds erupted in resounding applause when he called

for disarmament. He told a news conference, "A full-scale nuclear exchange, lasting less than 60 minutes with the weapons now in existence, could wipe out more than 300 million Americans, Europeans, and Russians, as well as untold numbers elsewhere. The survivors," he said, referencing what Khruschev reportedly told *Pravda* in July 1963, "would envy the dead." By August 12, congressional mail now only narrowly opposed the treaty by a 3-to-2 margin, which set off alarm bells in the panic-stricken military-industrial complex. "Will the Bottom Drop Out If Defense Spending Is Cut?" fretted a *U.S. News & World Report* headline. And by September the White House campaign had truly flipped public opinion, which now supported the treaty by 80 percent.

As the nation turned toward peace, the Pentagon's mahatmas made a last-ditch effort to save the day. At a September 12, 1963, National Security Council meeting, the Joint Chiefs of Staff again proposed a plan for a nuclear first strike, implying that the remaining months of 1963 might be most propitious for a sneak attack. We could kill 140 million Soviet citizens and disable Soviet retaliatory capacity, the generals promised Jack, with a maximum loss to the United States of only 12 million. Jack abruptly ended the discussion: "Pre-emption is not possible for us." On September 24, 1963, just as my baby sister Kerry was being born, the Senate ratified the Nuclear Test Ban Treaty 80–19, fourteen votes more than the two-thirds majority required for ratification. Jack considered its passage his greatest accomplishment as president. Two months later he would be dead.

By that fall, Jack had chosen his reelection campaign theme: "The road to peace." He would contrast America's idealism and moral authority with the warmongering belligerence of his expected opponent, Barry Goldwater. Jack had seen the message resonate, even in conservative enclaves. Americans, he felt sure, were weary of the Cold War. On a September 26 campaign stop in Utah, the Mormon Tabernacle Assembly shocked a cynical Washington press corps by greeting his vision with a prolonged standing ovation. America, Jack told them, must learn to live in "a world of diversity," where not even the world's top superpower could dictate world affairs. "We must first of all recognize that we cannot remake the world simply by our own command. When we cannot even bring all of our own people into full citizenship without acts of violence, we can understand how much harder it is to control events beyond our borders."

Jack went to Dallas in order to condemn the right-wing notion that "peace is a sign of weakness." He intended to argue that the best demonstration of American strength was not threats and destructive weapons, but rather a modeling of noble democratic ideals by "practicing what [we] preach" about equal rights and social justice, and by striving toward peace instead of being distracted by "aggressive ambition."

Immediately after signing the atmospheric test ban treaty, Khrushchev sent Jack the last of his personal letters, on October 10, 1963. In that missive, Khrushchev proposed the next steps for de-escalating and ending the Cold War. He recommended a non-aggression treaty between NATO and the Warsaw Pact's Soviet satellite nations, creating nuclear-free regions across the world, prohibiting the spread of nuclear weapons to other countries, banning nuclear weapons in space, taking specific measures for preventing surprise attacks, and other steps toward de-escalation. Reflecting on the period, Khrushchev's son and confidant, Sergei, told me of his father, "He was enthusiastic about his relationship with your uncle, and he was prepared to end the Cold War." Jack responded ten days later, saying that after reading Khrushchev's letter he was now "convinced that the possibilities for an improvement in the international situation are real." However, due to an alleged clerical error, the State Department never sent the president's letter and discovered it only after his assassination a month later. Meanwhile, Khrushchev had already secretly proposed to his own government radical reductions in the Soviet military, including the conversion of missile plants to peaceful purposes and dramatic spending cuts for defense. After Jack's death, that seemed to Khrushchev's Kremlin defense establishment a proposal for unilateral disarmament.

Less than a year after Dallas, as he pressed forward with this plan, Khrushchev's generals removed him from power. Sergei Khrushchev told me in 2017 that his father had also made a decision to accept Jack's invitation that our two nations should collaborate and go to the moon together. At the time of his death, Jack was planning his own trip to the Soviet Union, a gesture that would have decisively ended the Cold War. Forty years later, Sergei Khrushchev summarized the aborted days of détente this way:

> A great deal changed after the [Cuban missile] crisis: A direct communication link between Moscow and Washington was established, nuclear testing (except for underground tests) was banned, and the confrontation over Berlin was ended. But there was much that President Kennedy and my father did not succeed in seeing through to the end. I am convinced that if history had allowed them another six years, they would have brought the cold war to a close before the end of the 1960's.
>
> I say this with good reason, because in 1963 my father made an official announcement to a session of the U.S.S.R. Defense Council that he intended to sharply reduce Soviet armed forces from 2.5 million men to half a million and to stop the production of tanks and other offensive weapons. He thought that 200 to 300 intercontinental nuclear missiles made an attack on the Soviet Union impossible, while the money freed up by reducing the size of the army would be put to better use in agriculture and housing construction.
>
> But fate decreed otherwise, and the window of opportunity, barely cracked open, closed at once. In 1963 President Kennedy was killed, and a year later, in October 1964, my father was removed from power. The cold war continued for another quarter of a century.

A Soviet official told Pierre Salinger that Khrushchev wept openly when he learned of Jack's assassination, then withdrew into a shell. "He just wandered around his office for several days, like he was in a daze."

Rapprochement with Castro

When I first met Fidel Castro, in 1999, nearly four decades after the Cuban Missile Crisis, he acknowledged his own recklessness in inviting Soviet nukes into Cuba. "It was a mistake," he told me, "to risk such grave dangers for the world." But in 1963 he was still furious at the Russians for ordering the withdrawal of the missiles without consulting him. After the missile crisis, Khrushchev invited the embittered Fidel to Russia to smooth over the Cuban leader's anger. For the next six weeks, as the two leaders traveled across the Soviet Union, Khrushchev continually badgered Fidel to seek détente and pursue peace with Uncle Jack. Khrushchev wanted to convince Castro that JFK was trustworthy. Castro himself recalled how "For hours [Khrushchev] read many messages to me, messages from President Kennedy, messages sometimes delivered through Robert Kennedy." All of this was taking place under the CIA's watchful eye. In a top-secret January 5, 1963, memo to his fellow spooks, Richard Helms warned ominously that "at the request of Khrushchev, Castro was returning to Cuba with the intention of adopting a conciliatory policy toward the Kennedy administration 'for the time being.'"

Following Castro's release of the Bay of Pigs prisoners, Jack was open to such advances. Despite anger from some Cuban refugees for Jack's refusal to order U.S. military support during the Bay of Pigs invasion, the plight of the 1,150 Cuban prisoners held in Castro's Isla de Pinos prison had filled Uncle Jack with terrible remorse. He put my dad in charge of freeing them.

Fidel offered to trade the prisoners for 500 bulldozers and $60 million in baby food and medical supplies. Richard Nixon and Barry Goldwater condemned the exchange as caving to blackmail. A sanctimonious Nixon scolded, "Human beings are not to be bartered," and called for an invasion of Cuba. Undeterred, my father organized a bipartisan committee to raise the bulk of the funds. He got the last million dollars from Cardinal Cushing after a single phone call to the prelate from Grandpa. A liaison for the Cardinal delivered a cash-stuffed briefcase to my father at the Justice Department a few hours later. And the Cuban prisoners were freed.

After their release, veterans of Cuba Brigade 2506 flooded our home. My dad worked tirelessly to find jobs, housing, and legal help for the refugees and their families. He found homes for some of them near Hickory Hill. He placed a family in Mr. Zemo's former house and cleared the way for brigade members to join the U.S. military, and to obtain veteran's benefits. He helped find schools for their children. We carpooled with a number of families who enrolled in Our Lady of Victory. He used our morning horseback rides to visit and check in on brigade members in their houses. Refugee leaders visited us frequently at Hickory Hill. They were particularly fond of my mother because of her perpetual solicitousness toward the veterans and their families. She bought them groceries and found doctors for their children when they were ill. She asked them about their imprisonment, and one night, when two dozen of them came to dinner, she openly needled my father for leaving them on the beach. They were shocked and touched by her empathy and attention. They asked us to gather with them on the patio one day, where they presented my mother with a hand-made diamond and ruby brooch in the shape of the Brigada de Asalto 2506 logo.

Our most frequent visitor included legendary hero Erneido Oliva, the brigade's only black commander, whom my father greatly admired. My father made him tell us the story of how he stood down a Soviet tank by shoulder-firing a 50mm cannon. Che Guevara had come to visit him in prison to ask him why a black man would join the brigade. My parents' favorite was Enrique "Harry" Ruiz-Williams, a mining engineer who had provided explosives training to the brigade. Harry, a loud, swashbuckling rogue was tough, brave and truly funny. A borderline socialist, Harry was fervently anti-Communist. We all looked forward to his visits. At my dad's prompting, we heard from Harry how he had been blown into the sky during the invasion by a mortar shell. He had twenty pieces of shrapnel in his body. His Cuban army captors took him to a field hospital with both legs shattered and large puncture wounds through his lungs and neck. When Castro visited the hospital, Harry lifted a concealed .45 from his boot and pulled the trigger. The pin fell on an empty chamber. Castro asked him, "Are you trying to kill me?" Harry replied, "That's why I came here. I've been trying to kill you for the last three days!" Harry stayed with us frequently at Palm Beach and Cape Cod, and accompanied us during our Christmas ski trips. Another refugee, Manuel Artime, the brigade's civilian commander, had fought beside Castro and then turned against him when Fidel aligned with the Soviets. During one visit to Hickory Hill Artime borrowed Joe's horse, Shiloh, and took a spill while racing Daddy in a madcap gallop down the roadbed of the George Washington Parkway, then a dirt strip along the Potomac adjacent to the CIA. A lathered Shiloh cantered back to the barn alone with his thick leather reins severed by his hoof. Joe complained that the horse was never right again.

In the process of negotiating the release of these men, in the autumn of 1962, my father deployed his aide and friend John Nolan and James Donovan, a New York attorney and former OSS commando, to represent the American side of the discussions. (Tom Hanks later portrayed Donovan in another of his daring escapades in *Bridge of Spies*.) My father's two emissaries developed a close friendship with Castro. Following his return from Russia, Fidel asked Donovan how to go about normalizing relations with the United States. Donovan replied, famously, "The way porcupines make love—very carefully." Castro told Donovan that his ideal government was not "Soviet oriented," and that he was willing to end Cuba's guerrilla adventures in other Latin American countries. He asked for a parlay, assuring Donovan complete secrecy, pointedly adding that Che—an advocate for exporting revolution—would be excluded.

My father and Jack were intensely curious about Castro and demanded detailed, highly personal descriptions of the Cuban leader from both Donovan and John Nolan. "We went down there nearly every weekend for a year, so we spent a lot of time with Fidel," Nolan told me. The U.S. press had caricatured Fidel as an alcohol-sodden bully, but Nolan had a very different impression. "The Castro we met was handsome, charming, worldly, clean, and impeccably groomed, and an engaging conversationalist." Nolan recounted that he and Donovan had come to genuinely like Fidel.

The charismatic Cuban leader was reciprocally inquisitive about the Kennedy brothers. Nolan and Donovan spent many late nights smoking Cohibas and engaging in wide-ranging discussions. They traveled across the country, attending baseball games with Castro, who drove in an open Jeep with a security staff that was small "but very well organized," recalled Nolan. He and Donovan quickly came to understand that, despite the CIA's claims to the contrary, Fidel was immensely popular. "When we showed up at a baseball stadium it was always a surprise to the crowd and they would stand and applaud him spontaneously. There was nothing staged or compelled about it. Cubans adored him!" Castro took them to the Bay of Pigs and gave them a blow-by-blow account of the battle. He inquired about JFK's personality and opinions. For his part, Jack asked Nolan, "What do you think? Can we deal with this fellow?" Nolan told me that he enthusiastically encouraged Jack to move forward with Fidel.

In September 1963, Uncle Jack asked William Attwood, a former journalist, speechwriter, and U.S. diplomat attached at the time to the United Nations, to open secret negotiations with Castro. Attwood knew Castro since 1959, when he covered him for *Look* magazine before Castro had turned against the United States. Later that month, my father told Attwood to find a secure location to conduct a secret parlay with Fidel, and in October Castro began arranging for Attwood to fly surreptitiously to a remote airstrip in Cuba to begin negotiations on détente. On November 18, 1963, four days before Dallas, Castro agreed to an agenda for the meeting, and that same day Jack prepared the path for rapprochement with a clear public message. Speaking to the Inter-American Press Association in the heart of Cuba's exile community in Miami, Jack declared that

U.S. policy was to "not dictate to any nation how to organize its economic life." He added that Cuba's support for violent revolution across Latin America was the only obstacle to normalizing relations. "This, and this alone, divides us. As long as this is true, nothing is possible. Without it, everything is possible."

A month earlier Jack had opened another secret channel to Castro via French journalist Jean Daniel, a friend of William Attwood's. On his way to interview Fidel in Cuba on October 24, 1963, Daniel visited the White House, where Jack talked to him about U.S.-Cuba relations. In a message meant for Castro's ears, Jack criticized Castro sharply for precipitating the missile crisis but then changed tone, expressing the same empathy toward Cuba that he had evinced toward the Russian people in his American University speech. "There are few subjects," Jack told Daniel, speaking of the Cuban revolution, "to which I have devoted more painstaking attention. . . . Here is what I believe. I believe that there is no country in the world, including all the African regions, including any and all the countries under colonial domination, where economic colonialization, humiliation, and exploitation were worse than in Cuba, in part owing to my country's policies during the Batista regime. . . . I approved the proclamation which Fidel Castro made in the Sierra Maestra, when he justifiably called for justice and especially yearned to rid Cuba of corruption. I will go even further: to some extent it is as though Batista was the incarnation of a number of sins on the part of the United States. Now we shall have to pay for those sins. In the matter of the Batista regime, I am in agreement with the first Cuban revolutionaries. That is perfectly clear."

Between November 19 and 22, Castro conducted his own series of interviews with Daniel, carefully debriefing the Frenchman about every nuance of his meeting with Jack, particularly Jack's assertion that a corrupt Batista deserved to be overthrown. Three times Castro asked Daniel to repeat Jack's statements about Batista and then said to himself aloud, "He has come to understand many things over the past few months." Then Castro sat in thoughtful silence, composing a careful reply that he knew JFK was awaiting.

Finally he spoke slowly. "I believe Kennedy is sincere," he began. "I also believe that today the expression of this sincerity could have political significance." Fidel followed this with a detailed critique of the Kennedy and Eisenhower administrations, which had attacked his Cuban revolution "long before there was the pretext and alibi of communism." But, he continued, "I feel that [Kennedy] inherited a difficult situation; I don't think a President of the United States is ever really free, and I believe Kennedy is at present feeling the impact of this lack of freedom. I also believe he now understands the extent to which he has been misled, especially, for example, on Cuban reaction at the time of the attempted Bay of Pigs invasion."

Castro also suggested that Jack's challenge to his national security apparatus put him in peril. "Suddenly a President arrives on the scene," he said, "who tries to support the interests of another class (which has no access to any of the levers of power) to give the various Latin American countries the impression that the United States no longer stands behind the dictators, and so there is no more need to start Castro-type revolutions. What happens then? The trusts see that their interests are being a little compromised (just barely, but still compromised); the Pentagon thinks the strategic bases are in danger; the powerful oligarchies in all the Latin American countries alert their American friends; they sabotage the new policy; and in short, Kennedy has everyone against him. . . .

"I cannot help hoping, therefore, that a leader will come to the fore in North America (why not Kennedy, there are things in his favor!) who will be willing to brave unpopularity, fight the trusts, tell the truth and, most important, let the various nations act as they see fit. Kennedy could still be this man. He still has the possibility of becoming, in the eyes of history, the greatest President of the United States, the leader who may at last understand that there can be coexistence between capitalists and socialists, even in the Americas. He would then be an even greater President than Lincoln." With a broad grin he joked, "If you see him again, you can tell him that I'm willing to declare Goldwater my friend if that will guarantee Kennedy's re-election!"

Jack intended his back-channel negotiations to outflank Foggy Bottom and the spooks at Langley, but the CIA knew of the contacts and worked to sabotage Jack with cloak-and-dagger mischief. In April 1963, CIA officials secretly sprinkled deadly poison in a wetsuit intended as a gift for Castro from Jack's emissaries, John Nolan and James Donovan, hoping to murder Castro and presumably pin the deed on JFK, to discredit Jack's peace efforts. "If the CIA's scheme had worked, we would have been dead on the spot," Nolan told me, shaking his head in disgust. The Agency also delivered a poison pen to Castro's turncoat lieutenant Rolando Cubela in Paris, with instructions that he use it to murder Fidel. Historian Larry Hancock shows that Jack's peace feelers with Fidel and his November speech in Miami no doubt seriously alarmed the CIA, as well as those Cuban exiles whose entire lives were devoted to killing Castro and restoring their expropriated wealth. For that cadre, which included many spies, gangsters, and highly trained assassins, Jack's efforts at reconciliation with Castro was part of a betrayal that began when Jack refused to invade at the Bay of Pigs, and continued with his peaceful resolution of the missile crisis.

As the White House shut down the exiles' Miami sabotage operations like Alpha 66 and edged toward détente with Castro, even those Cubans who loved my dad stopped visiting Hickory Hill. My mother's jeweled Brigade 2506 brooch was stolen from her locked jewelry case, the lock expertly picked. My mother believes brigade members who had free rein Hickory Hill broke into her room to remove it. Many exiled leaders who had been great admirers of the Kennedys and personal friends of the family openly expressed their disgust with what they saw as White House treachery. On April 18, 1963, Bay of Pigs commander José Miró Cardona, chair of the Cuban Revolutionary Council, resigned in a fusillade of furious denouncements aimed at Jack and my father. In parting Miró Cardona promised, "There is only one route left to follow and we will follow it: violence." Hundreds of Cuban exiles in Miami neighborhoods expressed their discontent with the White House by hanging black crepe from their homes. In the spring of 1963, Cuban exiles passed around a pamphlet extolling Jack's assassination. "Only one development," the broadside declared, would lead to the return of their beloved country. "If an inspired act of God should place the White House within weeks a Texan known to be a friend of all Latin America." Howard Hunt told his son that he met with the CIA's leading anti-Castro operatives Frank Sturgis and David Morales in a CIA safe house in Miami to discuss the "big event" shortly beforehand—a plan to kill Jack.

On the day Uncle Jack was assassinated, Fidel Castro was meeting with Jean Daniel at his summer presidential palace in Varadero Beach. At one p.m. they received a phone call with news that Jack had been shot. *"Es una mala noticia,"* Castro said to himself. Then, turning to Daniel, "There is the end to your mission of peace. Everything is going to change." When the news came twenty minutes later that Jack was dead, Castro called it "a catastrophe." Then he asked Daniel: "Who is Lyndon Johnson? What authority does he have over the CIA?" Hearing that American authorities were in hot pursuit of a suspect, Castro told Daniel, "You watch and see—I know them—they will try to put the blame on us for this thing." And he was right. According to investigators on the House Select Committee on Assassinations, immediately following Jack's assassination, operatives in the CIA's Western Hemisphere Division promoted evidence—later proven false—suggesting that Castro had orchestrated President Kennedy's assassination. The Senate's Church Committee, which investigated the assassination for two years from 1975 to '77, concluded that Cuba had nothing to do with Jack's murder. Dan Hardway, an attorney who served as investigator for the House Committee, told me that the source of virtually every story blaming Castro was connected to the CIA's Western Hemisphere chief and propaganda guru, David Atlee Phillips.

Immediately after Jack's death, my father had sent his friend Bill Walton on a secret mission to Moscow to reassure the Kremlin leadership that, despite the CIA disinformation campaign, our family knew that neither Castro nor Russia had had anything to do with Jack's murder. Walton told Khrushchev via the spy Georgi Bolshakov that the culprits were "domestic plotters."

Fidel Castro, whose crack intelligence agency had thoroughly penetrated the CIA's Cuba operations, told my mother that he knew my father and Jack were unaware of the CIA's attempts to assassinate him in the early 1960s. Nearly four decades later, in 1999, Castro told me, "If your uncle had lived, the relationship between our countries would have been very different. He was a great president, an unusual man with love for children and a powerful understanding of the military and large corporations that run your country. We were on a road to peace."

CHAPTER 8

● ● ● ● ●

A Farewell to Camelot

IT SEEMED ODD TO ME HOW A DARKNESS CLUNG TO ANY MENTION OF Dallas in our home after Jack's death; Los Angeles never bore that stigma after my father's assassination. There was something mad about that Texas city—a poisonous hatred approaching zaniness that seemed to have set the stage for the murder. Notoriously corrupt, Dallas was a mecca for wing-nut racists, oil billionaires, and the fringe lunatics of the John Birch Society. The Ku Klux Klan sited its headquarters as Dallas in the 1920s, and the Dallas police department was still lousy with KKK members; its officers mingled also with local gangsters at venues like Jack Ruby's Carousel Club, buying drinks for so-called B-girls provided, at times, by the Mafia overlord Carlos Marcello. During the Warren Commission investigation, witnesses told the FBI, "Jack Ruby was well acquainted with virtually every officer of the Dallas Police Department," and that he was the mob's "pay off man for the Dallas Police."

Along with the Mafia, the CIA was a shadowy but potent force among Dallas's power elites and the city's law enforcement arm. CIA records released in 2017, pursuant to the JFK Assassination

Records Act, show that Dallas's then mayor, Earle Cabell, was a CIA agent active since 1956. JFK had fired Mayor Cabell's brother, General Charles Cabell, the CIA's deputy director, after the Bay of Pigs. Mayor Cabell was the commander of the Dallas Police Department, which, according to historian and journalist Russell Baker, included over a hundred officers with intelligence-agency affiliations. The city's large Cuban refugee community was a ferment of anti-Castro activism and seething anger toward the Kennedys.

Dallas's outskirts, clustered with military bases and oil fields, made the city a confluence of Wall Street interests, oil tycoons, and defense contractors, who saw the Kennedys as their mortal enemies. The reactionary petroleum and military oligarchs who ruled Dallas thrived on a dark symbiosis with a legion of fire-breathing fundamentalist preachers and hate radio and television bigots, virtuosos at wielding tribal rage, religious fervor, and zealous nationalism. Three weeks before Jack's murder, hecklers pelted United Nations ambassador Adlai Stevenson with stones and rotten fruit, spat on him, and beat him with picket signs at a UN Day speech in Dallas. "Mad things happened," reported historian William Manchester. "Huge billboards screamed 'Impeach Earl Warren.' Jewish stores were smeared with crude swastikas. Fanatical young matrons swayed in public to the chant 'Stevenson's going to die— his heart will stop, stop, stop and he will burn, burn burn!'" Public schools distributed radical right broadsides to their pupils; students booed the Kennedy name in classrooms; corporate CEOs fired and blackballed junior executives who refused to attend right-wing seminars. When Dallas's public school PA systems announced Jack's assassination, children as young as fourth-graders applauded. There was hatred elsewhere in the Old Confederacy, too; white children in Montgomery and Birmingham classrooms also cheered. A Birmingham radio caller declared that "any white man who did what he did for niggers *should* be shot." But in Dallas the malignancy had metastasized.

JFK's domestic and foreign policy thrusts seemed intended to vex every Dallas power center. The Dallas plutocracy viewed

JFK's firing of Allen Dulles as a declaration of war against the Wall Street/oil/intelligence nexus that had dominated U.S. foreign policy during the Eisenhower era. Dallas oil magnates had bet heavily on the Cuban-Venezuelan Oil Company (CVOC), a former Wall Street darling that owned development rights to three million acres on the island. In their view, those investments, thanks to Kennedy's "appeasement" of Castro—now had meager hope of redemption. Kennedy's peace efforts threatened Dallas's military contractors. The Nuclear Test Ban Treaty was already hurting uranium pricing—once a bonanza for the Dallas oil industry, which had made Texas the uranium mining capital since the 1920s. Furthermore, John Kennedy had torpedoed mineral and oil development in the resource-rich Amazon basin upon which Dallas oil potentates had placed another big bet. Worst of all, in their eyes, JFK was crusading against the Oil Depletion Allowance. A generous welfare program for Dallas's oil pashas, the allowance gave petroleum tycoons an automatic deduction, regardless of costs, for diminishing oil production at their wells. My father forced a horrified J. Edgar Hoover to deploy agents to collect information on an industry with strong ties to both the FBI and CIA. My dad had ordered Hoover's G-men to issue humiliating questionnaires to oilmen, for specific production, sales, and cost data. The *Oil & Gas Journal* accused Robert Kennedy of creating a "battleground on which business and government will collide." Dallas's leading oil magnates Clint Murchison, Sid Richardson, John Mecom, and H. L. Hunt (then the world's richest man) stood to lose hundreds of millions of dollars. Russ Baker, author of *Family of Secrets*, wrote, "The Kennedy administration struck at the heart of the southern establishments' growing wealth and power. Not only did it attack the oil depletion allowance, but its support of civil rights threatened to undermine the cheap labor that supported southern industry."

Many Kennedy family friends, including Ann Brinkley, David's wife, and Democratic National Committee member Byron Skelton, warned Jack to steer clear of Dallas in the days before the visit. Jack seemed to have his own sense of foreboding as he prepared for the trip. On Wednesday, November 20, after my father's birthday party, my mother found Jack distant and brooding at a dinner for the Supreme Court justices. Jack was fond of my mother and always effusive toward her; but that night, she recalls, he looked through her, as if she weren't there. The weekend before, Jack had made an unusual spontaneous trip to Palm Beach to say goodbye to his father. That night, Jack sang a haunting version of Maxwell Anderson's "September Song" to a group of friends after dinner at the White House: "Oh, the days dwindle down / to a precious few. / September . . . November! / And these few precious days I'll spend with you. . . ." Dave Powers found it chilling. Friends and advisers voiced ominous warnings about the trip, and Jack's own conversations drifted repeatedly to the subject of assassination, including on the morning of November 22, when Jack and Jackie saw the black-bordered front page of the *Dallas Morning News* accusing Jack of treason. Oil baron Nelson Bunker Hunt and the John Birch Society had financed the ad. "We're headed into nut country today," Jack told her. When she expressed anguish about his safety in Dallas he said, "But, Jackie, if somebody wants to shoot me from a window with a rifle, nobody can stop it. So why worry about it?"

The purpose of the Dallas trip was to heal a rift between Texas's right-wing Democratic governor, John Connally, and its liberal Democratic senator, Ralph Yarborough. But the underlying issue was civil rights. Jack had won Texas in 1960 by putting LBJ on the ticket, but the White House's push for racial justice had robbed the vice president of his mojo in the state, which now threatened to shift to the Republican column in 1964. The Yarborough–Connally feud was splintering what remained of the party in Texas. On Thursday, November 21, Jack left for Houston, and the following day he flew to Dallas to deliver a speech that was to have been a fusillade against right-wing militarism. He would condemn as "nonsense" the conservative mantra that "peace is a sign of weakness" and argue that the most convincing way to signal American strength is not by rattling sabers but by "practicing what it preaches about equal rights and social justice and pursuing peace instead of aggressive ambitions." Riding in the black Lincoln convertible beside John Connally, Jack and Jackie found exuberant supporters packing the parade route five deep. Probably due to Jackie's extraordinary popularity, the Texas crowds were over double the size predicted, and their enthusiasm was exultant. But sprinkled among Kennedy Democrats were familiar icons of hatred: Confederate flags, placards accusing Jack of being a Communist or a dictator, and wall posters showing Jack under the inscription "Wanted for Treason."

On Friday, November 22, 1963, my father was lunching on the pool patio at Hickory Hill with my mother, Bob Morgenthau, and his Rackets Division investigator, Sylvia Molano. They were discussing ongoing cases against various mobsters, including Carlos Marcello, the notoriously violent New Orleans and Dallas Mafia boss, who at that moment was awaiting a jury verdict in a deportation case. It was the second time my father's Justice Department had moved to deport him. My father was eating a tuna sandwich when my mother called from the pool house to say that J. Edgar Hoover was on the phone. My father rushed to take the call. Hoover never called Hickory Hill. Hoover told my dad that Jack had been shot and was en route to Dallas's Parkland Hospital. An hour later, Hoover called back. My mom and dad were in his den across from my bedroom. CIA director John McCone was with them. McCone had raced over from CIA headquarters, which was less than a mile from our house. Hoover informed my father flatly, "The president is dead." My father later told historian William Manchester that Hoover's tone was "not quite as excited as if he were reporting the fact that he'd found a Communist on the faculty of Howard University." Hoover never expressed his condolences. He spoke to my father again only once—when they happened to enter the Justice Department at the same time.

My dad immediately suspected that the CIA had killed Uncle Jack. After Hoover's call, he called a yet-unidentified CIA official and asked point-blank, "Did your outfit have anything to do with this horror?" Then my father invited McCone to take a walk with him in the yard. Out of earshot of my mother, he asked McCone whether the CIA had killed his brother. McCone denied it. According to historian David Talbot, McCone himself later concluded there had been more than one gunman.

A half hour later, walking with his Justice Department press secretary, Ed Guthman, my father said, "There's so much violence. I thought they would get me." Later that day, my father called Enrique "Harry" Ruiz-Williams, the last of the CIA's anti-Castro Cuban exile leaders who remained close to our family. Harry was at a Washington, D.C., hotel with Haynes Johnson, a journalist who had embedded himself in Brigade 2506 to write a book about the Bay of Pigs. Johnson years later confirmed that my father said to him, "One of your guys did it."

My mother raced to pick us up early from school—Joe and Kathleen were already in the car when I slid into the backseat in the Sidwell Friends driveway. Driving home from the District, we saw flags already at half-staff. Our mother told us that "A bad man shot Uncle Jack" and that he was in heaven. My father's friend Dean Markham picked up my brother David. "Why did they kill Uncle Jack?" David asked. Choking back his tears, Dean couldn't answer. When we got home, my father was alone under the white walnut tree with Brumus, our giant Newfoundland, Rusty, the Irish setter, and our flat-headed Lab, Battle Star. We ran and hugged him. We were all crying. He told us, "He had the most wonderful life. He never had a sad day."

David and I walked the cordon of U.S. marshals forming around our home to keep the growing crowd of reporters and photographers outside the metal cattle guards. My father had no trust in the FBI or CIA and little faith in the Secret Service, which he'd tried to move from Treasury to the Justice Department. But he loved the U.S. marshals. By default, they became his frontline troopers in his civil rights and organized crime battles. My father trusted Chief Marshal Jim McShane like a brother. And the marshals loved our family. That afternoon all ninety U.S. marshals based in D.C. volunteered for the job of guarding our home. Dean Markham and his wife, Susie, made their way past the marshals, accompanied by Supreme Court Justice Byron White and his wife, Marion, and my father's childhood friend Dave Hackett. Then Dave, Dean, and Susie bundled us into the station wagon and took us to the Markham house on nearby Turkey Run Road, where we watched the news coverage on TV and played with the Markham children, Dean and Marion.

The next day, my father walked Arlington National Cemetery with Bob McNamara, his sisters, Jean and Pat, and a gaggle of generals in a downpour as they chose Jack's gravesite. In opting for Arlington, my dad and Jackie were overruling Jack's Boston friends who wanted to bring him home to the family burial site in Brookline. Jackie and my father understood that Jack belonged to the nation now, and to history. I would visit Jack's grave at Arlington hundreds of times over the years, both alone and with my family and friends. Before it became the National Cemetery, Arlington had been the private plantation of Confederate general Robert E. Lee. On our first visit, my mother explained that Lee's estate had become the National Cemetery after the Battle of Bull Run, when Stonewall Jackson had whipped an overconfident Union Army just south of Hickory Hill. When General McClellan's brigadier asked Lincoln where to bury the Union dead, the angry president answered, "Bury them at Lee's feet." Exactly two weeks before his death, Jack had visited the site for Veterans Day and remarked, as he stood at Lee's mansion overlooking the Potomac and all the marble monuments of the American democracy, "I could stay here forever."

At ten a.m. we assembled around Jack's coffin in the White House's East Room for a small mass with family and friends, including Jack's Boston pals Kenny O'Donnell, Dave Powers, and Larry O'Brien. Treasury Secretary Douglas Dillon and the *Washington Post*'s editor Ben Bradlee were also there. Lincoln's body had rested in the same spot a century earlier. Pastors from all faiths, including, to our fascination, a hirsute Greek Orthodox prelate, convened with the family to pray for Jack. It was the kind of ecumenical gathering JFK had envisioned in his campaign address on Catholicism. Knowing Jack's love for the Special Forces, my father had asked the Green Berets to come up from Fort Bragg to lead Jack's last watch. During mass, I snuck glances at those warriors standing at attention with the hulking giant Major Francis Ruddy, whom all of us children knew well, behind the coffin and along the wall. Jack had created the Navy SEALs in 1962 and intervened with the Pentagon to allow the Special Forces to wear the green beret, and they considered him their godfather. Some of Ruddy's men had teary eyes. After Dallas, the elite unit would add a black band to its headgear in perpetual mourning for President Kennedy.

Most of that gray and cloudy day we stayed at the White House with my cousins, aunts, and uncles in a kind of informal wake. Following a small family mass on a jerry-rigged altar in the dining room, we had dinner with Jack's two best friends, Lem Billings and Red Fay, along with Dave Powers, the O'Donnells, Bob McNamara, Ted Sorensen, David and Sissy Ormsby-Gore, and the Greek shipping magnate Aristotle Onassis. My father ate with Jackie in the sitting room. Afterward, Jackie and my parents went into the East Room and prayed next to Jack's coffin. Then she summoned us to join her on the chairs and couches surrounding Jack's coffin, where we listened to funny stories and laughed a lot. My dad made Ari Onassis sign a contract promising to donate half his fortune to Latin America's poor. Red Fay, Jack's fellow PT boat skipper (who spoke always of himself in the third person as "The Red Head" and his wife, Anita, as "The Bride"), alternated with Dave Powers telling us tales about Jack during the war and in early political campaigns. Jackie smiled as Powers recounted how she had charmed West Virginia housewives, visiting them in ramshackle houses on remote hollows and by talking about her husband over supermarket PA systems while they shopped, and how during the 1952 Senate race she had, single-handedly, stolen the French-Canadian vote from Henry Cabot Lodge by spending an hour speaking French with a popular Chicopee priest. The priest, a former Lodge confidant, afterward ordered his congregants at each mass to vote for Jack Kennedy.

The next morning, Sunday, November 24, the weather broke. It was a clear, crisp day as we drove in from Hickory Hill with a long police escort, and watched, amazed, at the multitude choking the District. Washington was an ocean of mourners. They crowded every street and sidewalk, turning the District into a parking lot. Following mass in the East Room we assembled on the North Portico steps of the White House with President Johnson and Lady Bird. As we gathered for the funeral procession, Jackie and my father went back into the East Room to pray alone. They took a lock of Jack's hair, which my dad would keep until he died. He slipped a cutting of his own hair, a PT boat tie clip, and my mother's silver rosary into Jack's hands. Afterward, they wheeled the casket to the North Portico, where the caisson was waiting. Green Berets and soldiers from each branch of the armed forces strained to heft the five-hundred-pound mahogany coffin onto the gun carriage. Then they draped it with a flag. I was riveted by the sight of a caisson with its six matched horses as it pulled away from the North Portico, down the White House driveway, through parallel rows of soldiers and flags of every state, leading the procession onto Pennsylvania Avenue.

We rode behind the gun carriage. Uniformed soldiers and sailors, with fixed bayonets, marched slowly before the caisson as rows of drummers pounded a mournful beat. The grim sea of people on Pennsylvania Avenue was utterly silent but for the occasional wail of anguish. My father, Jackie, Lyndon, Lady Bird, Caroline, and John were in the first limo, and I was in the second with Uncle Sarge, my mother, and my cousin Bobby Shriver, with a Secret Service agent driving. Behind us was a long file of black limousines carrying all the other members of our clan, Kennedys, Shrivers, Lawfords, Smiths, Auchinclosses, and Fitzgeralds, followed by limos hauling the weeping White House staff, a bedraggled White House Press Corps, and police. Behind them the sidewalk crowds stood nine deep all the way from the White House to the Capitol Rotunda, where Uncle Jack would lie in state.

Mesmerized, I studied the riderless ebony funeral horse that followed the caisson with empty boots mounted backward in the stirrups—the symbol of a fallen prince looking back over his truncated life. The gelding's name, I learned, was Black Jack, after General Pershing. He was tall, with the fiery temper of a thoroughbred stallion, and his rioting against the poor soldier who struggled to cling to his bridle earned the horse national attention over the next two days. Black Jack never stopped bucking, furiously throwing his head and dancing wildly as his sleek coat shimmered black in the cold November sunlight. I could see the Army brand on his neck.

I had a million questions. I wondered aloud what would happen to Caroline and John. Where would they live? Would they still go to the White House school? What about the Secret Service "Kiddie Detail"—Mr. Foster, Mr. Walsh, Mr. Meredith? Would they still be at the Cape? What about Pierre Salinger and all the other people in the White House who worked for Uncle Jack? How would all these people find jobs? Would my dad still be attorney general? Would Uncle Sarge still run the Peace Corps? What would happen to our country? Our mother told us that Jack was in heaven with God, and with God's help he would watch over us and take care of the rest of the country.

Sarge distracted us from these anxieties by explaining the history of the government buildings and the equestrian statuary on Pennsylvania Avenue. He told us that a rearing horse in a statue meant the hero had died in battle, and that a single leg raised indicated that the rider was wounded in war. He told us about Lincoln's funeral, upon which Jack's was modeled, and how, dating back to the funeral of Genghis Khan, the riderless horse had symbolized a fallen hero who would ride no more. In front of us, crowds overflowed from sidewalks into the street, held back by soldiers in parade dress lining the funeral route. I marveled at how they stood still so long. My mother told us how the whole world was mourning Uncle Jack because he tried to help the poor and make peace.

Even the grown-ups were stunned by the crowds, their size, and their silence. Our mother told us to open the windows to wave our thanks. It seemed to break their hearts. We heard loud keening, agonized groans, and incoherent shouts of anguish. "Pray for us," people called out again and again. Their plaintive cries rose above the slow, muffled drumbeat, the jangle of Black Jack's tack, and the clatter of his hooves as he struggled frantically. They seemed to be pleading for some consolation that was beyond our power to give.

In front of the Justice Department a hundred thousand mourners surged in behind the casket and followed us toward the Capitol plaza, where hundreds of thousands more were already waiting to parade past the coffin. At the bottom of the Capitol steps soldiers fired a twenty-one-gun salute, and the military band played "Hail to the Chief." Nearly a million mourners watched us climb the steps. Everyone seemed to be crying, and their weeping made me want to cry. I watched the uniformed soldiers carry Jack's heavy coffin up the steep Capitol steps and place him on the bier where he would lie in state. Lincoln had reposed beneath the same frescoed dome.

The week following Jack's funeral, fifty thousand Americans sent to Jackie poems they had written that night. But Jack had touched people's hearts beyond America's shores, as I would frequently be reminded during my wanderings over the next five decades. The desolation evoked by his death was global. In nearly every nation, highways, businesses, factories, schools, and colleges closed. Radio and TV stations worldwide suspended advertising. Airplanes were grounded. Professional sports teams canceled contests. Sixty thousand Germans carrying torches flooded the square beside Rathaus Schöneberg, the former city hall for West Berlin where Jack had given his "Ich bin ein Berliner" speech. Three days later, the German government renamed it John-F-Kennedy-Platz. In Iran, every school, government office, and business shut down for days as the nation mourned. William Manchester reported that a Kalahari bushman walked ten miles to the U.S. embassy in Botswana and told the consul general, "I've lost a friend." In Kenya, Masai warriors held a tribal feast of mourning. The British prime minister described his nation as enduring convulsions of pain, and London teamsters who had never seen Jack roamed the streets in tears. In Greece all traffic stopped during the mass; Germany banned cars from the Autobahn. Traffic in New York also came to a halt, and trains stopped on their tracks across the country. Greyhound buses pulled over. The Soviet Union's deputy premier Anastas Mikoyan, the ruthless Bolshevik, burst into tears when Jackie said to him, after the funeral, "Please tell Mr. Chairman President that I know he and my husband worked together for a peaceful world and now he and you must carry out my husband's work."

At the services in the Capitol Rotunda where Uncle Jack lay in state, Speaker John McCormick, Senate Majority Leader Mike Mansfield, and Chief Justice Earl Warren spoke as we gathered around Jack's closed casket. Mansfield, who loved Jack, condemned "the bigotry, the hatred, prejudice and arrogance" that had laid the groundwork for the crime. After the service, Jackie and Caroline knelt and hugged the coffin. I looked up at Caroline's Dominican nanny, Providenci "Provi" Paredes. She was crying dolefully. Spasms of grief racked her body. The sight made me lose control. Suddenly I could no longer restrain my tears. I tried to wipe them away inconspicuously. Joe and Kathleen stood with my cousins nearby, also crying quietly.

As we came out of the Capitol, Washington was a sea of people stretching toward the horizon in every direction. We lingered for a moment at the top of the Capitol steps and gazed in wonder at the vast host. We would later learn that cars on New York Avenue were backed up to Baltimore, thirty miles to the north. Traffic had closed down every bridge leading to the District. By midnight, a hundred thousand mourners had paraded past the bier, and the line was still three miles long and five abreast. William Manchester described it as "the kindest most democratic crowd in the nation's history." Heavyweight champion prizefighter Jersey Joe Walcott waited in the creeping line among a group of African Americans who finally passed the bier together at 2:45 a.m. Native Americans in tribal regalia, nuns, priests, rabbis, doctors, nurses, uniformed soldiers, patients from local hospitals, embassy personnel from every nation, Boy Scouts, political leaders, and miners from West Virginia stood patiently. Universities across the eastern United States emptied as college students poured into Washington to mourn their president. In the queue were young men and women who had abandoned their cars on highway roadsides and walked for miles. Manchester saw a group of seamen practicing the salute they would execute on reaching the catafalque, where, as one of them put it, "There lay in honored glory a sailor well known to God." According to William Manchester, "All around people were praying, or singing 'We Shall Overcome.'" He describes a black man in a shiny suit "far too thin for the frigid weather," singing a stirring rendition of an Ed McCurdy ballad describing his dream about a giant room filled with a gathering of men assembled to put an end to war.

We left the Capitol around two-thirty and went back to White House. President Johnson came into the Blue Room and told us that Lee Harvey Oswald had been murdered. I didn't know how to take that news. I expected the grown-ups to be happy. They weren't. They just all reacted as if Oswald's murder only compounded the tragedy. They seemed in despair about the epidemic of violence in America. The man who killed Oswald at the police station in front of dozens of eyewitnesses and TV cameras was now in police custody. His name was Jack Ruby. I asked the grown-ups: "Did he love Uncle Jack? Did he love our family?" There was no answer. No one explained it.

The crowds lined up to bid farewell to Jack continued to build all night. By nine a.m. the next morning, November 25, when the District police finally closed the Capitol doors, thousands more were still waiting in a line that had continued to grow throughout the evening as new arrivals crowded into D.C. Rather than leave, the throng remained and knelt on the streets to pray.

We assembled in the White House early that morning and traveled by caravan to Capitol Hill. When we entered the Capitol at around eleven am, Aunt Jackie, my father, and Teddy went to Jack's flag-draped coffin and knelt, and we knelt behind them. After we had prayed, we stood aside and watched as the honor guard carried Jack's coffin back down the Capitol steps for the trip to the funeral mass. We got back in the limousines, and once again we followed Black Jack, with his shining silver sword and scabbard. In front of the White House, Jackie left her car and led an impromptu march to St. Matthew's Cathedral, walking with my father and Teddy, Lyndon Johnson, and virtually all the leaders of the free world, twelve abreast. To this day it stands as the largest gathering of heads of state and foreign dignitaries in American history. Jackie led 220 dignitaries from ninety-two nations and almost every member of Congress, and nearly every state governor. The Marine Band followed them. Our limousine was right behind the band. Seven days before, I had attended a concert by bagpipers from the Scottish Black Watch, in the White House Rose Garden. I was proud because Jackie had asked me to be an usher and help seat the crowd. It was the last time I saw Uncle Jack alive. Now I watched as the same pipers followed the Marine Band directly behind Jackie. With them were the Irish Cadets who had accompanied President Éamon de Valera from the Emerald Isle. One million people lined the route from the Capitol back to the White House.

While Jackie led the procession on foot into St. Matthew's, Jim McShane, my dad's favorite U.S. marshal, whisked Kathleen, Joe, David, Courtney, Michael, Kerry, and me to the church. We saw Cardinal Cushing waiting outside St. Matthew's in his red vestments. The sixty-eight-year-old prelate had known for years that he himself was dying. Jackie loved Cardinal Cushing, and earlier she had presented him with Uncle Jack's Navy dog tags, which he now wore under his vestments. Inside, Marshal McShane found us a pew from which we watched all the great figures of modern history parade before us, including Presidents Eisenhower and Truman, and the French president and resistance hero General Charles de Gaulle, who seemed to me a slouching giant. We saw Dr. Martin

Luther King, Jr., Richard Nixon, and Ireland's president, Éamon de Valera. To distract us, Sarge Shriver quietly identified colorfully clad heads of state: the king of Belgium, the queen of Greece, and Prince Philip of England, who wedged themselves with their swords and medals into overcrowded pews beside our favorite, the elegant, though tiny, Emperor Haile Selassie—five feet tall and all bespangled with medals, looking elflike beside the six-foot-five-inch de Gaulle.

Haile Selassie, who claimed to be a direct descendant of King Solomon and the Queen of Sheba, had led Ethiopia's heroic resistance against Mussolini's poison gas attacks before World War II. A year before, he had come to dinner at the White House and given Jackie a leopard-skin cloak, and a giant silver sword in a diamond-studded scabbard and ivory carvings for my cousins, John and Caroline. Another night he dined at Hickory Hill with Supreme Court Justice Earl Warren, and I managed to get the unlikely pair to sign a napkin for me. Emperor Selassie signed his name in the magical Ethiopian Ge'ez script. That evening Emperor Selassie had invited us all to his palace in Addis Ababa, where, he said, he had hunting cheetahs that wandered the royal hallways like friendly dogs. Selassie told us how he used the cats to run down small antelope called dik-diks.

I noticed Selassie was wiping his eyes. Tearfully, he whispered to Uncle Sarge, "President Kennedy has three hundred of his children helping my country." He was referring to young American Peace Corps volunteers. Behind us, to our right, we could see Khrushchev's deputy premier, Alexei Kosygin. The ruthless Russian prime minister under Stalin during World War II was also crying.

Jackie sat with John and Caroline in the first row with Grandma, my father, and Teddy. I sat in the second row with my brothers and sisters, while Lawford, Smith, Kennedy, and Shriver cousins were jammed in like sardines all around us. Governor George Wallace was wandering up and down the aisles, looking for a seat. No one tried to help him. Cardinal Cushing delivered the eulogy in his rasping voice and placed the Host on our tongues at the communion rail. Somebody read Jack's Inaugural Address and the poem from Ecclesiastes that was later memorialized in "Turn! Turn! Turn!" by the Byrds—"A time to be born, a time to die . . ." Then we sang Agnus Dei, and Jackie sobbed for the first time. Virtually every American watched the service, along with hundreds of millions around the globe, as it was broadcast worldwide, including behind the Iron Curtain, where the Soviets suspended their customary censorship. The Russian TV commentators said, "The grief of the Soviet peoples mingles with the grief of the American people." After dousing Jack's casket with holy water, Cardinal Cushing hugged it with both hands and, abandoning the Latin liturgy, blurted out in agony, "Ah, dear Jack!" That really made me want to cry some more.

Following mass, as the soldiers carried Jack's coffin out of St. Matthew's and down the massive steps, my three-year-old Cousin John, in his baby-blue suit, saluted his father, and a photo of that moment would become one of the iconic images of the era. Back in the limousine, we followed the gun carriage south to Constitution Avenue, past the Lincoln Memorial, between the giant gold lions on Memorial Bridge, and up to Arlington, passing the more than a million people who lined the six-mile route. At Arlington the buglers, pipers, cannoneers, artillery teams, Irish Cadets, details representing each of the armed services, and an honor guard of Green Beret commandos convened around Jack's grave on a knoll just below General Lee's old mansion. Kids climbed up into trees for a better view as the crowd surged around us on the hillside. William Manchester recalled seeing a tearful black woman who knelt and waved at Jack's coffin. "That's all right," she called to him. "You done your best. It's all over now." Bagpipes played. Fifty Air Force fighter jets screamed overhead in the missing man formation, and then Air Force One passed over, all alone, at low altitude, dipping her wings as she roared over us and disappeared, leaving only a vapor trail in the crystalline sky.

William Manchester wrote, "It was ironic that John Kennedy,

whom the world knew as a man of peace, and whose proudest achievements had been the Test Ban Treaty and the successful conclusion of the Cuban confrontation without bloodshed, should be buried as a warrior, but there really was no other way: if he must go in glory, and clearly he must, the troops were indispensable. There were no splendid traditions, no magnificent farewells for a hero of peace." Then Jack's flag-draped coffin sank and the bugler blew taps, breaking down in the middle to sob. "Battle Hymn of the Republic," Lincoln's funeral hymn, blared, and all my cousins were fighting back tears as Cardinal Cushing gave a final benediction. I watched my mother flinch at each volley as a line of riflemen in dress uniforms fired a twenty-one-gun salute. Jackie lit the eternal flame, a naked gas jet that would later be framed by a giant millstone of New England granite. Then it was over.

After the funeral, we all returned to the White House and swarmed around until late on our last night in the presidential mansion, laughing at Dave Powers's tread-worn stories about Boston's colorful political bosses: "Onions" Burke, from whom Jack had wrested control of the state party; Mother Galvin, the tough Charlestown czar; Curley's Henchman "Up Up Kelly"; the 300-pound "Knocko" McCormack, who always led South Boston's St. Patrick's Day parade riding his white mare; and Martin Lomasy, the West End boss who never put anything in writing and never spoke if he could nod. We celebrated—as best we could—our cousin John Kennedy, Jr.'s, birthday with presents, fried chicken, cake, and ice cream. Then Dave led us in a sing-along of Irish classics. He and Lem whistled "The Boys of Wexford." The RFK clan were the last to leave. My mother drove us home in the station wagon at close to eleven, while my dad stayed and returned with Jackie to Arlington at midnight, where they found Major Ruddy's beret on the pine boughs upon Jack's grave. They placed a bouquet by the flame and said a final prayer. That night my dad wrote us each a letter urging us to "remember all the things that Jack started—be kind to others that are less fortunate than we—and love our country."

Two days later, LBJ asked Congress to pass the Civil Rights Act in memory of Jack. His hope was to transform one act of hatred into unity and justice for our country: "No memorial oration or eulogy could more eloquently honor President Kennedy's memory than the earliest possible passage of the civil rights bill for which he fought so long. We have talked long enough in this country about equal rights. We have talked for one hundred years or more. It is time now to write the next chapter, and to write it in the books of law." He denounced the forces that contributed to the tragic event in Dallas, saying, "So let us put an end to the teaching and the preaching of hate and evil and violence. Let us turn away from the fanatics of the far left and the far right, from the apostles of bitterness and bigotry, from those defiant of law, and those who pour venom into our Nation's bloodstream."

IT'S HARD TO OBJECTIVELY GAUGE THE SUCCESS OF A PRESIDENCY, but if our measure is the goodwill that the president generated for our country abroad, Jack Kennedy wins hands down, with Franklin Roosevelt a distant second. Even during his lifetime, Jack's popularity beyond our shores was unmatched among American presidents in 150 years. German chancellor Konrad Adenauer told my uncle that the crowd of cheering Germans he attracted in Berlin was greater in both number and enthusiasm than the greatest crowds drawn by Hitler during his heyday. But Jack inspired the German people with a message of idealism rather than the easy, formulaic alchemies of demagoguery: fear, hatred and bigotry. After his death, nations across the world named roads, boulevards, hospitals, schools, and parks after my uncle. One hundred thirty-five nations issued commemorative postal stamps bearing his likeness.

Following President Ronald Reagan's death, anti-tax guru Grover Norquist initiated a national campaign funded by wealthy Republicans to name landmarks after President Reagan in all 3,067 American counties. It was a shrewd step in a purposeful effort to elevate President Reagan as an icon of the conservative cause. But in contrast to that crusade, the movement to name landmarks for President Kennedy was, both in the United States and abroad, a spontaneous outpouring.

John Kennedy's most impressive memorials, however, were in the example he inspired in a generation of progressive leaders and activists worldwide. Almost every day people relate to me how John Kennedy touched their lives in a positive way. Perhaps inspired by Jack Kennedy, they signed up for the Peace Corps, entered politics, decided to teach, went into nursing or public health, or joined the military. In South Hadley, Massachusetts, the Northeast Golden Gloves champion Carello DeSantes told me that in his home neighborhood, "every house has a picture of your dad and uncle over the mantel." I've heard and seen that in Mexican communities in California and New Mexico; on American Indian reservations; in African American neighborhoods in Watts, Brooklyn, Harlem; in Alabama and the Mississippi Delta; in union halls; and in homes of Appalachia coal miners. It's inspiring to me how many lives were moved in such positive ways by Jack's life.

Even more extraordinary, he left the same powerful legacy abroad. On every continent, I've seen his likeness—in huts in Africa and Latin America; on wallpaper, plates, and woven blankets hanging over mantels in Italy and Ireland; and in the offices of local politicians and heads of state. Anti-American mobs ran Vice President Richard Nixon out of Latin America during his goodwill tour in 1958. Likewise, in 1958, President Eisenhower had to cancel his trip to Asia because of anti-American rioting. All that changed during Jack's presidency. When I visited Poland and Germany in 1964 with my father and mother, the crowds, numbering in the hundreds of thousands, waved American flags. They were hungry for American leadership because they knew that John F. Kennedy understood the difference between leadership and bullying, and they trusted America, looking to it for moral authority, security, and inspiration.

Jack liked to observe, "There are children in Africa called Thomas Jefferson. There are none called Lenin or Trotsky or

Stalin." It would have pleased him that after his death, thousands of parents around the globe named their children for John Kennedy. I often meet adults in Kenya and in Latin America who are close to me in age whose first name is Kennedy.

In 2015, Ernst Simpson, a physicist from Jamaica, who is chief engineer at Terracycle and first cousin to Jamaica's prime minister, Portia Simpson, told me that when Jack died, every school in Jamaica closed and kids were sent home. "Around here, we talked about him as if he were one of us—as if we all knew him. There was no world leader of any nation in modern history that had that kind of influence on little people all over the globe. He gave us hope."

On August 5, 1996, while I was having dinner in Quito, Ecuador, with President-Elect Abdalá Bucaram five days before his inauguration, he became emotional and then began to cry as he described his love for my father and uncle.

In Tokyo in 2013, I met the president of the John F. Kennedy Breakfast Club, which has three thousand members and seventy affiliates. The club meets once a week to discuss international affairs. It was founded on November 23, 1963.

A recent global Pew poll found that when people across the planet are asked which nations they regard as the most dangerous, the overwhelming response is not Iran or North Korea. It is . . . the United States. Our country today is particularly loathed throughout the Arab world, but Jack Kennedy is still extraordinarily popular among the Arab nations. It's not hard to guess why standing up for self-determination, giving away milk to children, and trying to end poverty bring far higher returns on investment than drone strikes, torture, black prisons, weapons, and wars.

"If your uncle had not been killed or if your father had become president," Professor Hamid Arabbzadeh of the University of California, Irvine, College of Medicine told me in October 2014, "we would not have seen the rise of Islamic fundamentalism or the global anger against the United States." An Iranian expatriate, he recalled Jack's efforts to have the shah loosen controls on Iran. "He worked to usher in democracy; he hosted the shah in the United States for two weeks, and when the shah returned to Iran, he enacted land reforms, educational reforms, and more freedom for women." Professor Arabbzadeh continued, "Your father met with radical students all over the world—Communists and leftists—to say that, 'We want to include you. The United States wants the same things that you do for your country—self-rule, economic justice, social justice, democracy.' So many of us around the world think of JFK as our president. He did not just belong to the United States, and he made the United States popular around the globe."

A 2011 study in the *Journal of Peace Research* found that the perpetration of Islamic extremism was associated more significantly with poverty than with variables such as religiosity, lack of education or income, or dissatisfaction. John Kennedy understood that America's best approach to achieving security at home was to try to fight poverty and injustice abroad. This is the sort of vital strategic intelligence that our highly militarized intelligence agencies are unlikely to convey to America's political leaders. God forbid that we should consider whether investing billions of dollars in drones and smart bombs may simply aggravate violence and make us less safe, and that we would achieve far better national security gains investing in basic services such as health care and education for the poor. John Kennedy understood this calculus intuitively. His foreign policy approach, in Justice William O. Douglas's words, was to "show other nations the warm heart and clear conscience of America."

Arthur Schlesinger wrote in his daybook that "JFK accomplished an Americanization of the world far deeper and subtler than anything JFD [John Foster Dulles] ever dreamed of—not a world Americanized in the sense of adopting the platitudes and pomposities of free enterprise—but a world Americanized in the perceptions and rhythms of life. JFK conquered the [dreams] of [youth]; he penetrated the world as jazz penetrated it, as Bogart and J. D.

Salinger and Faulkner penetrated it; not the world of the chancel-
leries, but the underground world of fantasy and hope." Schlesinger
added ominously, "But if Kennedy's presidency gave rise to dreams,
it also triggered fear and reaction. To the Cold War establishment
and other bastions of the old guard, JFK was not a charismatic symbol of change, he was a stark threat."

CHAPTER 9

• • • • •

Senator Robert F. Kennedy

I cannot rest from travel. I will drink life to the lees.
Some work of noble note, may yet be done, not unbecoming
men who strove with Gods.
                    —ALFRED, LORD TENNYSON, *ULYSSES*

JACK'S DEATH BROUGHT OUR WORLD TO A HALT. DURING THAT GRAY autumn that followed, my father tried to keep us from the vortex of his anguish by taking us on hikes and engaging us in contests, sports, and games; but we felt his desolation. When we did not occupy his attention, he took long walks alone with the dogs. He wore Jack's leather flight jacket with the presidential seal, carried Uncle Jack's PT-109 tie clip, and kept a lock of Jack's hair in a tiny green jade frame in his dressing room. He looked haunted. My mother said he was like a man who had lost both his arms.

To avoid the press, my dad worked at the home of his friend Dean Markham, who would die two years later in a plane crash with my Uncle George Skakel. My father had lived for his brother Jack, striving only for Jack's success. Now he faced the prospect of finding his own destiny. My dad was the most dogmatic of all his brothers, putting his faith in an orderly universe where good defeats evil, brings order from chaos, and steers all affairs toward divine justice. But Jack's death made him question the basic assumptions of his cosmology. "The innocent suffer," he wrote in his journal. "How can that be possible and God be just?" Searching for understanding and coherence, he began edging toward a new intellectual and moral maturity. He never lost his Catholic faith, but he also looked elsewhere for meaning and insight.

My dad found particular solace in Shakespeare, in the Greek tragedies, in Emerson and Thoreau, and in the poets, particularly in the heroic musings of Keats and Tennyson. Jackie gave him Edith Hamilton's *The Greek Way*, which he consumed. In Schlesinger's words, "Hamilton's small classic, then more than thirty years old, opened for him a world of suffering and exaltation—a world in which man's destiny was to set himself against the gods and, even while knowing the futility of the quest, to press on to meet his tragic fate." My father carried *The Greek Way* with him, and had us read it, too. It lit a passion in me for ancient mythology, prompting me to read the plays of Sophocles and Aeschylus. I also scoured the library at Sidwell Friends for Edith Hamilton's other works, and read every book I could find on Greek, Roman, and Norse mythology, making myself a youthful encyclopedia on these subjects.

One day he came into my bedroom and handed me a hardcover copy of Camus's *The Plague*. "I want you to read this," he said with particular urgency. It was the story of a doctor trapped in a quarantined North African city while a raging epidemic devastates its citizenry; the physician's small acts of service, while ineffective against the larger tragedy, give meaning to his own life, and, somehow, to the larger universe. I spent a lot of time thinking about that book over the years, and why my father gave it to me. I believe it was the key to a door that he himself was then unlocking.

In the vernacular of the Greeks, Camus was a Stoic. That philosophy argued that, in an absurd world, the acceptance of pain, if accompanied by the commitment to struggle, transforms even the most common men into heroes and provides the most tragic hero peace and contentment. The hero, Sisyphus, condemned by the gods to roll a rock uphill for eternity, only to see it tumble back down, was ultimately a happy man. Even recognizing and accepting the futility of his task, he could find nobility in his struggle. It is neither our position nor our circumstances that define us, according to the Stoics, but our response to those circumstances; when destiny crushes us, small heroic gestures of courage and service can bring us peace and fulfillment. In applying our shoulder to the stone, we give order to a chaotic universe. Of the many wonderful things my father left me, this philosophical truth was perhaps the most useful. In many ways, it has defined my life, and has allowed me to find serenity and purpose even in the most trying and tragic circumstances. "Men are not made for safe havens," Edith Hamilton wrote, in one of my father's favorite passages. "The fullness of life is in the hazards of life. . . . To the heroic, desperate odds fling a challenge."

My mother was crucial to my father's recovery, gently pushing him toward reengagement. To give him a stable platform upon which to rebuild his life, she kept a sense of order and normalcy at Hickory Hill. She helped him recover through the example of her faith that Jack was in heaven, by ensuring that the children were cared for, and by her vigilant effort to rid our house of any evidence of sadness or mourning. The mention of Dallas was taboo. Magazines with cover photos of the assassination quickly vanished, and the TV was shut off whenever the film clips of Jack's murder appeared.

My mom pushed my dad to return to participating in public events. Three weeks after Jack's death, on December 20, we all went to our annual Justice Department Christmas party for foster children from Washington's poorest neighborhoods. It was my dad's first public appearance. In the commotion of our arrival, a tiny boy suddenly ran toward him shouting, "Your brother is dead!" The room went silent. Sensing he had done something wrong, the little boy began to cry. My dad swept him up in a warm hug and reassured him gently, "That's all right, I have another."

My mom urged him to go to New York for the renaming of

Idlewild Airport for Uncle Jack, and then to give his first public speech, a St. Patrick's Day talk in Scranton, on March 17, 1964. The outpouring of love by the Scranton Hibernians was beyond anything my father had previously experienced. The entire rust belt city emerged to greet him. Thousands lined the airport road in a snowstorm to catch a glimpse of him, and those who couldn't fit into the 1,100-seat hall listened outside to speakers set up in the streets, or on the radio.

Against the counsel of his advisers, who feared he couldn't speak the lines without breaking down, my father closed with a quote from the ballad "Lament for the Death of Owen Roe O'Neill" by Thomas Davis:

> Sagest in the council was he, kindest in the hall,
> Sure we never won a battle—'twas Owen won them all.
> Had he lived—had he lived—our dear country had been free;
> But he's dead, but he's dead, and 'tis slaves we'll ever be. . . .
>
> Sheep without a shepherd, when the snow shuts out the sky—
> O! why did you leave us, Owen? Why did you die?
>
> Soft as woman's was your voice, O'Neill! Bright was your eye,
> O! why did you leave us, Owen? Why did you die?
> Your troubles are all over. You're at rest with God on high,
> But we're slaves, and we're orphans, Owen!—Why did you die?

Somehow his voice held, though the newspapers declared that "There was not a dry eye among the stalwart sons of Erin." My father was regaining his strength.

With Jack's death—and Grandpa's stroke in 1961—my father was now the male head of our Kennedy clan. As he battled his own doubts, despair, and loneliness, he somehow found strength to hold our family together. He took special care of Aunt Jackie and her children, eating dinner with them in Georgetown several times each week, and, later, arranging for them to move to New York. We now spent our vacations with Jackie, Caroline, and John, as well as with the Lawford children, my Aunt Pat and her husband having separated. They accompanied us on river trips and ski vacations, as my father threw himself—and us—into challenging physical endeavors in the outdoors. "In the presence of nature," Emerson observed, "a wild delight runs through the man, in spite of real sorrows." I remember my father on those outdoor retreats increasingly encouraging, loving, and laughing.

In 1964, the Canadian government honored Uncle Jack by naming North America's tallest unclimbed peak "Mt. Kennedy." In March 1965 my father enlisted mountaineer Jim Whittaker to organize an expedition to the Yukon Territory to climb Mt. Kennedy. My dad first met Whittaker—the first American to summit the world's highest peak—at a White House ceremony celebrating his Mt. Everest climb. My dad, who didn't particularly like heights and had no time to train for the grueling Mt. Kennedy climb, joked with us that he would prepare for the trip by running up and down the stairs and screaming "Help." As he left for the expedition, Grandma Rose bade him goodbye, saying, "Don't slip, dearie." Afterward, he told us the climb had been the hardest thing he'd ever done. He showed me a letter he opened after returning. It was from a little girl. "I understand you are going to climb a mountain. I don't know why. I asked my Daddy why. He doesn't know why either."

Jim Whittaker became a cherished friend to my father. Jim, six foot five and superbly fit (his identical twin, Lou, was also a world-class mountaineer) had done such a bang-up job outfitting and leading the Mt. Kennedy summit that my father drafted him to organize wilderness adventures for our families. For me those trips were pure revelry. That summer, we embarked on our first whitewater trip, a family expedition down the Colorado River through the Grand Canyon. We were among the first few hundred people to run the Colorado—this was before the Glen Canyon Dam transformed the river into a giant plumbing conduit. There were still wide beaches on which to camp, and the Colorado's warm, muddy waters were home to large schools of eight native fishes, four of which have since gone extinct, with the other four on the way.

Our retinue included Lem Billings; crooner Andy Williams and his wife, Claudine Longet; humorist Art Buchwald; astronaut John Glenn and his wife, Annie; writer and sportsman George Plimpton; U.S. Olympic ski team coach Willy Schaeffler; Interior Secretary Stewart Udall and his wife, Lee; *Los Angeles Times* publisher Otis Chandler (who would later appear on Nixon's enemies list) and his wife, Missy; designer Oleg Cassini; my aunts Pat Lawford and Eunice Shriver; and sixteen Kennedy cousins. Our guides were the Hatch brothers, the famed western whitewater oarsmen who would outfit many of our subsequent expeditions. To this day I stay in touch with many of these wilderness boatmen. As mine did, their experience with rivers has helped turn their lives toward environmental activism. One of them, Tim Means, fought beside me on the front lines in our successful campaign to stop Mitsubishi from building the world's largest salt mine in Laguna San Ignacio, the greatest gray whale nursery on Mexico's Baja Peninsula. Tim went on to lead the battle against reckless development along the Sea of Cortez. Another guide on that Colorado trip, Bill Hedden, is one of the most effective advocates for the Grand Canyon itself.

At the time of our trip, the Colorado River run was still regarded as expeditionary; the inclusion of so many children was unprecedented. The guides were alarmed by my father's inclination to ride the rapids on an inflatable sleeping mattress, or in a life jacket, and by our practice of leaping in to join him. Wherever the river narrowed and deepened, we eddied out the rafts and climbed the canyon walls for big air jumps into the river. Each night, Art Buchwald led the grown-ups in campfire poetry reading. Then we laid out our sleeping bags on broad beaches, gazing at the glittering heavens as they revolved above the canyon walls, tracing the constellations that Grandma had taught us. After so much despair, it seemed that joy and laughter were returning to our lives.

Having caught the climbing bug, my father surveyed the horizon each morning for a high peak to summit before breaking camp. Our toughest climb on the Grand Canyon trip was Jack Ass Ridge, but my exhaustion disappeared on the descent when I spotted a small rattlesnake beside the steep trail. I caught him bare-handed, and slipped him into a sock I hastily stripped from my foot. It seemed providential. To me, rattlesnakes were the apogee of western romance, and I had dreamed of seeing one on this trip. Now I had one in my sock! I was desperately keen to keep this prize, but, sadly, the guides sabotaged my plans to take the rattler home; federal law forbade its removal from the park. I suspected the guides had cited the law because none wanted to have the snake on his boat.

As consolation, I skinned a small deer that I found dead on the riverbank and returned home with the pelt, which I treasured for many years. That afternoon, my alert sister Kerry spotted a terrified Chihuahua stranded on a rock in the fierce center of the Lava Falls rapid, and my dad rescued the trembling canine with a quick grab from our passing raft. There being no rules against removing stray dogs from the park, Kerry got to take it home. I suspect my father paid a bribe to its owners, who contacted us after recognizing their lost pup in news stories about our trip.

On the last day, we ascended the seven-mile South Kaibab Trail out of the Grand Canyon in baking 115-degree heat. It took six hours. Sweating and wheezing, Lem was characteristically unselfconscious in a broad, flowered cotton sunhat moistened with canteen water. Out of consideration for his asthma, the guides offered Lem a chopper evacuation, but he declined this "ride of shame," opting instead for a tiny burro, which carried the giant man with undisguised dismay. As if divining Lem's debilitating acrophobia, the mule took revenge by hewing to the edge of the lethally steep trail. Lem's audible anxiety—his nonstop expressions of anger, dread, and pleading with the mule—was a source of great hilarity to the children. I followed closely to eavesdrop and laugh, and the trip cemented a friendship with a man who became something of a surrogate father to me in later years. Other less fit grown-ups followed Lem's example and rode mules. We children struggled up the steep trail on foot. During a particularly challenging section, my father gathered us in a shady spot and inspired us by reciting the St. Crispin's Day speech from Shakespeare's *Henry V.*

After the Colorado, my father got hold of two white-and-red fiberglass Klepper kayaks, learned to Eskimo roll, and taught us the maneuver in the swimming pool. From then on, the kayak became his preferred whitewater vessel. The next summer on the Middle Fork of the Salmon River in Idaho, his combat roll still imperfect, my dad tumbled from his kayak and swam half a mile in frigid water through a boneyard of jagged boulders in the notorious Warm Springs rapid before pulling himself, shivering, into an eddy. In subsequent years my father and Jim took us on whitewater trips on Idaho's Snake River, and the Green and Yampa rivers in Utah. Ten years earlier, David Brower and the Sierra Club had saved the Yampa from a proposed dam at Echo Canyon. Senator Joe McCarthy had subpoenaed our guides, Don and Ted Hatch, who had joined Brower's campaign, for a grilling before the House Un-American Activities Committee (HUAC). McCarthy and others then considered opposition to hydropower Communistic. In 1965, we kayaked the Upper Hudson in New York's Adirondack Mountains, where my father rallied opposition to another proposed dam. Those whitewater trips were pure magic. I loved being with my family in the rough-and-tumble rapids in America's backcountry.

My father saw wilderness as the furnace of America's character. He prodded our guides to tell us about the region, its geology and natural history. We learned about the explorers who first ran the rivers—the one-armed Civil War veteran John Wesley Powell, Lewis and Clark, and John Frémont—the "Great Pathfinder" who had scouted the Green and Snake rivers, fought in the Mexican and Civil wars, and had run for president as an abolitionist. My father also introduced us to the Native Americans who had lived there long before the Europeans. He had read widely about Native American history and customs, and the guides pointed out ancient Hopi petroglyphs during our hikes. On the Colorado trip we stopped at a Navajo reservation near Red Rock, where we visited a medical clinic. I saw the crushing poverty—and how grateful people were for my father's attention. Even the children my own age seemed empty-eyed and hopeless. After we left the reservation my dad was quiet for a while, then he talked about how ashamed all of us should be—every American—that the Navajo lived in such desperation.

My father had a conservative nineteenth-century sensibility that children should be toughened by constant exposure to the elements and physical challenge. Wilderness and adventure, he believed, would imbue us with character as well as beef-jerky toughness; it would awaken our souls and instill in us the range of virtues that European Romantics associated with the American woods—self-reliance, Spartan courage, and humility. Both our parents urged us to challenge ourselves and take risks. Like his father, my dad didn't want his children to become indolent or diminished by privilege. He expected us to endure the exhausting climbs and occasional long days on a cold river in frigid rain with appropriate stoicism. Yet he also had a way of making the suffering seem fun— that we were all part of some noble adventure!

The Whittaker brothers were great skiers, and their families began joining us in the large crowds that comprised our annual ski trips. Goaded by my mother one night in Idaho, Jim impressed hotel guests by scaling the thirty-foot fieldstone fireplace in the Sun Valley Lodge's central ballroom. My dad rallied us for first tracks on the early-morning "milk run," and forbade us from quitting until the lifts closed at four. Then, after a sauna with my father and Uncle Teddy, we'd go out in our birthday suits and roll in the snow.

Every western ski trip included backcountry adventures. My father broke trail, on long climbs, through deep powder far outside the ski area boundaries, and we followed along, carrying our little wooden skis with their cable spring bindings on our shoulders, hiking steep, narrow ridges toward distant summits. One magical day, after a long hike, we were poised at the top of a stark head-wall behind Baldy Mountain in Sun Valley. Surveying the distant valley floor between the toes of my ski boots, there didn't seem to be a prudent way down. Our dad reminded us of Emerson's adage,

"Always do what you are afraid to do." And then he added: "Let's see you go." We dropped in, one after the other, over the cornice, into hip-deep powder, then worked our way through alpine meadows and thick woods, finally making it back to an unfamiliar road after dark. We hitchhiked back to the lodge.

My father sailed with a similar controlled recklessness. We grew up knowing the Kennedys came from Wexford, the seafaring Irish port town. Every summer morning, regardless of the weather, we were in the pounding surf, often watching my grandmother's bathing cap bobbing atop great rollers in the chilly water far offshore. "You have saltwater in your blood," our parents told us. "You should never be scared of the sea." Then, as today, there were nearly always Kennedy and Shriver sailboats out on Nantucket Sound, off Great Island, even when small-craft warning flags were battering the flagpole at the end of the Hyannisport pier and there were no other boats on the water. On breezy days and in small gales we'd help my dad rig the tiny Sunfish he kept above the tidal windrows on the family beach. He encouraged us to practice tipping it over and righting it before it turtled, so we would never lose our heads in a crisis.

The grown-ups relished maritime adventure. On impulse, my dad and Uncle Ted would load us in a sailboat and navigate to the Vineyard or Nantucket, or even up the coast of Maine to Penobscot Bay. More than once we used roadmaps for sailing charts. Soupy fogs or high seas were not an excuse to avoid our daily sail on the twenty-six-foot *Victura*, whose shallow draft and heavy keel were ideal for the stiff breezes and shallow shoals of the south Cape. On moonlit nights I loved to lie on the stern and watch bioluminescent jellyfish glow in the gurgling wake and try to cup them in my hands. In a stiff wind or a squall, my dad would heel the boat over on her side so we could walk up the mast, and practically stand on the keel. Leaving my mother or brother Joe at the tiller, my father would dive overboard and catch hold of the towline. One after the other, we would each jump into the water and drag behind as the boat headed out into Nantucket Sound.

Sailing, river trips, outdoor adventures, and sports at Hickory Hill were all opportunities to demonstrate personal gallantry. My father surrounded himself—and us—with real-life heroes like Jack Fallon, the rugged America's Cup champion, the bullfighter El Cordobés, mountaineer Jim Whittaker, and astronaut John Glenn., We also got to know Olympic gold medal decathlon champ Rafer Johnson (my brother Max's godfather) and boxing champ José Torres, and my dad made sure we understood that they both overcame racism and poverty to triumph at the highest levels of their sports. We all had the sense that in our adventures, we were being prepared for some larger role—I don't mean a hereditary post in politics, but that we were being schooled in gallantry and courage, prepared to take risks and make sacrifices to accomplish something for humanity.

My father had a special weakness for war heroes. Every weekend during the White House years, Gerald Tremblay, an ace fighter pilot in both World War II and Korea, came to stay at our home. Willy Schaeffler, the U.S. ski team coach who accompanied us on nearly every ski trip, had survived imprisonment by Hitler and a stint in a German penal battalion on the Russian front; he was also one of the few soldiers who lived through the German retreat from Moscow in 1942. Kenny O'Donnell, my dad's college roommate, had been shot down over Belgium and escaped a German POW camp; he was awarded the Distinguished Flying Cross for his heroism. My dad's closest friends and aides, his "band of brothers," which included Frank Mankiewicz, Byron White, Dean Markham, Dave Hackett, Ed Guthman, John Seigenthaler, Nick Katzenbach, Jim Symington and John Doar, were all combat veterans.

I understood that all these efforts to imbue us with pluck were boot camp for the ultimate virtue—moral courage. Despite his high regard for physical bravery, my father told us that moral courage is the rarer and more valuable commodity. In 1965, he wrote in *Life* magazine after reaching the summit of Mt. Kennedy that it was not "blind, inexplicable, meaningless courage" that he admired. "It is courage with ability, brains, and tenacity, and purpose. Winston

Churchill called this kind of courage 'the finest of human qualities' because it is the quality that guarantees all others."

Despite his grief, my dad continued to take an active interest in our sports, hobbies, homework, and school. My academic performance must have caused him anguish, because his standards were high. While he was alive, school remained a mystery to me, and homework an ordeal I dodged by heading for the woods as soon as the final bell sounded. Woods and barn beckoned; I needed to check my traps, fly my hawks, and see if the pigeons had paired up or if their eggs had hatched. These evasions must have irked him. Occasionally he brought me into his office and talked gently about the importance of working hard and being more organized. He never managed to work up a convincing rage about my shirking. Years later, I realized that his patience was probably tempered by memories of his own struggles in elementary and middle school and his own experience with a loving and supportive dad.

Despite my poor school performance, I felt his admiration in other realms. He took a keen interest in my hawks, and frequently came down to the barn to watch me fly them. He asked me questions about wildlife and animals, and made me feel that my knowledge and commitment impressed him. His interest and admiration in my extracurricular successes buttressed my confidence and self-esteem.

I'm happy my father supported me in this way. I wasn't a kid who needed scolding—I was hard enough on myself, and painfully conscious of my shortcomings. I didn't need to be told by another adult that my schoolwork wasn't measuring up. If my father had simply piled on, it might have crushed my soul. Unconditional support and love, the gifts my grandfather gave his children, probably allowed them to achieve so much. My dad gave these to each of his own children, too.

Even though the house was frequently filled with famous people, our father always made us feel that we were the center of his universe. In weekend touch football games he quarterbacked, usually opposite Dave Hackett or Kenny O'Donnell, and the teams commonly included professional athletes, football greats like Bobby Mitchell, Sam Huff, Frank Gifford, Don Meredith, and Rosey Grier, or baseball star Stan Musial. He often deployed those legends to decoy, or to throw a block, so that the younger kids—Michael, or David, or Courtney—could catch the pass or run with the ball.

My father taught us by example that we should approach the world humbly but with confidence. We should be able to laugh at our foibles, admit our mistakes, and never take ourselves too seriously. At the same time, we should be serious about the problems of the world. At dinner, he led us in debates about current issues and world affairs, such as the arms race, communism, and Vietnam. During the White House years, he told us what was happening at the Justice Department, in his battles with Jimmy Hoffa, and about the civil rights fights in the South. Later, when he was in the Senate, he talked about violence in the ghettos, and the other problems of poverty. Over a breakfast of Cream of Wheat, eggs, and Rice Krispies, he talked about kids he'd seen in Appalachia and Mississippi with rickets and swollen bellies, and asked us to imagine what it would be like to go to bed hungry, or to have no education and no hope of a good job. We heard about his work in Bedford-Stuyvesant, the Brooklyn neighborhood where he was spending so much time, and about the decrepit schools in Spanish Harlem, and his battle to close Willowbrook Institution on Staten Island, a medieval prison for children with intellectual disabilities.

Although he had devoted his own life to politics and often urged us to "do something for our country," my father did not feel his children had to follow in his footsteps into public life. But even as he encouraged us to pursue our own passions, it was clear that he would have been distressed if we did not involve ourselves in some way in the problems of humanity. In a 1967 television interview, he described a conversation he'd had with me: "I talked to one of my sons about [going into politics] one time, and he said that he wanted—he was twelve years old—he wanted to make a contribution. And he said, 'But I don't want to get involved in political life.' And I said, 'Well, what do you want to do?' He loves animals, and he said, 'I want to make a contribution like Darwin and Audubon did.' And so I think people work out their own lives . . . just as long as they understand that in the last analysis, what is important [is] that they give something to others and not just turn in on themselves."

As our father reengaged, our house filled again with laughter. He roughhoused with us, and organized games and competitions, from swim races to rope climbing. Even when we misbehaved he rarely went farther than to make us hand-copy famous poems. Occasionally, however, my behavior required an escalation of discipline.

In 1967, construction began on the Dolley Madison Highway, connecting Washington to northern Virginia's woodland and farm country across the Potomac. Since the new roadbed skirted our neighbor's farm, we had a front-row view as highway workers clear-cut the forests, leveled hillsides, buried streams, and uprooted trails to the pond where David, Kerry, Michael, and I caught tadpoles, frogs, and snakes. In the afternoons, we children stashed our books and climbed across corrals and pastures of cedar split-rail fences, hiked past the barns, through Mr. Ornstein's raspberry patch and the Neal family's cornfields and sunflower groves, until we reached the construction site, where machines had recently rerouted Pimmit Run through a concrete culvert. To my twelve-year-old brother David and me (thirteen), this seemed like a criminal act. We watched angrily as our woodland paradise gave way before giant bulldozers and bullying construction companies, advance shock troops of the rising suburban tide that would soon bury northern Virginia under concrete.

Neither of us was inclined to simply stew. "Aren't we men of action?" David goaded me. So one afternoon we pulled up surveyor's stakes and dislodged a large pile of stacked highway culvert pipes, sending them rolling down an embankment to shatter in smithereens. We were caught, and our parents punished us. After three weeks on lock-down, we had to work all summer helping to build and sweep a beach boardwalk to offset the costs of our sabotage.

But a year later, when the traffic started coming, we returned for another sally. This time we climbed a freshly cut hill to pelt passing cars on the new highway with mudballs. David got caught by the police, prompting my father to pull him out of school and take him on the road during his presidential campaign. Always reliable in a jam, David never ratted me out. Still, under a cloud of suspicion, I remained in hot water with my mom. Not long afterward, I went off to Jesuit boarding school at Georgetown Prep.

Of all of us children, David was the closest to my father, and the most vulnerable and dependent upon him. My dad had a special bond with David. They were very much alike: shy, vulnerable, tough, fearless, kind, loyal, and principled. David had a lisp when he was young, and our father amused us by asking David to say, "My sister Courtney blows her whistle." Then he picked David up and hugged him as we all laughed—including David. Of all his children, my dad reserved a special love for David, and David thrived on his affection. The night my father died in L.A., he left David alone in the hotel room with the television on, promising to return immediately after his victory speech. During the confusion following the shooting, it was hours before John Glenn found David, seated in front of the TV, tears streaming down his face, watching our dad's assassination play and replay. David never got over that loss. He died of a drug overdose in 1984, at age twenty-eight.

In June 1964, as my dad was beginning to emerge from his shell, General Maxwell Taylor, who parachuted into France on D-Day, invited our family to Normandy to commemorate the twentieth anniversary of the D-Day invasion (my mother was then pregnant with my brother Max). We flew in an Air Force cargo plane, stopping to refuel in Iceland, sleeping aboard the plane on bunks behind canvas webbing. We landed first in Normandy, where we paid our respects among the neat rows of white gravestones marking the resting place of the thousands of American men my father's age who died on Normandy's beaches. We explored German bunkers, and the hedgerows where Americans and Germans battled in close combat. We climbed through concrete trenches and pillboxes atop the cliffs overlooking the D-Day beaches—Utah, Omaha, Juno—where we could still see Nazi tank traps bristling like rusted teeth from the beating waves. In Paris, my mother scolded Joe, David, and me for brushing our teeth in the hotel bidet, a device with which, like most Americans, we were unfamiliar. My mother lacked the words to describe the purpose of the contraption. When we learned the truth, my brothers and I were at once deeply amused, grossed out, and astonished by the depravity of the French. Afterward, David, Michael, and I climbed to the very top of a tall lamppost outside the ambassador's residence.

Europe was still recovering from the war, and I was surprised by the Continent's poverty. The French rode bicycles, and drove rickety Renaults; an American car was a symbol of extreme opulence. My brothers and I were intensely entertained to discover telephone books, cut in neat quarters, substituting for toilet paper in the roadside lavatories.

Legions of reporters and photographers with noisy, blinding flashbulbs trailed us through the Louvre. Cheering crowds materialized every time we left our hotel, the French people's exuberance for Jack and Jackie spilling over to my father and his family. My favorite museum was the Jeu de Paume, where I saw the work of the great Impressionists and Post-Impressionists, Monet and Degas, Cézanne and Matisse. Many of these masterpieces had been recently returned to the museum, after being looted by Goebbels and Hitler during the war.

Even as a ten-year-old boy, it was evident to me that Europe was looking to America for leadership. Thousands of French, many waving tiny American flags, thronged our family with a warmth approaching delirium. They showered us with flowers, and spontaneously filled the giant plaza around the Arc de Triomphe.

Afterward, our parents took us to Berlin to dedicate a memorial to Uncle Jack. We visited Willy Brandt, the mayor of Berlin who had fought Hitler in the German Resistance and later as an Allied army officer. Brandt was a handsome giant, with an expansive forehead, determined mouth, and square jaw, who had single-handedly averted World War III in 1956 when he turned back a West German mob seeking to storm Communist East Germany at the Brandenburg Gate. In a heartrending speech that no other German could have convincingly delivered, he reminded the German people that their divided nation was the penance for having embraced Hitler. Together with Brandt and West German chancellor Konrad Adenauer we visited the Berlin Zoo, and saw the American bald eagle that President Kennedy had presented to them during Jack's *"Ich bin ein Berliner"* trip to Berlin in June 1963. I couldn't help but be struck by the irony of this symbol of American freedom being kept in a very small cage.

Mayor Brandt took us to the Berlin Wall, where I looked up at rolls of barbed wire spanning menacing guard towers, from which Communist sharpshooters studied us with shark eyes. We visited Checkpoint Charlie, the site of confrontations between Soviet and U.S. tank squadrons during Uncle Jack's Berlin Crisis. At the places marked with crosses, wreaths, and flower corsages we paid homage to the men and women fleeing totalitarianism who had been gunned down by East German guards, or had died leaping to their deaths for a chance at freedom.

Once again, giant crowds gathered and engulfed us wherever our family appeared. Seventy thousand Germans jammed the plaza beneath the Rathaus Schöneberg during the dedication of Jack's memorial, showing their affection for America and their hope for U.S. leadership. When reporters asked my father whether America was prepared to use its armies and weapons to reunify the country, he told them that the U.S. would stand to protect West Berlin from invasion, as our treaties required, but he warned against provocation that might lead to war. He didn't believe America should become the world's policeman.

Afterward we went to Poland, uninvited. Poland's Communist government had, in fact, forcefully asked us not to come. We ignored that request, and added insult to injury by bringing along my

Aunt Jackie's sister, Princess Lee Radziwill, whose husband, Stanislaw "Stas" Radziwill, was an emigré Polish prince, deposed during the Communist occupation. Poland's government suppressed media coverage of our visit. Government operatives even removed children from an orphanage just before we arrived bearing gifts for them.

I was excited to take my first glimpse behind the Iron Curtain. Embassy diplomats had warned us that Stasi and KGB spies would secretly monitor and record our every move, so Joe, Kathleen, David, and I took the precaution of searching our hotel room for listening devices. We tore the place apart. We lifted the rugs and piled up furniture in makeshift scaffolds to reach the chandeliers. We dismantled the crude and surprisingly sturdy black rotary phones, heavy as barbells, searching for concealed bugs. We ran the water and made loud noises—as we had seen James Bond do—to befuddle prying ears.

Despite the official hostility and media blackout, news of our arrival somehow spread. For five days mobs numbering in the tens of thousands surged around us across Poland, from Częstochowa to Kraków. At Częstochowa we visited the Catholic shrine of the Black Madonna. I drew an iguana next to my name in the visitors' book, where we also found Kennedy signatures from 1956, when my father and Teddy had visited. Teddy had signed: "Sincerly [sic], Edward Kennedy." Right next to Teddy's inscription, my dad had written: "That's not how you spell it, Teddy."

In the market square outside Kraków's Cloth Hall, crowds stalled our convoy, pulled my father from the car, and carried him around on their shoulders, finally depositing him on the roof of the embassy car, a Soviet-made Volga. We clambered up to join him. Thousands of ecstatic Poles reached out to touch my parents and embrace us, chanting, "May you live one hundred years," and singing "Sto Lat," meaning "good wishes and long life." It was the ancient song Poles traditionally sang to their legendary heroes, kings, and national leaders. Their chorus, the ambassador told us, was a subtle act of rebellion; the song had not been sung in Poland since the Communist takeover. As they sang, the crowds affectionately rocked the car, to our great delight.

In Warsaw the throngs were even larger. The multitude looked like a stormy ocean. Despite government warnings not to do so, we visited the Polish Primate, Stefan Cardinal Wyszyński (now a Catholic saint). As we waited for our audience with the prelate, a handsome priest ushered us into the kitchen, where he made ham-and-cheese sandwiches for us. He was animated and friendly. He smiled and laughed a lot and spoke halting English with spirited charm. As he worked on our lunch, he talked to us about communism with disdain and soccer and skiing with great enthusiasm. He told us he had dropped out of university when the Nazis invaded and sent his professors to concentration camps. His name was Father Karol Wojtyla, and forty years later he would be anointed Pope John Paul II.

Cardinal Wyszyński told us, "The best thing that happened to the Catholic Church in Poland was that it has been deprived of its wealth. This has brought the priests and bishops much closer to the people. The Communist officials have become the new oligarchy, and the Church has become the protector of the Polish people." My father later wrote of Wyszyński, "Without question, he is the most impressive Catholic clergyman I have ever met." The Polish people considered him a national hero. They credit him with saving the Polish Church during the Communist repression, despite his own three-year imprisonment.

While we attended mass at the Warsaw Cathedral, a hundred thousand Poles filled the square and adjacent streets, waiting for us to emerge. When the service ended, they surrounded our motorcade. Soon it became impossible to move, and my father again climbed atop the American ambassador's limousine. We scrambled up behind him, to wave to the exultant crowd. It looked to me like the multitude went on forever—there were people on rooftops and lampposts, waving from windows. Cheering Poles jammed every street leading into the central square. We waved and signed autographs. When they sang the now familiar "Sto Lat," we responded with a tone-deaf rendition of "When *Polish* eyes are smiling." It was a sappy performance—personally embarrassing, but they were delighted, and very forgiving!

Alarmed by the crowd's fervor and the suffocating crush, the American ambassador, John Moors Cabot, a stuffy Boston Yankee, lowered a window slightly and called up to us: "I say, would you tell the attorney general that the roof is caving in." My father told him, with feigned bewilderment, "But this is how we always come home from mass." We went on to demolish the roofs of cars across Europe, and I remember some complaints from the family's comptroller, my Uncle Steve Smith, when the embassies sent their bills for the damaged vehicles.

Ambassador Cabot proclaimed himself annoyed by our visit, and the Polish government, he insisted, was "livid with anger and shock." Yet he acknowledged in his dispatch to Lyndon Johnson's equally disapproving State Department that the beneficial impact of the trip on America's image abroad could not be underestimated, and he termed our visit an "undoubted success."

On our way home we stopped in England to visit Runnymede, the famous battlefield where the British people forced King John to sign the Magna Carta in 1215. The document was the precursor to our own Constitution. The British considered it the perfect spot to honor America's slain president. Queen Elizabeth, the following year, would lay a stone monument at the center of Britain's most hallowed soil, declaring on a plaque, "This acre of English ground was given to the United States of America by the people of Britain in memory of John F. Kennedy." The ceremony somehow reminded me of Camelot.

In France and Germany my father had briefly been drawn out of his mourning to marvel at the seismic outpouring of grief from people who'd never seen or met his brother. I saw my parents' wonder at equally fervent multitudes in a Communist country where the press blocked any good news about America. The yearning in those crowds has remained a vivid memory for me for more than half a century. I think it affected my father the same way. After

Jack's death, my dad told me that he intended to leave public life. Perhaps he would become a college president, or teach history and public policy. I am certain that trip abroad in 1964 helped change my father's mind, and, together with my mother's steady confidence, encouraged him to gather his strength and continue Jack's legacy.

My father often recalled Jack's love of Thomas Paine's words, "The cause of America is in a great measure the cause of all mankind." The yearning of those crowds was unmistakable. The Europeans were counting on American leadership. Afterward, my dad told me that this love we were witnessing for our country made it especially important for us to perfect our democracy and make it an inspiring template for the rest of the world. America's challenge was to prove to the world that government by the people can work, and to avoid the slide back into feudalism or totalitarianism. America as an exemplary nation with a sacred mission to uplift the world was still an ennobling idea. Today, it has devolved into a messianic justification for American imperialism. But my father gave us to understand that exceptionalism must be based on values, not on wealth or military power. The world wanted our leadership, not our bullying—and Europeans knew the difference.

The Rift with LBJ

In 1965, President Johnson sent me a long letter after I swung through a window, severing the major tendons in my left leg and shaving off a couple of toes. The repair and reattachment required 147 stitches and kept me in a cast for two months, and in physical therapy for a year. The president's note was cornpone but really sweet, characteristic of the multitude of kindnesses that LBJ showed my family in the wake of Uncle Jack's death. Teddy and Johnson genuinely liked each other, and Teddy worked hard in the Senate to pass LBJ's ambitious domestic agenda. After Jack's death, LBJ reached out in many little ways to comfort my father. The president encouraged him to undertake a peace mission to Indonesia, to negotiate a peace treaty between Sukarno and the Dutch. Out of compassion, Johnson was purposely trying to encourage my father to reengage, to rescue him from his despair. But we were all also aware of the growing tension between LBJ and my dad as the press and public increasingly focused on my father as a potential presidential contender.

My father was good-natured about the rivalry and even kidded about it with LBJ. When a little girl sent my dad a letter asking for a copy of his inaugural address, he sent it along to the president with a cover note: "Dear Mr. President, I thought you would want to see a typical example of my mail." My father was guarded in expressing any distaste for the president. My mother, characteristically, was unconstrained. From her we heard that Johnson was cynical, vulgar, and self-interested. She openly expressed my father's quiet anxiety that LBJ was torpedoing many of Jack's programs, and undermining America's role in the world. LBJ put Thomas C. Mann, a free-enterprise Texan, ardent colonialist, U.S. business booster, and CIA shill, in charge of the State Department's Latin American Division. Arthur Schlesinger called that appointment "a declaration of aggression against the Kennedys." Under Mann's guidance, LBJ quickly embraced the right-wing junta that deposed President João Goulart's government in Brazil, sent 20,000 troops to invade the Dominican Republic, and ended U.S. aid to the starving poor of Peru in retaliation for government threats to nationalize Standard Oil's Peruvian oil holdings.

In December 1963, less than one month after Jack's death, my father urged Secretary of State Dean Rusk to end the ban on travel by American citizens to Cuba. Travel bans had always irked him, but that recommendation may also have been my father's way of dismissing the CIA's disinformation campaign that had tried to pin Jack's murder on Castro. My dad argued that the ban violated the rights of Americans to travel—one of the freedoms he was sworn to protect as attorney general—and that lifting the ban would "contrast with such things as the Berlin Wall and Communist controls on travel." When the appeal fell on deaf ears, he pushed harder, urging the State Department to consider ending the Cuban embargo. He said it made America look like a bully and Castro a hero for standing up to us. The embargo undoubtedly contributed to Fidel's relentless hold on power, but my father's suggestion was heresy at Foggy Bottom, and afterward the State Department excluded him from all national security discussions.

LBJ was intensely suspicious of my father. Although he kept him on as attorney general, LBJ systematically clipped his wings. He effectively removed my father's authority over the FBI by encouraging a smug J. Edgar Hoover to report directly to the White House. Hoover was so sure of gaining a direct line to LBJ that, on the day Jack died, Hoover ordered FBI agents to stop meeting my father at airports when he landed, a courtesy the FBI had always extended to the attorney general.

With Johnson and Hoover now in cahoots, the FBI began spying on my father. Hoover sent thirty agents to the Atlantic City Democratic Convention to snoop out the relationship between my dad and Dr. King. Hoover was also conducting aggressive surveillance of Uncle Teddy. During FDR's presidency, Hoover had spied on Eleanor Roosevelt, whom he suspected of being a Communist, but prior to LBJ no American president had ever ordered the FBI to spy on a potential political opponent. Those poison seeds would later flower at Watergate.

The Democratic convention in August 1964 fortified Johnson's worst suspicions. LBJ, the garish Texan, saw a calculated personal slight in my father's use of a quote from *Romeo and Juliet*, recommended by Aunt Jackie, to close his tribute to Uncle Jack: "When he shall die, / Take him and cut him out in little stars, / And he will make the face of heaven so fine / That all the world will be in love with night, / And pay no worship to the garish sun." The delegates' twenty-two-minute ovation for my father only rubbed salt in Johnson's wounded pride.

As the estrangement grew, Johnson kept close track of my dad's movements and discouraged Cabinet members from carrying on friendships with him. Johnson was intensely suspicious of any of his appointees who had been close to the Kennedys. He often chided General Maxwell Taylor, the new ambassador to South Vietnam, about his namesake, my brother Max. "How is that Kennedy boy named after you?" When White House friends or senators visited Hickory Hill or Hyannisport, the president would often phone them there to let them know he was keeping score. Bob McNamara, Dick Goodwin, Averell Harriman, Senator Fred Harris, all received unsettling calls from LBJ at Hyannisport. Presumably, Hoover had our homes under surveillance.

One night, not long after Jack's death, our parents took us to the McNamaras' for dinner. In the middle of the meal, a fuming President Johnson barged into the house unannounced, accompanied by his Secret Service entourage. McNamara appeared rattled as he fetched a chair for his uninvited guest. My mother bundled us up as if a tornado were coming and cleared us from the room, leaving my dad and a befuddled Bob and Margie McNamara alone with the president. She rushed us home to Hickory Hill for an improvised dinner of cold cereal. I remember thinking how huge Johnson was next to my father, and how caustic his manner. It was my opinion that the dinner was going to be no fun.

My father had entertained thoughts of becoming Johnson's vice president. Dick Goodwin told him, "If Johnson had to choose between you and Ho Chi Minh as a running mate, he'd pick Ho in a minute." But on July 10, 1964, Johnson preemptively announced that no one from his Cabinet would be selected for the job. My dad joked that he was "sorry he'd brought so many good men down with him." Johnson initially offered the vice-presidential slot to my uncle Sarge Shriver, hoping, it seems, to divide Kennedy supporters and diminish my father. After consulting with my father, Sarge refused. Two days later Johnson chose Senator Hubert Humphrey of Minnesota.

Hamstrung by the White House, my father needed an independent power base. Averell Harriman and other friends urged him to run for the Senate in New York, where my dad had grown up. My uncle Steve Smith began making preparations.

The Run for the Senate

By 1964, we were teasing my father that his hair was starting to gray around the temples, and we had started counting the wrinkles on his brow. One day he sat down at the table and told us, "Well, children, how would you like it if Daddy runs for senator?" Our objection was unanimous—we didn't want to live in New York. When I asked, "Could we still live at the Cape and Hickory Hill?" he promised that we could.

On September 1, my mother, pregnant as usual, flew the seven of us down from Cape Cod to New York in the Skakel DC-3. We all sat in the balcony of the 71st Regiment Armory on Manhattan's East Side, wearing our campaign buttons saying, "Let's Put Kennedy to Work for New York," and watched the New York State Democratic Convention nominate my father for senator on the first ballot. When a reporter asked him if he was simply using New York as a stepping-stone, my father assured her that he had no presidential ambitions—"and neither does my wife, Ethel-bird."

My father was far from a shoo-in for the seat then occupied by the popular Republican Kenneth Keating (whom I last saw in the Hickory Hill pool, swimming in his dress suit during my mom's party for John Glenn). Keating, also Irish Catholic, enjoyed a strong lead in the polls. Following its usual pattern of hostility toward my dad, the New York Times endorsed Keating and denounced my father, expressing an "uneasiness that is no less real because it is elusive and difficult to define." The liberal establishment also lined up against my father. Salon liberals Gore Vidal, I. F. Stone, the Nation magazine, James Bolder, Richard Hofstadter, Edward Keating, and the clubhouse columnists at the Village Voice, still bitter about his crusade against the Teamsters and his abbreviated engagement with Joe McCarthy, all condemned his candidacy.

I was always surprised that, despite the negativity in the mainstream press, regular people still seemed to love my father. All over the state we found exuberant crowds. We traveled with him to

Rochester, Rye, Syracuse, Schenectady, Binghamton, Buffalo, and Albany. We campaigned in Long Island, and upstate around the Finger Lakes and the Adirondacks. He took us skiing at Hunter Mountain in the Catskills and to Lake Placid in the Adirondacks, where we rode the bobsled run and skied with my cousins Caroline and John in frigid conditions. I got frostbite on my cheek that has dogged me for life, and my eyeballs froze as I skied. My brothers and I marveled that our spit would crackle before it struck the ground; we did a lot of spitting. That night we had already put on our pajamas and assembled for bedtime prayers at the Lake Placid Club when my father learned that it was "restricted." This, we children were surprised to learn, meant no Jews were allowed. Furious, my dad moved us out in the middle of the night, in a convoy of snowmobiles across the frozen lake, to a tiny motel on the far shore.

My mother, with her characteristic embrace of theories soon to be stripped of scientific credence, believed that frigid air and direct sunlight greatly improved the health of children and imbued them with vigor. She insisted we sleep year-round with windows wide open and electric fans blasting. Even on winter trips she was adamant that we make no concessions to arctic weather. Portaging the fans, numerous and large, across the lake required several snowmobile trips. Our cousins John and Caroline, who had accompanied us, were also provided fans and made to understand the advantages of these health treatments. Near morning, through my open window I saw Jack Walsh, the Secret Service detail chief assigned to protect John, shivering outside on the hotel porch. It was still dark when I greeted him through the screen. "What's up, Mr. Walsh?" I asked. Replying through chattering teeth, he told me in his South Boston brogue, "I came out heah to get waam."

My father took us to the New York World's Fair near LaGuardia in Queens to campaign in the giant crowds. It was common knowledge in our family that my dad did not like to speak publicly, nor was he very good at it, at first. But at the urging of Charles Evers, my father reluctantly stopped using notes and began speaking extemporaneously, and from his heart. He was at his best advocating for America's most vulnerable and disenfranchised, and during this campaign he made himself their champion.

I especially loved campaigning in Harlem and Brooklyn, where the crowds were most demonstrative. At every stop, gangs of kids ran beside our motorcade, chasing our convertible and calling out, "Kennedy, Kennedy, you my main man!" My father always had a football that we threw around on breaks between speeches, and sometimes he threw it back and forth with the kids as they jogged beside our motorcade. We loved it when he lifted them into the car and we all rode together for a few blocks. He continually met with moderate and militant black leaders and listened as they poured out their thoughts, complaints, and frustrations. As John Glenn later observed, "Many of our national leaders would not have been physically safe in those areas, but Bob came to them, learned from them, and they from him, with mutual understanding and dignity." After emerging from one East Harlem tenement, my father told me he'd seen a Puerto Rican baby with cats in its crib to keep the infant safe from rats.

With Jack's memory still fresh, my father's popularity extended even to upstate New York's Republican strongholds. When he arrived in the Hudson River town of Glens Falls at one a.m., five hours late for his speech, 4,000 people were waiting with their children in their pajamas. Thirty years later, while I argued a prolonged lawsuit in Glens Falls against the General Electric Company for polluting the Hudson, dozens of residents, young and old, approached me in diners, on the streets, and in municipal buildings to share their memories of that night. One man told me he'd waited six hours perched on his father's shoulders for a momentary wave at my dad. In Buffalo, a hundred thousand people lined the streets from the airport into downtown. Once, we arrived several hours late at a rally in Lower Manhattan where actor Tony Randall had kept the crowd warm for four hours by leading them in a sing-along.

On Election Day, my dad took us to the Bronx Zoo, which we toured with a horde of reporters. I fed an Indian elephant and got my first glimpse of swallow-tailed kites. I also got interviewed, and probably came off not much worse than my father, who was too thoughtful, honest, and awkward to make a good sound-bite politician. He was miserable at bantering. That night, as the votes were tallied, my big sister Kathleen reminded him of all the work we children had done to ensure his victory. "Daddy, you better win or else!" He defeated Kenneth Keating by 700,000 votes.

That Christmas we all went skiing in Aspen with Olympian Tom Corcoran, the McNamaras, and Jackie, Caroline, and John. While we were on the hill, the ski patrol notified us that Bob McNamara's father had died. We all raced to accompany him to the bottom. I crashed and knocked out my front tooth. The day after we got home, on January 4, 1965, my father was sworn in as U.S. senator. His first speech before the Senate, in June 1965, was a jeremiad against the spread of nuclear arms. He called for disarmament before atomic weapons spread to terrorists, military dictators, and "arrogant madmen" who would hold civilization for ransom with threats of nuclear apocalypse. With proliferation, he warned, every minor conflict "might well become the last crisis for all mankind."

The problem of pollution was also high on his agenda. He spoke out against the contamination of Lake Erie, which the New York State Conservation Department had recently declared officially dead. He expressed alarm at paper plants that had poisoned Lake Champlain with dioxin, and at the destruction of Long Island's wetlands. From our new apartment at United Nations Plaza we could see two towering brick smokestacks from Con Ed's East River power plant spewing greasy smoke over the neighborhoods of lower Manhattan. My father was furious that state officials weren't doing anything about it. Soon after his election, he made a trip up the Hudson to inspect the pollution. Dead for twenty-mile stretches north of New York City and south of Albany, the river caught fire regularly and turned whatever color they were currently painting trucks at the GM plant in Tarrytown. Today, in my office at Waterkeeper Alliance, I have two photos of my dad examining dead fish along the riverbanks near Newburgh, New York.

In March 1965, when I was eleven years old, my father took us to the Adirondack Mountains to paddle the Whitewater Derby on the upper Hudson. My dad hoped the trip would help bring attention to the Hudson's spectacular headwaters, and help to derail a proposed dam on the Hudson River Gorge. Snow was falling when we arrived at the put-in and a glassine layer of ice coated the near shore shallows. We were in kayaks and canoes, and the downdraft from the press helicopters flattened the waves, making the rapids impossible to read, so we kept capsizing. My brothers and I ended up swimming several formidable rapids; spilling from his Klepper kayak, my father swam five times. Snow was piled on the bank and clusters of broken pack ice floated in the river. To this day it was the coldest I have ever been—colder even than Lake Placid. John and Caroline felt the same way. Perhaps they wondered if their relatives were trying to freeze them to death in the Adirondacks. That was my first experience on the Hudson, and I remember how surprised I was when our guides told me not to drink the river water. During all those trips out West our guides had carried tin cups that we dipped in the river around to slake our thirst, so I took it for granted that whenever you did whitewater, you got to drink the river. But these outfitters said the Hudson water was poison. Even at that age it occurred to me: "There's a theft involved here," and I took it personally. The outrage stayed with me, and I spent my career trying to redress the crime.

Champion of the Underdog

My father did not suddenly become compassionate after Dallas, as some biographers have suggested. He had always stood up to bullies on behalf of the underdog. But after his brother's death, he couldn't bear to see suffering of any kind. He made me get rid of a pet boa constrictor because it ate live mice. I recall discovering him seated one afternoon on the grass beside the tennis court, striving, with tweezers, to straighten the naturally curved bill of an evening grosbeak, believing it to be the result of an injury. My dad's treasured Revolutionary War–era pistols disappeared from the house, as did our BB guns.

Concern for the poor had been a central theme of the Kennedy administration—Head Start, the Peace Corps, the Alliance for Progress, and the focus on civil rights were its offspring—but whereas Uncle Jack had the ability to put himself in the other person's shoes while still keeping himself apart, my father lacked that capacity for emotional detachment. He experienced the suffering directly. With the miseries of the poor and alienated hounding him, my father turned his public efforts to battles against poverty and injustice. As Arthur Schlesinger would write, "He made himself in the Senate the particular champion of those who in the past had been the constituents of no one." His approach was direct engagement. During a Saturday drive through Southeast Washington, with us in his convertible, he spotted a high school swimming pool that had been closed for nine years. As soon as he returned home, my dad got on the telephone, called government officials and private donors, and arranged for the pool to reopen. He regularly visited public schools in the poorest neighborhoods, often with us in tow, just to show the young people that someone in high office cared about them.

My father immersed himself in the problems of poverty in every region of the country, and his concerns became our central preoccupation as well. He toured West Virginia and eastern Kentucky, and our home was often filled with coal miners and experts on the impacts of hunger, like Dr. Robert Coles, of Harvard. He returned from a tour of the Mississippi Delta with Marion Wright Edelman and Charles Evers and told us he had seen families living in shacks smaller than our dining room, and that their children had only one meal a day and went to bed hungry. He said he wanted us to do something to help those families when we grew up. As a result of that trip, he wrote, and Congress passed, the Special Supplemental Nutrition Program for Women, Infants, and Children (WIC), guaranteeing nutrition for infants from poor families across America.

He took us repeatedly to Bedford-Stuyvesant in Brooklyn, the toughest ghetto in America, which became a second home for him.

César Chávez

In 1966, my father took up the cause of César Chávez in California, flying with his Senate Subcommittee on Migratory Labor to Delano, California, to conduct hearings on farm labor in the fruit-packing industry. A local sheriff explained to the astounded committee members that he had jailed twenty-five of Chávez's strikers because the scabs and thugs employed by the growers threatened to cut them to pieces with bowie knives and machetes. "So rather than let them get cut," the sheriff explained, "we removed the cause." Characteristically caustic with public officials who were abusing their power or neglecting the public trust on behalf of greedy corporations, my dad told him brusquely, "This is the most interesting concept—you go arrest somebody that hasn't violated the law. Can I just suggest that the sheriff reconsider his procedures in connection with these matters? Can I suggest during the lunch period that the sheriff and district attorney read the Constitution of the United States?" When reporters asked my dad if the farmworkers were Communists, he told them indignantly, "No, they are not Communist, they're struggling for their rights." He explained to the growers and the press that the farmworkers had a right to form a union. Then he took a place with Chávez's workers in the picket line. "He was amazing," UFW cofounder Dolores Huerta told me. "He not only endorsed us but joined us. We were frightened that he was going too far in supporting us, and he might harm himself."

My father would have loved César under any circumstances. They had much in common. They were both short of stature, tough, stoic, and shy. They detested bullying and felt called to give their lives to a higher purpose. My dad saw César Chávez as the Martin Luther King of the Hispanic battle for equality. He shared my father's Catholic faith, hatred of injustice, and abhorrence of violence. From that first meeting they became both friends and allies. My dad told his staff to do everything they could to help César, who was the first person outside the family my father told when he decided to run for president. In 1968 a group of younger farmworkers began arguing for direct action, including violence. César vowed to stop eating until all United Farm Workers (UFW) members renounced violence. When the issue was resolved, my dad traveled back to Delano to bring communion to César, breaking his thirty-day fast. My mother both participated in and broke his subsequent fasts, and my brother Joe played that role in 1986. My brother David worked in Delano the summer after my father's death, and I worked with César, Dolores Huerta, and César's son-in-law, UFW president Arturo Rodriguez, in César's final campaign against pesticides.

Joe, Kerry, and I were pallbearers at César's funeral, in 1993. We helped lead a crowd of 40,000 farmworkers for four miles in the blazing sun, and laid César to rest in a pine box handmade by his brother and adorned with the UFW emblem. My mother walked in front of the casket hand-in-hand with César's wife, Helena, and Dolores Huerta. To this day Arturo Rodriguez and I continue to work together on issues of labor, pesticides, and environmental justice.

Native Americans

My father saw the plight of Native Americans as our nation's original sin and greatest shame. Several of us children would accompany my dad on most of his campaign trips, and wherever there was a reservation nearby he took us with him to visit with Hopis, Navajo, Sioux, Shoshone, Apache, Cherokee, Mohawks, and the Choctaw, who, my father recalled, had famously raised money for Irish relief during the 1848 potato famine. Many of us were given Native American names by tribal leaders in naming ceremonies; my Sioux name, Wambli Gleska, means Spotted Eagle. The Sioux earlier honored my Uncle Jack with a naming ceremony. He promised them,

"From here on in, when we're looking at westerns on TV, I'll be rooting for *our* side!" Native American or Eskimo leaders often visited the house, bringing gifts from their great chiefs, including peace pipes and eagle-feather headdresses, and beaded moccasins for us children, objects we treasured.

My dad admired Native Americans for their courage and toughness, but the conditions in which most had to live offended his sense of justice. Of the approximately seventy campaign appearances he made during the first month of his presidential campaign, ten were on reservations or at Native American schools. He told us that on the Pine Ridge Reservation he had seen an entire Sioux family living in the burned-out hulk of an abandoned car. Tim Giago, founder of the *Lakota Times* and founder and editor of *Indian Country Today*, the nation's largest Native American publication, had been with my father that day. He told me that the sight had caused my dad to cry, a sight that I'd never seen—even after Jack died. On the day of the primary election, the last day of my father's life, he won 99 percent of the vote on that reservation, and carried the state of South Dakota. Tim told me, "It was the first time we ever had a national political leader who was our champion. We knew he was thinking all the time about how to solve our problems. His door was always open. When he died, we felt like we had lost a family member."

On May 16, 2001, I attended a conference in Rapid City, South Dakota, with 150 American Indian leaders. Prior to my speech, Tim Giago asked the audience how many of them had grown up in homes adorned with photos or wall hangings depicting John and Robert Kennedy. Nearly all of them raised their hands.

My own youthful experiences with Native Americans had a profound impact on me, and as a result I have spent a good deal of my career representing the interests of indigenous Americans in the United States, Canada, and Latin America in treaty negotiations and in courtrooms, facing off against governments, and against mining, timber, hydroelectric, and oil-industry forces endeavoring to steal their resources and destroy their lands and tribal culture.

Representing America Abroad

In June 1966 my parents visited South Africa, which was then governed according to apartheid, a system that excluded blacks from any meaningful participation in government or the economy, and narrowly restricted the places where they could live and work. My father went at the invitation of the National Union of South African Students (NUSAS) and its president, Ian Robertson. South Africa's prime minister and government officials objected to my father's visit and refused to meet with him. They placed Ian Robertson under house arrest for inviting him, and the government made every effort to discourage attendance at my dad's public appearances. Nevertheless, when he flew into Johannesburg airport at midnight, 1,500 people were waiting for him, tearing his shirt and popping off his cufflinks in their enthusiasm. The *Times* of London called his 1966 Johannesburg speech "the most stirring and memorable address ever to come from a foreigner in South Africa."

The next day, Afrikaner students at Stellenbosch University gave the government a shock. Stellenbosch was the alma mater of every prime minister in South African history and the cradle of apartheid. After his speech, my father engaged the Afrikaner students in a ferocious debate. "Why don't they [the blacks] take part in your elections? Why don't you allow them to worship in your churches?" Then he asked a question that shocked his devoutly religious pro-apartheid audience and rang across South Africa for many years thereafter. "What the hell would you do if you found out that God was black?" His candor won their hearts. Instead of booing him, the students paid him their highest tribute, pounding their spoons on the long wooden dining tables, an honor reserved, until that time, only for the demagogic racism of the nation's most beloved Afrikaner leaders. Stellenbosch's student body next shook Afrikanerdom to its depths by breaking from the Afrikaner student government association and entering talks to affiliate with

NUSAS, Ian Robertson's "damnable and detestable organization" that had brought my father to South Africa in the first place. Exultant crowds of record size of black Africans mobbed him in slums like Soweto and papered the walls of their shanties with newspaper accounts of his visit for years thereafter.

As a front-page editorial in *Rand Daily Mail* opined, while bidding my father farewell at the end of this trip, "Senator Robert Kennedy's visit is the best thing that has happened to South Africa for years. It is as if a window has been flung open and a gust of fresh air has swept into a room in which the atmosphere had become stale and fetid. Suddenly it is possible to breathe again without feeling choked. . . . Kennedy has taken the youth of the country by storm, through his message of confident, unashamed idealism." Nelson Mandela later told me that my father's visit lifted his badly beaten spirits in his prison cell on Robben Island, and brought hope to other anti-apartheid movement leaders and continued to inspire them for years.

My father traveled to Indonesia, Japan, and Latin America. His openness, candor, and willingness to debate helped defang anti-American feelings. In Jakarta, confronting a hostile crowd, my father grabbed a bullhorn and spoke extemporaneously for twenty minutes on the meaning of freedom:

> *Let me tell you what the United States stands for. We were born in revolution. We believe that government exists for the individual and the individual is not a tool of the state. We in America believe that we should have a divergence of views. We believe that everyone has a right to express himself. We believe that your people have a right to speak out and give their views and ideas.*

Communists in Chile and Indonesia, who attacked and heckled him on his arrival—bombarding him with missiles from rotten fruit to spittle—later cheered him openly.

We followed his trips with excitement, devouring the news

articles sent to us by my father's assistant, Angie Novello, and fighting over the phone to hear the stories when our parents called us
each evening before bedtime.

Our house saw a parade of visitors from Japan, Italy, Germany,
Indonesia, Morocco, and Kenya. Students from Iran visited to
thank my father for derailing their deportation—at the request of
the Iranian government—by the State Department. They told us
our dad had saved their lives. Returning home would have been a
death sentence.

My father, whom some represented as ruthless, devoted his
energies to those neglected and forgotten people who shared his
ideals about the role of America in the world. Our house hosted
a continual parade of men and women who shared his ideals and
placed their hopes in American values: heads of state from Asia,
Africa, and Latin America; a chauffeur from Jakarta; student leaders from India, Thailand, and Japan; Indians, Latin Americans, and
Chicanos—people who believed in America's promise and the values of our country.

CHAPTER 10

● ● ● ● ●

The Final Campaign

Come my friends, 'tis not too late to seek a newer world.
ALFRED, LORD TENNYSON, *ULYSSES*

FOR FIFTEEN MONTHS, PRESIDENT JOHNSON CLEAVED CLOSE TO JFK'S strategy in Vietnam—arming and training South Vietnam to fight its own war, avoiding the use of heavy bombs or the introduction of U.S. combat troops. But by the summer of 1964, administration hawks, including National Security Adviser McGeorge Bundy, were pushing harder than ever to escalate. The Pentagon and the CIA initiated a series of military probes in early August, hoping to lure North Vietnam into committing a provocative act that would justify direct U.S. involvement. On August 4, 1964, the spy ship USS *Maddox*, operating within North Vietnam's coastal waters, opened fire with what the *New York Times* then described as a defensive barrage against attacking North Vietnamese torpedo boats. In fact there were no enemy boats. The *Maddox* was firing "at flying fish," Johnson later confided, but he used the alleged attack to galvanize Congress into passing the Gulf of Tonkin Resolution, granting the president full authority to defend South Vietnam against

"Communist aggression." Johnson now had the green light to introduce conventional U.S. forces into the war on the North.

In March 1965, Johnson sent in American aircraft to carpet-bomb North Vietnam, and deployed U.S. combat forces to fight in the South. Suddenly Vietnam was an American war. Today, more than half a century later, with North Vietnam a thriving capitalist state at peace with its neighbors, friendly to the United States, and still bridling against political incursions by its traditional enemy, China, it's hard to give credence to the hawkish voices that warned against the so-called domino effect—the inevitable viral spread of communism were it allowed to infect Vietnam. But in 1965, my father's calls for disengagement were a lonesome voice in the wilderness.

In April of that year, distressed by the bombing, my dad met privately with LBJ to urge him to call a cease-fire and initiate negotiations. When a truculent Johnson made clear that he had no intention of parlaying with North Vietnam, my father broke openly with the president in a speech to CIA trainees at Washington's International Police Academy in July. The bombing and troop infusions, he argued, were reversals of President Kennedy's policies and counterproductive. After all, the South Vietnamese government could not claim legitimacy so long as it was complicit in the American bombing of its own people. He called for immediate de-escalation. He urged the administration to talk terms with the Vietcong, whom congressional hawks portrayed as "terrorists" and Sino-Soviet vassals, but whom my father had come to regard as representing a genuine freedom movement. During one of our family dinner-table debates, my father recounted that Ho Chi Minh had fallen in love with democracy when he worked as a youth as a baker at Parker House in Boston; that he had fought beside us in World War II against the Japanese, and that Ho had quoted Jefferson, not Mao, in his own inaugural address.

We launched the war, he told us, to forestall the 1956 election that we had agreed to in the Geneva Treaty because the CIA and its puppet, Ngo Dinh Diem, realized that Ho Chi Minh would win by a landslide. "We are torturing, burning, and shooting Vietnamese," my dad told us, "to prevent free elections. It's immoral and very un-American." Speaking before the Senate in January 1966, my father condemned the bombing as a moral abomination. The inadvertent killing of Vietnamese civilians, he said, was shattering America's moral authority in the world; and our policies, he argued, were pushing Ho Chi Minh closer to Mao and the Soviets. Going a step further, my dad committed the heresy of calling for rapprochement with the Red Chinese.

The press responded savagely. The *Chicago Tribune* ridiculed my father as "Ho Chi Kennedy!" A New York *Daily News* cartoon portrayed him wielding a hatchet emblazoned with "Appease the Vietcong." Similar derision echoed from the *Washington Post* and *New York Times*. Richard Nixon scourged my father, warning that anything short of complete victory for South Vietnam "would mean, ultimately, the destruction of freedom of speech for all men for all time, not only in Asia, but in the United States as well." Hubert Humphrey led Democrats in condemning my dad from the left. Our clan took temporary refuge from the whirlwind of political vitriol with a quick family ski trip to Stowe, Vermont. But even amid New England's tranquil winter landscapes we were all keenly aware that our dad was under assault from all sides.

Unbowed by the attacks, my father expressed horror, during a series of post-holiday speeches and interviews, at our military's use of Agent Orange to defoliate the Vietnamese countryside, and napalm to incinerate civilians. He railed against the Pentagon's practice of turning Vietcong prisoners over to the South Vietnamese, making America an accomplice in their torture. "This is wrong in itself, and it is also bad policy," he said, "because we can never hope to get them to come over to us if all they can expect is to be tortured." When he publicly observed that the South Vietnamese probably didn't care for our puppet, General Ky, any more than they did for the Vietcong, the *New York Times* ridiculed him. Nevertheless, the truth of his observation was self-evident. At the war's peak in 1968, we had 525,000 troops in Vietnam, and our ally General Ky's South Vietnamese army numbered 250,000. Together we were fighting a few thousand South Vietnamese rebel Vietcong and 20,000 North Vietnamese regulars sent south by Ho Chi Minh; yet Ky's notoriously corrupt government was so unpopular among the Vietnamese people that it was still losing the war! When he learned that it was costing America close to $100,000 for every Vietcong killed ($750,000 in today's dollars), my father suggested, sarcastically, over dinner, "Why don't we offer to buy each of them a house in Beverly Hills?" The fact that those resources might instead have been directed to fight the war on poverty at home further steeled his resolve. Daniel Ellsberg told me that of all the people he met in Washington during that era, "there was no one more intensely opposed to the Vietnam War than Bobby Kennedy."

Every evening, almost without exception, my father would call his pals, journalists Jack Newfield and Pete Hamill, to talk about the war. He would then call his old friend Bob McNamara, and browbeat him about getting out of Vietnam. "I don't recall a single night that he didn't call Bob. He urged him to publicly resign, tell America the truth, and condemn the war," my mother remembers. "It became a ritual." Despite the stern judgment of history against him, McNamara mostly agreed with my father. He desperately wanted to resign from his job as LBJ's secretary of defense, but he told my father that he was the only thing keeping LBJ and the Joint Chiefs from bombing the dikes and dams in North Vietnam—a strategy that would have drowned or starved a million civilians. His friends have recounted that a conscience-stricken McNamara would shut his office door and cry. LBJ was apparently concerned enough that he remarked, "We don't want him to go out the same way that [Defense Secretary James] Forrestal did." Forrestal leapt from a window.

In January 1966, while my father was visiting France, members of the French government approached him. Ho Chi Minh's representatives had asked the French officials to convey a peace feeler: if the United States halted its bombing, North Vietnam would negotiate a peace treaty. When he briefed the president about this offer, LBJ was enraged by what he considered my dad's foreign policy meddling. The president predicted we would win the war by July. "I'll destroy you and every one of your dove friends in six months," he shouted at my father. "You'll be dead politically." Back in his office, my incredulous father told his political adviser, Frank Mankiewicz, "Those guys are out of their minds. They think they are going to win a military victory in Vietnam by summer." By spring 1967, U.S. bombers had carpeted the tiny, impoverished North with more explosives then we dropped in all of World War II. And on March 2, 1967, my father delivered his biggest break with the president yet.

We were eating breakfast when my dad came downstairs before going to speak on Vietnam in the Senate. His advisers Adam Walinsky, Dick Goodwin, and Frank Mankiewicz had slept at our house, after laboring most of the night. While Adam Walinsky worked on a draft in my father's bedroom office, Dick Goodwin wrote a competing version in the attic. We were all in on the conspiracy to keep Dick hidden from Adam, who considered Goodwin both a rival and dangerously centrist. Mary Jo Kopechne was banging away at the typewriter, while Jinx Hack and Carol Gainor sprinted the staircases with papers and coffee.

Before leaving for school, I locked my coatimundi in the basement playroom, where I commonly let her wander during the day. My mother, seven months pregnant with my brother Douglas, went down with a scrum of reporters to show them my reptile collection; the coati, who never liked my mom, bushwhacked the group and laid into my mother's legs with her sharp teeth and ferocious claws. Journalist and broadcaster Dick Schaap lifted my mom onto the pool table to rescue her from harm's way, but the determined animal shinnied up a table leg and resumed the assault. After beating a retreat with the journalists, my mother had her legs bandaged at the Georgetown Hospital ER. The attack sent her into premature labor. I had to part with the coati, though I ducked additional trouble because of the preoccupation at Hickory Hill with the fallout from my father's Vietnam speech.

Later that day, my dad told a full Senate gallery, in words that

echo poignantly today as we pummel Afghans, Syrians, and Ye-
menis from drones, "The most powerful country the world has
known now turns its strength and will upon a small and primitive
land." He asked his fellow senators to imagine, "The vacant moment of amazed fear as a mother and child watch death by fire fall
from an improbable machine sent by a country that they barely
comprehend." Then he addressed the American people, saying that
the savagery was "not just a nation's responsibility but yours and
mine. It is we who live in abundance and send our young men out
to die. It is our chemicals that scorch the children and our bombs
that level the villages. We are all participants."

A trio of senators applauded my father's speech, William Ful-
bright, George McGovern, and Albert Gore, whose son, Al Gore, Jr.,
was then serving in Vietnam. The rest stewed. When Senator
Scoop Jackson, my dad's good friend, read on the floor a letter from
Lyndon Johnson excoriating my father for giving comfort to the
enemy and damaging U.S. military morale, Arthur Schlesinger advised my dad to send Scoop a present: "Why not the coatimundi?"
Spurred both by my father and his own conscience, six months
later, in August 1967, Robert McNamara publicly admitted to the
Senate that the bombings had been ineffective. Johnson responded
that November by naming him president of the World Bank without calling to inform him he'd been fired as defense secretary.

By this time the dogs of war were circling at home. Fueled by
returning Vietnam vets fresh from the rice paddies, schooled in
savagery, and abandoned by the administration as military expen-
ditures moved antipoverty efforts to the back burner, our nation's
ghettos became pressure cookers of discontent. Black militants ominously recalled the Negro spiritual, "God gave Noah the rainbow
sign. / No more water, the fire next time." Then, in the "long hot
summer" of 1967, the cities exploded and burned. A five-day riot
in Newark, New Jersey, killed twenty-seven. LBJ called up the National Guard and sent tanks to quell the violence. The nation was
splitting at its seams, but still pundits cited the domino theory to
justify our continued involvement in Vietnam.

During a November appearance by my father on *Face the Nation*, liberal journalist Tom Wicker recited the classic justification for American intervention still invoked today by armchair patriots to justify U.S. wars against Islamic rebels on the Asian continent. Wicker declared that, given the "great threat from Asian communism" to American security, shouldn't we "do as much as needs to be done" to achieve victory? Bridling, my father turned toward the camera and asked Americans to think through the moral implications of that response. "Now we're saying that we're going to fight them there so that we don't have to fight in Thailand? So that we don't have to fight on the West Coast of the United States? So that they won't move across the Rockies? . . . So we're going in there and we're killing South Vietnamese, we're killing children, we're killing women, we're killing innocent people because [the Communists are] 12,000 miles away [from us] and they might get to be 11,000 miles away."

One of my father's persistent preoccupations was the thinness of the veil between civilization and savagery. The example of Nazi Germany haunted him. What made us so sure we were different from the Germans? At dinner he reminded us that Germany had been the best-educated, most culturally advanced, most tolerant, most democratic nation in Europe when its people elected Hitler as their chancellor. What, then, were the barriers that stood between us and the barbarism of the Nazis? He was wary of our capacity for self-delusion and sanctimony, of our tendency to judge ourselves by our intentions rather than our actions, or by how God or the world sees us. My father asked Arthur Schlesinger, "Do you suppose that ten years from now we will all look back and wonder how the American people ever went so far with something so terrible?"

On January 31, 1968, the Vietcong launched the Tet Offensive, a massive surprise assault on military installations and cities across South Vietnam. Attackers even penetrated the U.S. embassy compound in Saigon. My cousin George Walter Skakel died that March in the battle at Quang Tri. Tet was a military disaster—with 50,000 soldiers killed—for the North Vietnamese, but an overwhelming political victory. Johnson had promised that our most recent troop surges had stabilized the country, crowing just a month earlier that "All the challenges have been met." Now everyone knew the administration had been lying. "Tet," my father said a week later, "finally shattered the mask of official illusion with which we have concealed our true circumstances, even from ourselves."

Predictably, the Joint Chiefs reacted to Tet by demanding 250,000 additional troops. The White House obligingly announced yet another surge and escalated its bombing campaign, while my father kept his relentless focus on the moral dimensions of America's foreign policy. "Are we like the God of the Old Testament," he asked his fellow senators, "that we can decide in Washington, D.C., what cities, what towns, what hamlets in Vietnam are going to be destroyed?" He told his trusted friend Tom Watson, Jr., president of IBM, "I'd get out of there in any possible way. I think it's an absolute disaster."

Within a month of Tet, my father had decided to run for president.

RFK for President

For the better part of a year my dad had been resisting calls from antiwar leaders like George McGovern and Allard Lowenstein urging him to throw his hat into the ring for president. Then, on November 30, 1967, Senator Eugene McCarthy of Wisconsin declared his own candidacy, and the antiwar movement rallied around him. My father was now even more reluctant to splinter his own party, which might give an advantage to Nixon, the probable Republican candidate. Furthermore, Robert Kennedy had little chance of winning; he would be running against an incumbent president who had the support of the unions and the Democratic Party leadership.

Although he doubted he could take the nomination, my father was certain he could beat Nixon in the general election. He had, after all, masterminded Nixon's defeat in 1960. A steady stream of friends and advisers came to Hickory Hill to discuss his prospects, and we children—I was fourteen years old—overheard many of those debates. My dad's most seasoned counselors and closest friends opposed his candidacy. Uncle Teddy, who enjoyed a cordial friendship with LBJ, was adamant, fearing the impact his brother's candidacy would have on the Democratic Party. Bob McNamara and Bill Moyers cautioned my father that he could not possibly win. Arthur Schlesinger, who initially opposed the adventure, soon migrated across the fence to join my dad's young turks, Dick Goodwin and Adam Walinsky, the only close advisers who were champing at their bits to see him run. Kenny O'Donnell wisely predicted that LBJ was a bully and would flee from the battlefield, but counseled only that my father should decide whether he had the stomach for a fistfight.

I don't recall being consulted about the decision, the way we children had been about my dad's Senate race in New York, but I clearly recall my feelings of dread. I had listened to the arguments and speculation of his advisers, and I read the papers daily. I was certain he would lose. No Kennedy had ever lost an election, and I didn't relish my father becoming the first.

I didn't see how my father could beat Johnson, and my feelings didn't change even after Eugene McCarthy wounded LBJ by making a good showing in New Hampshire; after all, Johnson still won decisively (48 percent to 42). Johnson had advanced Uncle Jack's civil rights agenda, garnering solid support within the black community. And the sizable majority of Americans, including most Democrats, were still strongly behind Johnson on Vietnam, whereas my father's antiwar pronouncements smacked of radicalism. His only reliable base was the young people who opposed the war and they were migrating to McCarthy. My dad could not even count on the Catholics who had voted for JFK. Business, labor, the media, and politicians were all arrayed against him. In January 1968, his pollster, Louis Harris, advised him he couldn't win.

My mother was the strongest advocate for my dad running. She knew he would never be happy on the sidelines. With the moral leadership of our country at stake, she realized he would come to regret having shied from the fray while our cities burned, while we fought an unjust war, and while we wandered far from the values that made him proud of our nation. Knowing his thoughts better than anyone, my mom goaded him with a piece written by his friend Jack Newfield in the *Village Voice*, which she carried in her purse for that purpose. "If Kennedy does not run in 1968, the best side of his character will die. He will kill it every time he butchers his conscience and makes a speech for Johnson next autumn. It will die every time some kid asks him, if he is so much against the Vietnam War, how come he is putting party above principle? It will die every time a stranger quotes his own words back to him on the value of courage."

For his part, my father was, as Schlesinger put it, "deeply fearful of what another Johnson term might do to the world." My dad knew he could not support LBJ, since that would require saying things that he didn't believe. Nor could he stomach the idea of supporting McCarthy, whom he considered vacillating and self-righteous. Having lived through the great crises of the Kennedy administration, my father questioned McCarthy's competence to be president. Even worse, Eugene McCarthy, who chaired the Senate Banking Committee, was in bed with the banks and insurance companies, which explained why he had no real commitment to address poverty and no plan for saving our cities.

My dad finally resolved his political dilemma by reducing it to an ethical one. America had lost her moral center. Our cities were in flames for the third summer in a row from the worst race riots since the Civil War. American soldiers were on nightly TV, burning straw huts in Vietnam villages with Zippo lighters and flamethrowers. At Georgetown Prep we saw even worse. Older boys showed us underground magazines depicting American soldiers raping Vietnamese women, and Jack's beloved Special Forces emerging from the jungle with strings of human ears as trophies. What had happened to our country? Here we were, committing the very atrocities we had long condemned in others. My dad came to see the presidential election as a struggle for the soul of our nation, and he could not sit idly by.

By early March my father had made his decision, but he decided to wait until after the New Hampshire primary to announce, so as not to dilute the opposition to LBJ. On March 10 he left on a trip to see César Chávez. My father's aides, knowing how he was leaning, warned him that the visit to Chávez would hurt him with California voters. "I know," he told them, "but I like César." On March 16, 1968, my nine brothers and sisters and I paraded behind my parents into the Senate Caucus Room and sat down in a long row of stiff chairs behind my mother, then pregnant with Rory, soon to be her eleventh. Squirming in my flannel suit, I heard my father say in his soft voice that he was running to "end the bloodshed in Vietnam and in our cities . . . to close the gaps that now exist between black and white, between rich and poor." I noticed that my father's hands were trembling as he spoke. While the reception and press inside the Caucus Room were tepid and skeptical, as we exited the Capitol and came out onto the street, crowds enveloped us, shouting, "Bobby! Bobby!"

Many people tell me today, "I voted for your father"; but what I remember most is the anger his announcement provoked. The young priests at Georgetown who opposed the war were furious at my dad for trying to usurp McCarthy. Mr. Joyce, a Jesuit brother whom I greatly respected, told me my father was an "opportunist." He was not assuaged when I pointed out to him that my father had been on record against Vietnam long before McCarthy. Some of the worst antipathy came from reformers and liberals, who nearly wore out the term "ruthless," the epithet originally applied by Jimmy Hoffa. On Friday, March 15, my dad told Jack Newfield and Haynes Johnson, "Not to run and pretend to be for McCarthy while trying to screw him behind his back, that's what would really be ruthless. I know I won't have much support. I understand I'm going it alone. It is a much more natural thing for me to run that not to run. When you start acting unnaturally, you're in trouble. . . . I'm trusting my instincts now and I feel freer . . . I know my brother thinks I'm a little nutty for doing this, but we all have to march to the beat of our own drummer."

On March 17, the day after his announcement, a largely Irish crowd booed my father as we walked in New York City's St. Patrick's Day Parade. His fellow senators also expressed outrage at his decision. With the exception of the UAW and the machinists, organized labor opposed him, its bigwigs still choking on his persecution of Hoffa. The big-city Democratic machine bosses, including Chicago's Mayor Daley—a longtime Kennedy stalwart—condemned my father for jeopardizing the party's hold on the White House. All three of my siblings' famous godparents, Maxwell Taylor, Douglas Dillon, and Averell Harriman, opposed my father for the same reason. My dad couldn't even count on the campuses; he complained, only half-jokingly, that McCarthy had all the A students and he had all the B students. It seemed like everyone was against him. The only Americans who seemed to welcome my father's candidacy were the poor, who traditionally didn't vote, and Republicans, who rejoiced that he was splitting his party.

When Uncle Jack ran for president he had a well-oiled machine and years of preparation, as well as a base within the party. He had formidable, well-financed organizations in all the primary states. My father, by contrast, had no campaign organization and no staff. The primary season was already well under way, and he had missed filing deadlines in Wisconsin, Pennsylvania, and Massachusetts. The remaining primaries were in Indiana and Washington, D.C., on May 7, followed by Nebraska on May 14, Oregon on May 28, California and South Dakota on June 4, and New York on June 18. There were only seven weeks before Indiana, and my dad had to learn to speak and inspire large crowds of Democrats, and to persuade party delegates, bosses, and leaders—already under the firm thumb of the White House—not to pledge their support to LBJ prior to the convention.

It Started with the Sunflower State

One day after running the gauntlet of Hibernian hecklers on Manhattan's Fifth Avenue, my father flew to Kansas to meet an earlier obligation to give the Alfred M. Landon Lecture at Kansas State University. It was not the venue he would have chosen to launch his presidential race. Even then, Kansas was among the most conservative states. KSU students didn't burn draft cards or march in antiwar protests or wear love beads. They were clean-cut, corn-fed, and reliably conservative, not the species of tie-dyed, long-haired hippies found on East Coast campuses. KSU's young men were hawkish enough that they didn't wait to be drafted; they dutifully volunteered for the war. Finally, Kansas didn't even have a primary! It was not, by anyone's estimates, a good launch platform for a national campaign.

My father's advisers and the press that filled his plane were shocked to find a record crowd of 19,000 cheering students waiting to greet him at the Topeka airport. Well-wishers reached through a chain-link fence to touch his face and hands. My dad told me afterward that a huge man gripped his hand and refused to let it go. Arriving at the KSU auditorium, my father looked out over a silent sea of scrubbed faces, neat haircuts, neckties, and proper calf-length skirts. Perhaps believing that his cause was lost in any case, he started by saying, "Some of you may not like what you're going to hear in a few minutes, but it's what I believe, and if I'm elected president, it's what I'm going to do."

He began his speech in the heartland of America with a pointed and still poignant critique of our country, lamenting how our gross national product counted air pollution, cigarette advertising, napalm, and nuclear warheads but not poetry, intelligence, or integrity. Then he turned to Vietnam. He started strategically by taking more responsibility for the Vietnam War than he needed to. "I was involved in many of the early decisions of Vietnam, decisions which helped set us on our present path." He said those efforts may have been "doomed from the start," since the Diem regime had been riddled with corruption. "If that is the case, as it may well be, then I am willing to bear my share of the responsibility before history and before my fellow citizens. But past error is no excuse for its own perpetuation. 'Tragedy is a tool for the living to gain wisdom, not a guide by which to live.' Now, as ever, we do ourselves best justice when we measure ourselves against ancient tests." Then he invoked his favorite Greek playwright, Aeschylus: "[A] good man yields when he knows his course is wrong and he repairs the evil. The only sin is pride.'"

He then attacked the moral and strategic rationales for the war. "I am concerned—as I believe most Americans are concerned— that the course we are following at the present time is deeply wrong. That we are acting as if no other nation existed, against the judgment and desires of neutrals and our historic allies alike." The White House's only strategy, he said, was "the ever expanding use of military force where military force has failed to solve anything." He asked them, "Can we ordain to ourselves the awful majesty of God—to decide who will live and who will die and who will join the refugees wandering in a desert of our own creation?" As he spoke, the audience, initially hostile and tentative, warmed to him. At each new idea, students leapt to their feet, until the entire audience was standing and erupting in waves of cheers as he continued gaining confidence. Student leader Kevin Rochat found himself in tears. "My God," he thought, "he's saying exactly what I've been thinking." Jack Newfield described the scene as the unleashed passions of these formerly subdued Kansas farm boys shook the hall. "The field house sounded as though it was inside Niagara Falls. It was like a sound truck gone haywire."

When my father finished, the students rushed the stage, toppling chairs, screaming and clapping and weeping and reaching out to touch him. Pandemonium overtook the hall as students cheered until their voices broke, and applauded till their hands ached. A few hours later, some 20,000 students at the University of Kansas repeated the wild eruption. On the plane back, my dad told Jimmy

Breslin, "You can hear the fabric ripping. If we don't get out of this war, I don't know what these young people are going to do. There's going to be no way to talk to them. It's very dangerous." Some of my father's political advisers warned him to tone down his rhetoric, concerned that the sight of those riotous crowds would frighten middle-class voters. Others, like Adam Walinsky, disagreed, arguing that it was critical politically for my father to show the Democratic Party bosses he could inspire intensity in the electorate. Recalling that moment, Walinsky told me, "Your dad's only hope was to make it impossible for them to deny him the nomination." It was his only hope, Jack Newfield said, to prevail against the "totalitarian arithmetic at the Democratic Convention."

Three days after his Kansas speeches, my dad went to another unlikely venue, the University of Alabama, in Tuscaloosa, the same school where he and his brother had forcibly registered black students five years earlier. He explained to the students that anyone who sought the presidency "must go before all Americans: not just those who agree with them, but those who disagree; recognizing that it is not just our supporters . . . who we must lead in the difficult years ahead." The prolonged ovation mainly by southern white students shocked the political pundits. A few weeks later he delivered the same message to an all-black audience in Watts. "And I tell you here in California the same thing I told those in Alabama with whom I talked. The gulf between our people will not be bridged by those who preach violence, or by those who burn or loot." The *Los Angeles Times* described the audience as "uproarious, shrieking, and frenzied." My father knew his chances were slim, but win or lose, he was at peace. For the next eleven weeks, his crowds, black and white, were consistently delirious, convulsive, and hysterical. They reached out maniacally, to touch and claim him. He lost so many sets of cufflinks that he switched to plastic ones. He wore clip-on ties to avoid being choked. And their love buoyed him. "So many people hate me," he told his friend, the comedian Alan King, "that I've got to give the people that love me a chance to get at me." He had joined the battle. He was speaking the truth. He would do his part to remind our citizens of the substance of American values. He was "like a dog finally let off the leash," said Peter Edelman. George Stevens compared him to "a volcano erupting."

After Kansas, my dad spoke in sixteen states in fifteen days. Rounding up his friends, he hastily assembled a campaign staff that included former Justice Department aides Ed Guthman and John Seigenthaler, Senate speechwriters Jeff Greenfield and Adam Walinsky, Pierre Salinger, Kenny O'Donnell, and Ted Sorensen. He was able to attract people of tremendous caliber—decorated war heroes, distinguished journalists, and academics willing to set aside their own careers for service, not for my father, but for America. Charles Evers signed on as a volunteer, even though it meant resigning from his desperately needed paying job at the NAACP.

My dad's staff and supporters were overwhelmingly young, and if he had won, my father would have been the youngest elected president in history, six months younger than Jack when he took his inaugural oath. Since my father loved to be around dogs, he brought my Irish cocker spaniel, Freckles (a gift to me from the Wexford, Ireland, clan of John Barry, the American Revolution war naval hero), on the campaign trail with him.

My father's security crew as he traveled included former G-man Bill Barry, family friend Jim Whittaker, decathlon champion Rafer Johnson, and a posse of NFL players, including Rosey Grier, Merlin Olson, Lamar Lundy, and Deacon Jones of the Los Angeles Rams' "Fearsome Foursome" defense. It sometimes took all four of them to hold my dad in position as he stood on automobiles, flatbed trucks, and podiums as fervent crowds mobbed him, often pulling him from the vehicles. Football great Sam Huff organized West Virginia; Olympic skiing medalist Tom Corcoran organized New Hampshire. My father's entourage included artists and entertainment stars like John Lennon, Buddy Hackett, Jonathan Winters, Tony Randall, the Smothers Brothers (whose popular TV show was canceled by CBS due to their criticism of the Vietnam War), and Alan King, whom my father thought one of the funniest men alive and who devised the campaign's unofficial slogan: "Sock It to 'Em,

Bobby." The folk and rock band the Kingston Trio accompanied him and played at every stop at hospitals and factories and children's homes, and on a whistle-stop tour across Indiana aboard the Wabash Cannonball. The atmosphere was electric. The Jefferson Airplane and the Grateful Dead endorsed him and performed concerts on his behalf. Psychedelic artist Peter Max designed the campaign poster, and everywhere my father went, people held signs proclaiming: "Bobby Is Groovy."

Even when it wasn't fun, it was fun. Jim Whittaker recalled a day campaigning in Oregon in a driving downpour when my dad ordered the driver to drop the top on their convertible so they could greet the crowds lining the sidewalks. Whittaker protested, "It's pouring, Bob." My father pointed to the crowds and answered, *"They're* getting wet!" Forty minutes later, as the car finally crept out of town, the top still down, the two men shared a soggy piece of chocolate cake a constituent had given them, which Whittaker realized would be their only lunch. My dad looked at Jim's grim face across the seat and remarked, "Name something you have done that was more fun than this!"

Freed from the shackles of political correctness by the hopelessness of his campaign, my father was able to speak truth to the American people, and they loved it. He violated every political rule. He openly criticized our country for falling short of its ideals. "Once we thought, with Jefferson, that we were the 'best hope' of all mankind," he wrote in the *New York Times*, "but now we seem to rely only on our wealth and power." He declared on *Meet the Press*, "I am dissatisfied with our society. I suppose I am dissatisfied with my country."

And he refused to pander. At Indiana University School of Medicine he spoke about the need for national health care. When an angry medical student asked who was going to pay for the program, he answered, "You are," and then proceeded to accuse the students of not living up to their community responsibilities. At Creighton University, in Omaha, shocked students initially booed him when he promised he would do away with their student deferments. He then asked them if they considered it unfair that the children of well-to-do parents could buy their way out of the war by attending college. When a hostile student argued that the draft was a way of getting black youth out of the ghetto, he responded: "Here at a Catholic university how can you say that we can deal with problems of the poor by sending them to Vietnam? Look around you. How many black faces do you see here, how many American Indians, how many Mexican-Americans? If you look at a regiment or division of paratroopers in Vietnam, forty-five percent of them are black. How do you accept this? You're the most exclusive minority in the world. Are you just going to sit on your duff and do nothing?" It was no problem, he added, for him to get all ten of his children into college somewhere in the United States, but other fathers with ten children weren't so fortunate.

A day after Dr. King's assassination, with violence still shaking Watts, RFK told a group of 2,200 white civic leaders that white Americans bore responsibility for the riots. Days later, he elaborated on that theme in a speech to privileged white leaders in Cleveland: "For there is another kind of violence, slower, but just as deadly, destructive as the shot or the bomb in the night. This is the violence of institutions; indifference and inaction and slow decay. This is the violence that afflicts the poor." Even in the bleakest black neighborhoods, he challenged vitriolic militants about their advocacy of violence. In a blistering meeting in Oakland, Black Panthers derided him savagely and denounced his close friend Rafer Johnson as an "Uncle Tom." Instead of backing down, my dad challenged the militants for romanticizing lawlessness. "I'm glad I went," he said afterward. "They had to know that somebody will listen." The following day the same Black Panthers who berated him at the Oakland meeting escorted his car through the ghetto. Generally Americans loved his frankness, and they swamped his speeches and cheered his motorcades with increasing fervor.

By March 23, the campaign had turned a corner. Gallup polls showed my father ahead of LBJ, 44 to 41 percent (with McCarthy distantly trailing both), and newspapers began predicting that

Democrats might dump LBJ from the ticket. On March 31, Johnson went on TV purportedly to give a speech on Vietnam, but instead he stunned the nation by announcing his withdrawal from the presidential race. By then I was boarding with the Jesuits at George-town Prep in Rockville, Maryland. There was a small television in the library annex and I'd gone in to watch the president's speech. I couldn't believe what I was hearing. "We're going to win," I said to myself for the first time. "The war will be over in January. Our soldiers are coming home. Instead of building million-dollar bomb-ers, our country will spend that money constructing schools and health centers and rebuilding our cities." All the things I had heard my father talk about were about to come true. He would restore America's moral standing, revitalize the cities, and make the poor a part of our democracy. Suddenly I believed it was all possible—and so did a lot of other people.

David went off to campaign in Indiana—on a whistle-stop tour aboard the old *Wabash Cannonball*. At every small town, the train whistle and the music of the Kingston Trio would summon the crowds. The campaign entourage of embedded journalists would pour from the press cars and listen to him deliver his stump speech from the caboose with my mother and a retinue of Hollywood and sports celebrities at his side. Then the whistle would blow "all aboard," and the train would be on to the next town. As it pro-gressed, the crowds—mostly white farmers—grew larger and more demonstrative. A victory in Indiana's May 7 primary, my father's first, would be a concrete sign that he might win the nomination. Indiana was a conservative state with a small minority population, a state dominated by agricultural issues. On April 4, my dad was scheduled to give a campaign speech in the roughest part of India-napolis's black ghetto, a place no white American politician would go, as voters were sparse and safety concerns manifold. Just before he headed into the inner-city neighborhood, he learned that an as-sassin had murdered Martin Luther King, Jr., in Memphis. My fa-ther staggered in anguish at the news. "Oh, God," he said, "when is this violence going to stop?" The Indianapolis police chief warned him not to enter the neighborhood. When my father said he was going in, the captain withdrew his police escort.

Most of the electrified crowd were unaware of the tragedy as he climbed on the flatbed to address them. The night was cold. "I have bad news for you, for all of our fellow citizens and people who love peace all over the world, and that is that Martin Luther King was shot and killed tonight." The crowd gave a collective gasp of horror, then my dad went on to speak directly for the first time about the circumstances of Jack's death. His brother "was killed by a white man." Speaking from the battlefield of his own psychic struggles, with a calm but breaking voice, he urged everyone not to be "filled with bitterness, with hatred, with a desire for revenge." We could respond with violence and polarization, he said, "or we can make an effort, as Martin Luther King did, to understand and to comprehend, and to replace that violence, that stain of bloodshed that has spread across our land, with an effort to understand with compassion and love."

My dad's Indianapolis speech was not an artful, rehearsed oratorical masterpiece. It was an anguished visceral expression of shared agony. He had never spoken about his brother's death like that before. The crowd sensed he was doing something unusual, and gave him their hearts. Indianapolis was one of the few cities with large black communities that did not explode in riots that night. One hundred and nineteen other cities were not so fortunate, and President Johnson deployed 75,000 soldiers to quell the violence. Casualties included 2,500 injured and 39 dead. From our boarding school in Rockville, Maryland, we watched smoke rise over the Capitol and troop trucks roll past all day. My father found armed soldiers wearing gas masks patrolling Washington's streets when he returned the following morning. Walking through the capital's slums, he stepped over broken glass and past wrecked stores and businesses. A middle-aged black woman, standing amid the ruin, watched him intently as he picked his way toward her down the tangled slopes of a smoking rubbish pile, "Oh it's you!" she said, surprised, and then, "I knew you would come!" The crowd that

gathered around him became so large that the National Guards-
men momentarily mistook it for a mob of looters. My dad spent
part of his day consoling Coretta Scott King and her children, and
arranging for Martin's body to be flown from Memphis to Atlanta.

Civil rights leaders like Andrew Young, John Lewis, and Ralph
Abernathy now focused on my father as their last hope for America.
Hosea Williams later recalled, "King's murder left us hopeless, very
desperate, dangerous men. I was so despondent and frustrated at
Dr. King's death, I had to seriously ask myself, 'Can this country be
saved?' I guess the thing that kept us going was that maybe Bobby
Kennedy would come up with some answers for the country. I remember telling him he had a chance to be a prophet. But prophets
get shot."

Looking back, there were so many people for whom my father
represented the last best hope. And they weren't all black, or Native
American, or Hispanic. Indiana was conservative, skeptical, and
white, but many Hoosiers came to believe that he was the only person who could heal the country. My father was able to accomplish
what historian and psychiatrist Dr. Robert Coles later called "The
miraculous"— his ability to attract the support of frightened, impoverished, desperate African Americans and their angry leaders,
as well as working-class white people, farmers, and ethnics who felt
abandoned by their government. He won 42 percent of the primary
vote in Indiana, with the state's favorite son, Governor Roger Bran-
igin, winning 31 percent; McCarthy trailed with 21 percent of the
vote. That same night my father beat Hubert Humphrey by nearly
thirty points in Washington, D.C. On May 14 the farmers gave him
Nebraska with 51.5 percent of the vote. His coalition of poor whites
and blacks held. My dad won 80 percent of blacks, nearly 100 percent of Hispanics, and 60 percent of low-income workers.

My dad's insistence on campaigning in the May 28 Oregon primary was evidence of both the fatalistic aspect of his crusade and
his determination never to back down. He could have ducked the
state, as Goldwater did when he faced defeat there in 1964. Defeat
for my father was a virtual certainty. My mom, fretting over polls

that showed him losing the state disastrously, explained to us that Oregon was only 1 percent black, and only 10 percent Hispanic. There were few Eastern European ethnics, and the Native Americans who might have supported him didn't vote. His advisers described the state as a giant suburb with high employment, no slums, no urban decay, no real issues of racial injustice, and practically no one you could call poor. Even white liberals were scarce, and they overwhelmingly supported McCarthy. "I do my best where people have problems," my father observed; and there were few people with serious problems in Oregon. Moreover, the Oregon Teamsters, still stinging from my father's 1959 investigation that ended with the jailing of Dave Beck, Oregon's most powerful labor leader, made a special effort to defeat him. The 50,000 members of the AFL-CIO launched an energetic campaign to defeat him.

In the dark days before the Oregon primary, when polls showed him trailing McCarthy, columnist Joseph Kraft asked my father if he thought the whole campaign was a mistake. My dad shook his head. "I don't regret it. I don't regret it because whatever happens, I've done everything I could. I've made the effort." He had integrated the philosophy of the Stoics. Still, it was a blow when he lost Oregon to McCarthy by six percentage points, becoming the first Kennedy ever to fall short in an election.

———————

HE CAME TO CALIFORNIA FEELING STRANGELY LIBERATED. CALIFORnia, our largest state, is also the most diverse—blacks, whites, Hispanics, Asians, large ranchers, small farmers, military bases, lumbermen, oil tycoons, car dealers, and film stars. Its neighborhoods ran the gamut from Beverly Hills to Disneyland and from Santa Barbara and San Bernadino to Watts and Compton; it was a true cross section of America. As usual, he refused to pander to hostile audiences, and he reveled in the adulation and hopes of the impoverished—touching the poor and being touched—throwing himself into the crowds.

The last stop in California was Los Angeles, and there he ran a gauntlet of tumultuous, exultant crowds—black, brown, and white all surging to embrace him in blizzards of confetti. He lifted children into his car to save them from being trampled. The "Fearsome Foursome" struggled to hold on to him as admirers repeatedly pulled him out of cars. Marveling at his candor with those who opposed him and the fervent hopes he excited among the poorest and most idealistic Americans, the reporters who covered him lost all objectivity. They became his biggest cheerleaders. New York's most hardened political journalists, Jimmy Breslin, Jack Newfield, and Pete Hamill—all of them a blend of cynicism and idealism—moved from embedded journalism to become active advisers and speechwriters for his campaign. That final week Dick Harwood, of the *Washington Post*, asked his editors to take him off the campaign. "I'm falling in love with this guy," he explained.

More Bullets Reach My World

As with Jack, there had been many omens. Shortly after my father announced his candidacy, Jackie confided to Arthur Schlesinger: "Do you know what I think will happen to Bobby? The same thing that happened to Jack. There is so much hatred in this country, and more people hate Bobby than hated Jack." That same week, the influential conservative guru Westbrook Pegler vented his aspiration "that some white patriot of the Southern tier will spatter [Robert Kennedy's] spoonful of brains in public premises before the snow flies." In 1968, J. Edgar Hoover's deputy and companion, Clyde Tolson, was less poetic: "I hope that someone shoots and kills the son of a bitch." At a hotel lounge in Indiana a group of reporters covering the campaign debated whether my father could win the race for the White House. "Of course he has the stuff to go all the way," said *Newsweek* reporter John J. Lindsay. "But he's not going to go all the way. Somebody is going to shoot him." A stunned silence followed that remark, but no one disputed it.

Like most children, I saw my father as invincible, but having lived through Jack's assassination and other violent deaths, I knew he was at risk. Nevertheless, I understood that violence was not going to deter him from his course. We had been raised to live life fearlessly and to fight for our principles without regard to personal danger. I knew there were people who hated my dad, and death threats were a regular fact of life for our family. Many of them were credible, but they were so numerous and so routine that they had become banal.

I understood that every worthwhile endeavor includes an element of risk, and that this time the stakes were enormous. After all, men were dying every day in the rice paddies and rain forests of Vietnam. And back home in America, ghettos were burning, people were living in desperation and meeting violent deaths. Every day, children were lost in our cities from gang or random violence, from neglect or lack of health care, or as they gave up on their country and turned to crime. If there was any chance of mitigating these horrors, wasn't my father's decision to put himself in harm's way worth the hazard?

My father, as ever, was fatalistic about his own destiny. At that time, Secret Service protection was not available to presidential candidates (Congress changed the law following my father's assassination). My dad had refused J. Edgar Hoover's offer of FBI agents to protect him, knowing that Hoover would use the agents primarily to spy on him for the benefit of LBJ. Because of its reputation for brutality and racism, its ignominy within minority community, and the extreme right-wing disposition of its ultimate commander, Los Angeles mayor Sam Yorty, my father also rejected offers of protection by the LAPD. In any case, my father did not want to be surrounded by security. "That's not the way I want to run a campaign. It's not the United States of America. In some other countries a candidate may have to talk through some kind of a shield, but not here." He wanted to engage and touch the crowds. Often, former federal agent Bill Barry was alone with my dad on the back of a car. In hindsight, it was certainly reckless, given the power and determination of his many enemies. I suppose it was hubristic, too.

My father spent the day of the California primary in the setting he loved best, along the seashore with my mom and six of my younger siblings, at the Malibu Beach home of filmmaker John Frankenheimer. Standing in the swimming pool, he tossed Kerry and Michael high in the air to splash in the water. He buried coins in the sand for my three-year-old brother, Max, and together they kicked the rubbery kelp balls that lay in windrows at the tidal fringe. He talked about the pollution of the beautiful coast and the disappearance of the great old kelp beds. While swimming in the breakers, my brother David got caught in a powerful undertow and my father pulled him out, probably saving his life.

The polls closed early that day in South Dakota, a conservative farm state with a large white majority. South Dakota was Hubert Humphrey's birthplace. But Sioux Indians and white farmers, ordinarily fierce antagonists, came together to give my father a significant victory in America's most rural state. That evening my dad also won America's most urban state, California, bridging an abyss that still marks our most daunting national divisions. In the Golden State, voter turnout in ordinarily apathetic Watts exceeded that in affluent Beverly Hills. For the first time in history, thanks to César Chávez, there were lines around the block when the polls opened in South Central Los Angeles. My father had rebounded from his Oregon defeat. Before he left to give his victory speech at the Ambassador Hotel in Los Angeles, he took a call from Mayor Daley, who pledged his support at the Chicago convention. My dad's aides celebrated that call as a virtual guarantee that he would become the Democratic Party's candidate. My father closed his speech at the Ambassador Hotel saying, "So, my thanks to all of you—and on to Chicago. Let's win there." He flashed a peace sign.

Shortly after my father died, we found in his desk a prophetic excerpt from John Keats, written in my dad's cramped hand. "While we are laughing, the seed of some trouble is put into the wide arable land of events. While we are laughing it sprouts, it grows and suddenly bears a poison fruit which we must pluck." The bullets found him as he reached out to shake the hand of a seventy-five-dollar-a-week Mexican busboy, Juan Romero, in the kitchen of the Ambassador Hotel. Romero had purposefully traded with another worker to bring room service to my father earlier that evening. He had grown up in a small village where every house had a picture of President Kennedy. He was curious to meet JFK's brother. Paul Schrade from the UAW was wounded by that first bullet. My dad's last words were an expression of concern for Paul. "Is everyone okay? Is Paul okay?"

My father died surrounded by his fiercest supporters: my mother, his life's love; Dolores Huerta of the farmworkers' union; civil rights leader Charles Evers; radical black student firebrand John Lewis, who once led SNCC; and Budd Schulberg, the great screenwriter who had celebrated American workers in films like *On the Waterfront*, then had retired to run a writers' workshop in Watts. Jack Newfield of the *Village Voice* and two other hard-nosed, blue-collar reporters who never went to college, Pete Hamill and Jimmy Breslin, were both close by. Sitting on the curb outside the Ambassador Hotel, Charles Evers told Newfield, "God, they kill our friends and they kill our leaders."

I was fourteen years old and asleep in my Georgetown Prep dorm room when the school's disciplinarian, a broken-nosed former Golden Gloves champion, Father Dugan, woke me at six a.m. and told me gruffly that there was a car waiting outside to take me home. At Hickory Hill, Jinx Hack told me my father had been shot, but I was still thinking that he'd be okay. He was, after all, indestructible. But when I heard someone order all the campaign offices to close, then the bullets reached my world. I knew my dad was going to die.

Douglas was an infant. My mother was pregnant with my sister Rory. Courtney, Kerry, and my little brothers David and Michael were on the campaign trail. Kathleen, Joe, and I flew with Lem Billings to Los Angeles on Vice President Humphrey's plane, Air Force Two. There were thousands of supporters outside Good Samaritan Hospital when I arrived. People were holding "Pray for Bobby" signs. I heard sirens wailing as we drove past the silent crowd. The hospital floor was cleared of all but close friends, family, and a security force of U.S. marshals. Teddy, Jean, Pat, Steve Smith, and Jackie were there, as was Coretta Scott King. John Seigenthaler, John Glenn, Rafer Johnson, Andy Williams, and Jimmy Breslin were all standing glumly in the dark hallway. I walked past them into a recovery room.

My father lay on a gurney with his head bandaged and his face bruised, especially around the eyes. A priest had already administered last rites. My mother sat beside him, holding his hand. She stayed there all night. I sat down across the bed from her and took hold of his big wrestler's hand. I prayed and said goodbye to him, listening to the pumps that kept him breathing. Each of us children took turns sitting with him and praying opposite my mom. My dad died at 1:44 a.m., a few minutes after doctors removed his life support. My brother Joe came into the ward where all the children were lying down and told us, "He's gone." We all cried. My father was forty-two years old when he died. My mother, now his widow, was thirty-eight.

The following day, President Johnson sent Air Force One to pick us up. At Los Angeles International, a long procession of Highway Patrol motorcycles and patrol cars escorted us onto the runway. Behind the hearse were about a hundred civilian motorcycles, part of the squadron of bikers from clubs across California who had served as my father's informal escort during his campaign. It was a motley crew of beach boys, off-duty cops, biker gang members from San Bernardino, and even Black Panthers from Compton. Aiming to give my dad a final send-off, they raced the plane down the runway and I watched them fade behind us as the plane took off. Some of them, riding too close to the engines, were blown from their bikes by the jet's backdraft.

We flew my dad's body back to New York, taking turns to sit with him in the stern cabin, never letting him be alone. I stared at his coffin, thinking it seemed too small to contain him. At nine a.m. we landed at LaGuardia, and Teddy and my brother Joe and I helped to lift the casket onto a hydraulic lift, which lowered it off the plane. Hundreds of people met us at the airport, including the leading lights of New York politics: Governor Nelson Rockefeller, Mayor John Lindsay, and Senator Jacob Javits. We drove to Manhattan and lifted my father's casket from the hearse and carried it into the nave of St. Patrick's Cathedral. Grandma Rose came in, and we had a short family service. Jackie broke down and wept as she hadn't allowed herself to when we buried Uncle Jack.

On Friday, we gave my father a wake at St. Patrick's, and we took turns standing honor guard around his casket that night in half-hour shifts. Teddy stayed all night. I stood vigil near the closed casket as a red-eyed crowd in the tens of thousands filed quietly past, saying their goodbyes. Countless mourners spent long hours waiting in the sweltering heat for just a moment beside his flag-draped coffin. The line seemed undiminished on Saturday when the great cathedral opened for the requiem mass. For many blocks the crowd lined the sidewalks, eight to twelve deep, listening to the mass on loudspeakers. In St. Patrick's, separated only by a few pews, Chicago's law-and-order Mayor Richard Daley and radical activist Tom Hayden both wept uncontrollably. A few weeks later they'd be at war with each other on Chicago's streets, with Tom leading protests at the Democratic convention and Daley's bully-boy cops pummeling youthful demonstrators in a display that would split the party, frighten the nation, and usher in the election of Richard Nixon. My father had been the glue that kept those divergent groups at peace and common purpose.

The funeral was on June 8. There were hundreds of television cameras at St. Patrick's, with reporters occupying all the back pews, their wires running up the aisles. A hundred million people watched on TV and two thousand attended, including Lyndon and Lady Bird Johnson. We rode in a convoy to St. Patrick's, where Grandma was waiting. My mother and Jackie knelt by the coffin. Andy Williams sang "Ave Maria." Cardinal Cushing once again found himself leading the funeral mass for a young Kennedy, but this time, at my mother's insistence, all the most far-reaching changes from Vatican II were implemented. The spirit of Pope

John XXIII was present. My mother wanted a mass that was in English, participatory, and filled with joy. Latin, as Pope John XXIII recognized, was a tongue that Christ never spoke. It was the language of the Roman Empire, an oppressive, cruel, authoritarian, violent plutocracy.

Teddy, who had not slept for several days, delivered the eulogy in a breaking voice. "My brother need not be idealized, or enlarged in death beyond what he was in life," Teddy said. "He should be remembered simply as a good and decent man, who saw wrong and tried to right it, saw suffering and tried to heal it, saw war and tried to stop it." When Andy Williams sang "Battle Hymn of the Republic" a cappella, everybody broke down. Teddy and Jackie wept pitifully as we left the cathedral. A small group of black men and women standing on the sidewalk began singing "Glory, glory hallelujah . . ." when suddenly the booming voice of a large black woman coaxed everyone to join the singing. The weeping also spread. I watched a woman collapse, sobbing hysterically as we carried my father down the cathedral steps to the hearse that led our convoy to Penn Station for the train ride to Washington, D.C.

We placed my father's flag-draped casket in the last car, resting it on red velvet chairs in the dining section, high enough so that the crowds lining the tracks could view it as it passed. Black bunting and green pine boughs adorned that car. I sat with Michael and David in the car just before and watched as the train passed the waiting crowds. Two million people lined the tracks to bid my father farewell as we moved south through the ghettos of Newark, Philadelphia, Wilmington, and Baltimore—and through the rolling countryside of Pennsylvania and Maryland. The mourning multitudes slowed the train so that a two-and-a-half-hour trip stretched to seven hours. On board, the mourners included many who would have been in my father's Cabinet, or running government agencies. All of them believed Robert Kennedy would have ushered in an era of idealism, peace, and justice. Pierre Salinger said, "The passengers on that train could have run the most exciting government this nation has ever seen!"

Among the vast crowd that lined the tracks were whites, blacks, Orthodox Jews, and Hispanics. Uniformed soldiers and boy scouts saluted the passing train. Cops and firefighters stood at attention alongside long-haired hippies in tie-dye and adolescents in Catholic school uniforms. Black militants sporting giant Afros waved clenched fists. Catholic priests, brothers, and nuns stood with hands folded, including a half dozen sisters perched on the bed of a yellow pickup truck. A team of Little Leaguers halted its game to salute us. Others, including honor guards from high schools and VFW and American Legion halls held flowers or flags or signs reading "Goodbye, Bobby," "God bless the Kennedys," "We will miss you, Bobby," and "So long, Bobby." I could see onlookers crying, covering their faces, and kneeling with clasped hands. In the countryside, people held babies high in the air and shouted, "Pray for us, Bobby." A man in a suit stood alone in a field wiping his eyes. A wedding party awaited our train in a Delaware farm field; the bridesmaids in pink and green flung their bouquets at my father's car as it passed.

The train slowed in Newark and Trenton, and we crawled past large crowds. mainly black, gathered on the platforms, singing "Battle Hymn of the Republic." In the Philadelphia station, women held photos of my father. As we crossed the Ramapo River, the crew of a city fire boat stood, saluting us. Even in the smaller stations, we could hear people singing the "Battle Hymn of the Republic," often accompanied by high school bands. My brother Joe and, later, my mother, walked the entire length of the train's twenty-two cars, shaking hands with all 1,100 passengers. My mom took the time to sit beside and comfort first Pete Hamill and then light-heavyweight champion José Torres. She found both men weeping dolefully. In Philadelphia and the other stations, thousands of people waited, openly crying as my mother waved to them.

When we reached Washington's Union Station, President Johnson met our funeral train and escorted us up to Arlington. The mournful faces I saw that day were the same diverse cross section of Americans that I'd seen mobbing my uncle and father since I was a little boy. As our procession crept past the Mall, we saw the thousands of impoverished people—the so-called Poor People's Campaign—encamped in "Resurrection City," a crowded metropolis of shanties and tents of plastic sheets, a gathering conceived by my father and launched by Martin Luther King, Jr.—who himself had been assassinated just two months before. Earlier that spring my dad had told Marion Wright Edelman, the formidable civil rights leader, that government would not seriously address the issues of poverty until the poor claimed political power. My father had suggested organizing a poor people's march on Washington, much the way Dr. King had organized the Civil Rights March on Washington in 1963. Marion Wright Edelman had brought the idea to King, and my dad's office partnered with King to organize the "Poor People's Campaign of 1968."

Now all these homeless people came out to the curb and, holding their hats against their hearts, bowed their heads or waved forlornly as we passed on our way across the Potomac River to Arlington National Cemetery and up the hill to bury my father beside his brother, between two magnolias, beneath a small black stone simply engraved with his name and the years he lived. Joe helped my mom to the front seat, while I and the other pallbearers carried the casket from the hearse to the grave. By now it was nearly ten-thirty at night. Under a brilliant moon we said our final goodbye, as mourners with candles turned Arlington Cemetery into a city of lights.

When I got home that night, I went alone into my father's office and shut the door. He kept a single bed there for nights he had to work late, and I could smell the faint aroma of his favorite cologne. I lay down and looked at the framed pictures on the wall—Aunt Kick, Uncle Joe, Jack, his friend Dean Markham, all so young, and all dead now. Then I lay down on my dad's bed and wept until my tears soaked the pillow. After a while, Dave Hackett, my father's closest and oldest friend, came in and sat with me. I knew his heart was also broken. We stayed there for a long time in silence. Finally he said, "He was the best man I ever knew."

———————

ONE DAY IN MAY 2010, OVER FORTY YEARS AFTER MY DAD DIED, COLO-rado senator Tim Wirth told me as we hiked together in the Rocky Mountains that he had gotten into politics because of my dad. He'd been in the Senate caucus room when my father declared his candidacy for the presidency, and returned home himself to run a long-shot bid for the Senate. That same day in May, an African American woman passing me on a moving walkway in the Denver airport shouted "Kennedy!" with a broad smile, and clapped her hands. A flight attendant on the airplane back to New York stopped by my seat with brimming eyes, touched her heart, and said, "I loved your father."

The three people who spoke to me of my father that day were all part of a steady drumbeat of strangers, who have approached me, sometimes daily, for over four decades, all of them haunted by the emptiness of unfulfilled expectations. They stop me in airports, on street corners, at public functions and political rallies—doctors, lawyers, nurses, city mayors, labor leaders, politicians, civil rights workers, grassroots organizers, and volunteers in every imaginable cause for human enrichment and dignity—to tell me that they entered public service inspired by my father's example. I have seen pictures of him mounted prominently in the offices of mayors, senators, congressmen, journalists, and business people more times than I can count. Even fifty years after his death, people cry when they see my face because it reminds them of my father. Flight attendants smuggle me first-class meals in coach, restaurant waiters refuse to give me a check. Even toll collectors on highways, who will never see me again, refuse to take my money. A few random examples in my journal include testimonials from people at every level of our society.

In mid-October 1991, following a speech in Indianapolis, a pretty girl asked for my autograph. She introduced herself as Courtney, explaining that she was born in June 1968, just after my Dad died. Her parents named her after my sister. After that same speech, a local reporter told me that he had gone into a depression for three months when my dad was killed, and had to go to a counselor. He was crying, and he said that he'd read every book on my father and tried to model his life after Robert Kennedy's vision. He ducked away, wiping his flowing tears.

In May 1993, George McGovern told me that he had just come from the Civil Rights Convention in Birmingham. "Everybody remembered me, not because I had run for president, but because I was friends with your dad." McGovern told me, "I'm more proud of that friendship than anything I've done in my life."

On Wednesday, May 7, 1997, an old man approached me after a speech in Flint, Michigan, to say that other than the death of his parents, the death of my father had been the most devastating time of his life. He is one of many people who have told me the same thing.

Another man at that same event said, "The Kennedys were born with silver spoons in their mouths, and took them right out, and began feeding the poor."

Three days later, on May 10, 1997, a Saturday, the mayor of Nitro, West Virginia, found me in a gym on an exercise bike at my hotel— "I've got a picture of your uncle in my office, and the town clerk has a picture of your dad in hers."

Tuesday, March 10, 1998. After class, I went to sit shiva with Eric Breidel's parents at their home on Fifth Avenue. Former secretary of state Henry Kissinger pulled me aside. "I want to tell you again how much I admired your father. I once wrote your mother a letter comparing him to Mozart. I don't think she had any use for me after I went into public life, but I wanted her to know how priceless and irreplaceable I thought her husband." He paused, "This country would have been very very different. . . ."

One day in August 1993, in a taxi to the airport in Vancouver, British Columbia, the driver, a middle-aged woman named Julie, would not take my money, despite the twenty-dollar tab on the meter. I asked if she would at least have her picture taken with me. "Now I'm gonna cry," she said. And she did.

The great Greek general and statesman Pericles, who ruled Athens in its Golden Age and built both the Parthenon and the Acropolis, observed, "What you leave behind is not what is engraved in stone monuments but what is woven into the lives of others." The best tribute to my father's life is the love he wove through so many people's lives.

• • • • •

CHAPTER 11

Lem

To those people in the huts and villages of half the globe
struggling to break the bonds of mass misery, we pledge our
best efforts to help them help themselves, for whatever period
is required, not because the Communists may be doing it, not
because we seek their votes, but because it is right. If a free
society cannot help the many who are poor, it cannot save the
few who are rich.

—JOHN F. KENNEDY

THE SUMMER AFTER MY FATHER'S DEATH, HIS OLDER CHILDREN DIS-
persed to give my mother a chance to recoup. Kathleen went to
teach on a Navajo reservation in New Mexico, and later to an Inuit
village in Alaska. Joe moved to Washington State to work for Jim
Whittaker as a guide on Mt. Rainier, while David traveled with
our cousin Chris Lawford to La Paz, California, to work for César
Chávez. I spent the summer in East Africa, traveling with my Uncle
Jack's boyhood best friend, LeMoyne Billings.

Lem, who at the time of my father's death was fifty-three, had
been Jack's roommate at Choate, and for the one year Jack attended
Princeton. Lem shared Jack's love of art and culture, and his special

passion for Americana, including folk art. An engraved sperm-whale tooth—a gift from Lem—that inspired Jack to begin his famous scrimshaw collection, shared Jack's Oval Office desk with the etched coconut that saved his skin in the Solomon Islands. Jack so enjoyed Lem's sense of humor, his gift as a raconteur, and his deafening laugh that he assigned him a bedroom at the White House. Lem spent every weekend of Jack's presidency at the White House with Jack and Jackie, or with them at their Middleburg, Virginia, estate, Glen Ora. He often traveled with Jack on presidential trips. He was with Jack in Vienna during his meeting with Khrushchev, and he chose for Jack the wooden model of "Old Ironsides" that Jack presented to Khrushchev as a gift. Lem was also with Jack in Berlin in 1963 for JFK's *"Ich bin ein Berliner"* speech. While his friendship with our family began with Jack, Lem was as close to all the Kennedys as we were to each other.

Before she was killed in 1948, Lem had been equally close to Aunt Kick. After Uncle Jack's death, Lem and my father became best friends, Lem being one of the few people who suffered the president's loss as acutely as my dad. Teddy recalled the winter weekend trips he spent with Jack and Lem at Cape Cod in his early adolescence, sleeping on cots in the garage of the boarded-up summerhouse and cooking their own food as "one of the biggest thrills of my childhood." Aunt Eunice said of him, "I consider Lem my best friend, and I think there must be fifty to a hundred human beings who thought Lem was their best friend too."

Starting at six o'clock every morning until long after he went to bed, Lem was answering the phone and making people laugh at the other end. His war buddy Brook Cuddy called him "a genius at friendship." My earliest childhood memories include expeditions with Lem to hunt salamanders and crawdads in Pimmit Run near Hickory Hill. To this day, Lem was the most fun person I've ever met in my life. After he died, in 1981, I assembled and edited a collection of reflections from some two hundred of his "best friends," including every living member of the Kennedy family. I wanted the next generation of Kennedy children to have a record of Lem's extraordinary example of love.

On our way to Nairobi in 1968, Lem and I stopped in Paris to visit the Shrivers. Uncle Sarge, who had run the Peace Corps and served as architect of LBJ's War on Poverty, was by then serving as Johnson's ambassador to France. His wife, Eunice, my godmother, was Lem's soul mate; they spoke to each other every day on the phone. After a couple of days in the U.S. embassy, playing with my cousins Bobby and Maria, and touring Paris's art museums, Lem and I boarded a TWA flight to Nairobi.

The Kenyans welcomed us like visiting royalty. A delegation of Kenyan diplomats took us from the airport to a meeting with President Jomo Kenyatta, the anticolonialist leader whose imprisonment by the British had helped inspire the bloody Mau Mau uprising and the eventual British withdrawal from East Africa. I had read voraciously about the Mau Mau in school. Ms. Kooligan, my third-grade teacher, who equated Britain's loss of its empire with the decline of civilization, recounted bloodcurdling tales of Mau Mau leaders brutally murdering their white colonial masters and ritualistically devouring their still-beating hearts. That sort of story made me particularly enthusiastic about meeting Kenyatta, who had led the country since Kenya's independence in 1963. He did not disappoint. He greeted us at home in his gray dashiki and spoke with great fondness of my father and uncle, both of whom he knew, and of their support for tribal nationalism. Later, we saw him hold court in his traditional leopard robe, wearing a couple of pounds of gold rings and wielding the wildebeest-tail scepter that served as a symbol of his office. Unfortunately for his nation, he leaned toward authoritarian rule and shared power disproportionately with his fellow Kikuyus, reducing Kenya to a tribal kleptocracy that would become a dictatorship under his Kikuyu successor, Daniel arap Moi.

We also met another colonial liberator, President Julius Nyerere, the Socialist leader of Tanzania who had helped win its independence for Tanganyika and then united his nation with the island kingdom of Zanzibar. My father had visited Nyerere a year before. Nyerere was precisely the kind of developing-world leader he most admired. He was a non-aligned Socialist who worked to build a fair, just, and equitable system for his people. Thoughtful and eloquent, Nyerere shared my father's love for Shakespeare and made a hobby of translating the bard's plays into his own language. Nyerere told me that he loved my father and was not just saddened but deeply embittered by his death. People across Africa, he told me, had yearned for his presidency. When Nyerere died in 2001, he was one of Africa's most successful, humane, and incorruptible postcolonial presidents. He built Tanzania into a progressive socialist democracy that largely escaped the violence, corruption, and tribalism that afflicted its East African neighbors.

Lem and I toured East Africa's spectacular national parks with their new African managers. We accompanied park police as they arrested ivory and rhino horn traffickers and raided poacher camps in the Serengeti on Kenya's northern frontier, confiscating wire snares, wooden bows, and poison-tipped arrows. Most of the poachers were impoverished, hungry men clad in tattered burlap that hardly covered their bodies. Some had the handsome features and aquiline noses of the Sudanese. The uniformed park guards treated them roughly, and despite my low opinion of poaching, I felt sorry for them.

Everywhere we went we found a universal sense of loss from the deaths of President Kennedy and my father. It was touching for me to see that even in the most remote villages, people had invested enormous hope in Uncle Jack's presidency and my dad's candidacy. Jack's legacy, including the Peace Corps and other programs benefiting the poor, had left an indelible impression. From poverty-stricken mothers in Nairobi's sprawling slums to non-literate Masai, Kenyans expressed to me unbearable sadness. Passing Lem and me on the street, knowing only that we were American, strangers would smile and greet me saying, "Ah, Kennedy milk." The Food for Peace program had helped nourish a generation of the world's poorest people and inspired warm feelings for the United States in far-flung corners of the globe. I saw schools and hospitals named for Uncle Jack in urban areas and backwaters. Every city seemed to have a library or thoroughfare bearing his name. As part of their daily routine, schoolchildren in Kenya sang a patriotic song that referred to John Kennedy as their father. We found his photo inside many huts and homes.

Lem and I also spent considerable time with Kenya's charismatic labor leader and visionary Tom Mboya, one of the nation's founding fathers, and my father's favorite African leader. Widely expected to succeed Kenyatta as the young nation's president, Mboya was respected across the African continent as one of the most sophisticated, thoughtful, and articulate advocates for African nationalism. He had almost single-handedly built the trade union movement in Kenya, Uganda, Tanzania, and elsewhere. Mboya hailed from the Luo, a small tribe of bright, gentle, and industrious fishermen from the Lake Victoria region. He earnestly wanted Kenya to escape the violent gravitational pull of tribalism that tore apart so many African countries, stunted the potential of their people, and caused governments to devolve into plundering, family-run affairs. Mboya's heroes were Abraham Lincoln, Thomas Jefferson, and Mahatma Gandhi. His ambition was to make Kenya a true African democracy.

I first met Mboya eight years before, in 1960. Great Britain had promised to grant Kenya independence by 1964. A coleader, with Kenyatta, of Kenya's liberation movement, Mboya worried that Kenya had no black college graduates with the training to manage self-rule. He knew that a modern nation required an educated class. Employing his charm and diplomatic skills, he extracted commitments from two hundred American colleges to give full scholarships, beginning in the fall of 1961, to nearly three hundred of Kenya's smartest students, whom Mboya had identified. By summer 1960, he lacked only the money necessary to fly his students to the United States. He traveled to the United States to ask for help from Richard Nixon and the Eisenhower State Department, but they turned down his request.

At the suggestion of calypso singer Harry Belafonte, Mboya visited us on the Kennedy compound at Hyannisport at the end of July, seeking support for his plan. Uncle Jack, then a U.S. senator and chair of the African sub-committee, was impressed by this charming and brilliant man, and had arranged for our family to donate $100,000 on the condition that the contribution remain anonymous, to keep it from being politicized. Word of JFK's commitment leaked, however, to the Nixon camp, which made the donation public, apparently believing it would hurt Jack with Southern voters. Mboya's project became known as the "Kennedy Airlift," and it ultimately educated five hundred young Kenyans, including many of the country's modern democratic leaders—among them the 2004 Nobel Peace Prize winner, Wangari Maathai.

Mboya and my father had found kindred spirits in each other and developed a strong friendship. My dad saw in Tom Mboya a bright light of Africa's future. That summer of 1968, Lem and I spent time with Mboya, touring Nairobi, the Ngong Hills, and the famous Treetops Hotel where Princess Elizabeth became Queen Elizabeth II of England. A year later, government assassins shot and killed Mboya outside a Nairobi pharmacy.

In July 2004, while I was living on Martha's Vineyard with the comedian Larry David, the two of us attended the Democratic National Convention in Boston, where I spoke. Afterward, Larry and I watched a young, little-known U.S. senator from Illinois deliver a stirring keynote address. We spoke with Senator Barack Obama and learned that he was also on his way to Martha's Vineyard. The following night we joined him at a small dinner party on the island. During our conversation, I learned from Barack that his father was Kenyan. I asked him what tribe his father was from.

"My father was Luo," he told me.

"Have you heard of Tom Mboya?" I asked.

Obama replied, "Tom Mboya is the reason I'm in this country." Barack's father, he explained, had been one of the first Kenyan students to come to America via Mboya's "Airlift" program.

As we were preparing to leave Kenya, we received an invitation from Emperor Haile Selassie to visit his Addis Ababa palace, and another from President Gamal Abdel Nasser to visit Egypt. Nasser was the father of Arab nationalism. He was a secularist, but his independence worried the CIA, which tried to overthrow him in 1959. Nevertheless, he had great affection for Uncle Jack, for which reason he feted Lem and me. We climbed the Great Pyramid of Giza, and toured Karnak and the Valley of the Kings. Our guide on the expedition was the chief justice of Egypt's Supreme Court, whose most excellent judgment dictated that I accompany him and Lem on a late-night escapade to a rooftop nightclub to see a famous belly dancer. Being a fourteen-year-old with active hormones, that show nearly compensated for our missing the cheetah hunt with Emperor Selassie.

I would return to Africa frequently over the coming years. In 1975, I spent the summer there in tents and on horseback, again with Lem. This time I was making the pilot for a TV series on tribal culture and wildlife with a producer named Roger Ailes, who had created the *Mike Douglas Show* and had been Richard Nixon's communications adviser during the 1972 presidential campaign. Ailes was brilliant, charming, and hilariously funny. We formed a friendship that—thanks to our mutual love for Lem—survived Ailes's creation of Fox News, which has played such a sorry role in polarizing our country and poisoning public discourse.

———————

LEM BILLINGS WAS AN INSPIRING MENTOR, NOT JUST TO ME, BUT TO AN entire generation of fatherless Kennedy children for whom his home on East Eighty-Eighth Street in New York City became sanctuary, library, classroom, and museum. Historic portraits filled every wall not already occupied by bookshelves crammed with leather-bound classics, histories, biographies, and art books. Lem slept in an antique walnut bed beneath an "I Hate Barry" (i.e., Gold-water) button pinned to the upholstery. There he would lie with the earpiece of his glasses in his mouth, voraciously reading the latest biography of a European monarch, or of Ulysses S. Grant, or yet another life of George Washington, while his African basenji, Ptolemy, lay impatiently at his feet, awaiting the summons onto Lem's chest, where he slept each night. After reading each new book, Lem would discuss it with the legions of Kennedy kids who gathered at his house.

Lem didn't just know history. He seemed to be a part of it. His ancestor Elder Brewster had come to Plymouth on the *May-flower*. Lem's great-great-grandfather, Francis Julius LeMoyne, M.D., was a leading Abolitionist who founded LeMoyne College (now LeMoyne-Owen College) in Memphis, Tennessee, and man-aged a principal stop on the Underground Railroad. Lem's great-grandfather, Luther Guiteau Billings, fought in three American wars, including the Spanish-American War and World War I, in which he held the rank of rear admiral. During the Civil War he was captured three times and escaped twice. He led a prison revolt in Savannah and captured a prison train, killing his Confederate guard with a saber. In 1868, Luther's ship, the last flat-bottomed frigate in the American Navy, was the sole survivor of the famous earthquake and tidal wave that wiped out the town of Arica (then in Peru). Lem made me a gift of the German Mauser rifles that his grandfather captured in Cuba in 1898.

Lem Billings pronounced "Nazi" to rhyme with "nasty." Imme-diately after Pearl Harbor he enlisted in the American Field Ser-vice, a fire-eating paramilitary medical corps attached to the British Eighth Army, so he could join the brawl against Hitler without delay, despite the poor eyesight that had thwarted his efforts to join the U.S. armed services. He drove an ambulance across the Sahara behind General Montgomery, pursuing Rommel, the "Desert Fox," 1,500 miles from Egypt to Tunisia. He told us what it was like to be stung by a scorpion and almost die, and to be blown off the roof of his ambulance by cannon fire from a German Messerschmitt. (He carried shrapnel in his back for the rest of his life.) After Tunisia, Lem got hold of a contact lens prototype that allowed him to convincingly pass his military vision test, and Grandpa helped him get into the Navy in time for the Pacific war. At Iwo Jima a wounded Japanese solder tried to detonate a grenade hidden in his behind while Lem was toting his stretcher to surgery belowdecks. When it didn't explode, Lem carried the grenade three decks up and tossed it overboard.

Lem knew how to stop an Army mule from passing manure by whacking it upon the hindquarters with a shovel. He could name all the U.S. presidents and their wives, as well as the kings and queens of England, in chronological order. He remembered a time when the Ku Klux Klan was a regular and highly regarded presence in his hometown of Pittsburgh.

Many members of Lem's generation will regard these facts as less than extraordinary, but to me these were the enchanting details of a magical life. Lem was as close to Jack and my father as anyone in their lives, including their brothers and sisters. They were so close, in fact, that they sometimes rubbed off on each other in ways that met with Grandma's disapproval. "P.S. Jack," she wrote in a June 1944 missive, "Please say yes-ter-day not yes-day, and please say Mis-sus not Mis' Longworth. You acquired these at Choate School and from a certain young man in Baltimore, not in New England."

Prior to attending Choate, Lem had never met a Catholic, a deficiency that ended when he encountered Uncle Jack. "You don't look like a Catholic," he remarked to Jack at their first meeting, reflecting the prejudices of his Pittsburgh upbringing. Lem's lifelong friendship with Jack grew out of their early escapades at Choate, where both boys rebelled against a system of regimentation designed to crank out cookie-cutter graduates. Both had elder brothers who were star athletes and scholars, against whom they had little hope of measuring up. After the school's notoriously tyrannical headmaster, George St. John, promised in his weekly sermon to root out the "muckers"—those who did not fit the Choate mold— Jack and Lem assembled a select group of friends to inaugurate a

"Mucker's Club." Each member purchased a gold clam rake lapel pin engraved with their name, the phrase "Mucker's Club," and the title "president." The headmaster discovered the prank, dubbed them all "public enemies," and expelled the lot. He reversed the sentence only after pleading by other masters and visits to the school by Grandpa and other affected parents. St. George placed Jack and Lem on rigid probation. Their classmates responded by voting Jack "most likely to succeed."

Lem and Jack became inseparable, and Jack rarely came home without Lem. "I was twelve years old before it dawned on me that Lem wasn't my older brother," recalled Teddy. Lem always said that Jack Kennedy was the most fun person he'd ever met. Lem and Jack referred to each other as Billy and Johnny. In hundreds of letters to each other they also employed a legion of nicknames including Lem, Lemmer, Leem, Moynie, Jack, Ken, and Kenadosus. Lem kept nearly two hundred letters from Jack, which he left me to archive when he died. Tucked away in the shelves at Eighty-Eighth Street, among the dozens of scrapbooks, was a newspaper Jack presented to Lem at Penn Station when he met Lem's train arriving from Princeton. "Train Stops With Jerk," the headline read, "Billings Gets Off." It's hard to describe it as just friendship. Eunice Shriver said of Jack's friendship with Lem, "It was a complete liberation of the spirit. I think that's what Lem did for President Kennedy. President Kennedy was a completely liberated man when he was with Lem."

Lem stood six foot four, and never lost his strong arms, broad shoulders, and the tree-stump thighs that made him a wrestling and crew champion at Choate and Princeton. A lumbering giant, his athleticism lay in his brute strength, endurance, and imperviousness to pain. He was a great bear, slow on the move, and his poor eyesight and lack of agility did not serve him well in the ball sports—tennis, football, baseball, golf—favored on the Kennedy compound. Jack, by contrast, was lean and buoyant, with Zen-like timing and proficiency in a variety of sports. His long spiral passes were graceful, unhurried, and glamorous.

But Jack was sickly in stark contrast to Lem, who had an inde-structible constitution. My uncle stoically battled a procession of painful and life-threatening afflictions. He was in acute physical pain most of his days from back and stomach ailments, and re-ceived last rites three times. He spoke about his suffering to almost no one except Lem. When Jack had yet another near-fatal spinal surgery, in 1947, Lem took six weeks off work to stay with his friend in Palm Beach. As president, Jack told Lem that he would trade all his political successes for Lem's famous good health. Lem joked that if he ever wrote Jack's biography, he would call it "John F. Ken-nedy: A Medical History." With the exception of chronic asthma, Lem never seemed to get sick. When he was injured on a remote Peruvian river in 1975, and I stitched together a long gash in his knee with a sewing needle and dental floss, without anesthetic, he never flinched.

In the summer of 1937, Lem and Jack made a two-month tour of Europe, beginning at Le Havre, where Jack's Ford sedan convert-ible came off the boat from New York. Jack was anxious to get to the cabarets of Paris, but Lem, a Renaissance art major with a con-suming passion for architecture, made him stop at every cathedral town from Rouen to Beauvais to the bomb-blasted ruins of Reims before exploring Paris and the Loire Valley. At Reims, on July 8, Jack wrote: "My French improving a bit and Billings' breath getting French."

Their primary objective was to observe the Spanish Civil War, a prelude to the destruction and tyranny that fascism was about to wreak on the rest of Europe. At the Spanish border, heavily armed frontier guards turned them back, but Loyalist refugees fleeing Franco's bloody persecution gave them firsthand accounts of the fascist general's deliberate bombing of civilians from the air, con-duct that both boys considered unprecedented barbarism. At every stop, the two friends wrote detailed letters home filled with keen observations about culture, art, and politics.

Ironically, it was at Lourdes, the sacred springs where millions of ill and infirm Christian pilgrims bathe each year in hopes of miraculous cures, that a rare event occurred: Lem got sick. In his journal, Uncle Jack describes their next stop, Carcassonne, as "an old medieval town in perfect condition—which is more than can be said for Billings." Despite Jack's triumphs at the gambling tables in Monte Carlo, the peripatetic duo continued to stay in the cheaper hotels out of deference to Lem's limited means. "He was perfectly happy to live at places for forty cents a night, and we ate frightful food," Lem later recalled of Jack, "but he did it [because] that was the only way I could go with him." In those days only the wealthier Europeans traveled by automobile, so Jack and Lem parked their convertible on the outskirts of a village and hiked in to negotiate a better rate at the local hotel. In Austria, they found even cheaper accommodations. "Stayed at a Youth Hostel in Innsbruck," Jack wrote, "which caused 'her ladyship' [referring to Lem] much discomfiture. It was none too good as there were about 40 [of us] in a closet, and it was considered a disgrace to take a bath."

In Rome they attended a Mussolini rally before driving to Munich. Inside the Third Reich, they experienced widespread contempt for Americans. Instead of shaking hands, the Germans would raise an arm and shout "Heil Hitler!" expecting a return salute. Instead, Lem and Jack would casually throw back their hands and wave, clownishly saying, "Hiya, Hitler." On one occasion, two Nazi Brown Shirts greeted their gesture by spitting on them. In the attic of a Munich beer hall, they drank with a group of Nazi Black Shirts. One friendly trooper, Oxford educated and baby-faced, encouraged them to steal a couple of signature ceramic beer steins as souvenirs and escorted them to a back door. But the young Nazis had secretly alerted a house detective who arrested and detained Lem and Jack on the street outside, confiscating the steins and their passports. As they were hauled away, they saw the treacherous Black Shirt and his companions laughing triumphantly. Lem and Jack came to loathe the Nazis, but Lem always regretted their decision to leave Germany three days before Hitler's famous speech at Nuremberg.

Lem and Jack were together on the Sunday afternoon of December 7, 1941, when the radio announced the Japanese bombing of Pearl Harbor. "We knew then that our world had changed," Lem told me. In a letter of recommendation endorsing Lem's application to the American Field Service, Grandpa remarked, "He has been my second son, as Jack's closest friend for ten years." The letter worked. For his part, after a stint in naval intelligence, Jack attended the Midshipmen's school in Melville, Rhode Island, and shipped out, in March 1943, to the New Hebrides, in the South Pacific, as a PT boat skipper. Lem told me that Jack was glad for the autonomy of commanding his own vessel, since he bridled under authority. John Hersey's chronicle in the *New Yorker* of Jack's heroic conduct following the sinking of his PT-109 in November 1944, caused a national sensation. Jack received the Navy and Marine Corps medals and Purple Heart, but he was characteristically understated in his own account of the incident. At the end of a lengthy letter to Lem, Jack casually added, "We have been having a difficult time for the last two months. Lost our boat about a month ago, when a Jap cut us in two. Lost some of our boys. We had a bad time for a week on a Jap island, but finally got picked up—and have got another boat." Jack was always self-deprecating about the incident. "I didn't do much. They sank my boat," he would reply when people asked about his heroics. By that time, Lem had also finally gotten into the Navy, using experimental contact lenses he'd gotten from his brother, a doctor at Johns Hopkins University.

Two weeks after the Japanese surrender, in August 1945, Jack, fresh from his discharge, wrote Lem from Hyannisport. At this point, Lem was chief purser aboard the USS *Cecil* in the South Pacific, having barely begun his deployment. "As I was getting over VJ Day," wrote Jack, "your letter set me to thinking and wondering if it wasn't time you came home for a long rest. I have been figuring out your points [to calculate time in the Navy] and it comes to nine, as close as I can figure it, which leaves you forty more to go, which indicates eighty months more, which comes to, say, seven years. Frankly, Billings, are you happy in the Service?" The sardonic humor in that letter, and many others like it to Lem reflect the honed skepticism of military bureaucrats that would inoculate President John Kennedy against taking seriously the advice of warmongering generals like LeMay and Lemnitzer and Admiral Arleigh Burke when he became their commander in chief.

After the war, their military service complete, Jack pressured Lem to skip out of Harvard Business School to work for his 1946 congressional campaign in Cambridge. Lem protested, "I'm a Republican, a native of Pittsburgh, and an Episcopalian." Nevertheless he went on to work fifteen-hour days at the Kennedy campaign headquarters. When Campaign Director Dave Powers sent Lem into blue-collar Charlestown, his most important instruction was: "Whatever you do, Lem, don't tell anyone you went to Choate." Lem worked tirelessly in that election, and the 1960 presidential primaries, and, later, in the general election, campaigning around the clock in both Wisconsin and West Virginia.

President Kennedy offered Lem his choice of jobs in the new administration, including running the Peace Corps or an ambassadorship. Lem refused, wary that having Jack as his boss would change their relationship. Instead he kept his position as vice president of Lennen & Newell, a New York City public relations firm. He did, however, accept a $75/day commission as trustee of the National Cultural Center (later the Kennedy Center) that Jack hoped to build in Washington. Lem became the driving force behind that project. He also assisted Jackie in her famous remodel of the White House. Above all, he was Jack's perpetual companion during his presidency.

Jack treated Lem's lack of portfolio as a source of amusement for himself, but was at times nerve-racking for Lem, who always worried about how Jack would introduce him at official functions. Jack presented Lem to German chancellor Konrad Adenauer as "Mr. Billings, one of our top cultural people," to astronaut Alan Shepard as "Congressman Billings," and to a bevy of puzzled admirals at the launching of an aircraft carrier as "Lieutenant Junior Grade, Billings."

LEM BILLINGS WAS JACK KENNEDY'S ALTER EGO, BUT AFTER DALLAS HE became one of my father's closest friends. Lem gave me a precious perspective on my dad after he died that most children never get of their parents—as a child. Lem said that he loved my father from the first day he set eyes on him, when my dad was eight years old and Lem sixteen. Lem was instantly struck by my father's thoughtfulness, generosity, and sensitivity toward others. "He was the nicest little boy I had ever met." I learned from Lem the details of my father's journey from doctrinaire Catholicism to a more nuanced and complex view of life. Lem explained the stubborn loyalty and compassion that drove my dad to visit Joe McCarthy on his deathbed, when all the world considered him a pariah. Lem also told me how devastated my father felt during a difficult three-hour meeting in New York in 1963 when he was harshly, even viciously, rebuked by James Baldwin, Lena Horne, Harry Belafonte, and a room full of civil rights leaders, including Harlem psychologist Kenneth Clarke, Martin Luther King's lawyer Benjamin Jones, and the heroic and unspeakably courageous Freedom Rider Jerome Smith about his insensitivity to their struggle. Lem told me that a call from Belafonte the next day had brought him around. Belafonte told him, "You can't take this personally. You need to understand that everyone in that room is looking to you and your brother as the final hope for black people." That phone call became a turning point in my father's evolution. He understood that their pain was beyond anything he had ever experienced or conceived could occur in America. Instead of being angry he resolved to listen harder.

One evening Grandpa loaned Lem his brand-new Buick for a spin about town. Coming home early the next morning, exhilarated by his classy wheels and tipsy from a night at the Cuban Tea Room, Lem missed the ninety-degree turn on North Ocean Boulevard, careening straight into the parking lot of the Palm Beach

Country Club, where he wedged the car between a Rolls-Royce and a Cadillac. The Buick's running boards were twisted up against the neighboring cars' doors, forcing Lem to exit through the window. As he squirmed his hulking body through the constricted aperture between the two parked vehicles, Joe Timilty, Boston's police commissioner, and a crony of Grandpa's, bounced jauntily out of the club and waved cheerfully as he sauntered toward his own car, offering Lem neither assistance nor comment. After extricating himself, Lem hiked back to the house and told my father, who disappeared upstairs to explain the situation to Grandpa, while a shaken Lem awaited the verdict. He would have preferred to face a Japanese banzai charge at Iwo Jima to Grandpa's disapproval. My father returned with Grandpa's summons, and a terrified Lem then ventured into Grandpa's study. There Grandpa greeted him with a gentle smile and expressed only his relief that Lem wasn't injured. Grandpa never mentioned the incident again. Two years later, Grandpa wrote Lem an unusual letter that revealed his profound gratitude for Lem's friendship to the Kennedy family. "This is a good time to tell you that the Kennedy children from young Joe down should be only proud to be your friends, because year in, year out, you have given them what few people really enjoy: True Friendship. I'm glad we all know you."

In an enjoyable and highly sympathetic biography, *Jack and Lem: The Untold Story of an Extraordinary Friendship*, author David Pitts alleges that Lem Billings was gay. I never saw any evidence that Lem was gay. To the contrary, he almost always had a steady girl. Nor did I hear even the hint of a rumor to that effect. Pitts's claim—that Jack knew that Lem was gay, and accepted him, nonetheless—is speculative, but it does support the notion that Lem's sexual orientation, whatever it was, would have been irrelevant to Jack. Jack had numerous gay friends and confidants, including Gore Vidal, Joe Alsop, and Bill Walton. His cosmopolitan acceptance of their sexual preference was consistent with his tolerance, open-mindedness, and humanity in every other arena. If indeed Lem had a secret gay life, it occupied very little of his time and energies and, in any case, that fact would in no way have diminished our friendship or the intensity of the love that I and my entire extended family felt for him.

Lem's most obvious trait was his loyalty and love for his friends and family. I saw this in every aspect of my own relationship with him, and I also saw the unswerving devotion that he practiced in every important relationship. He wrote his brother every week. He visited his drove of nieces and nephews often, dividing the holidays among them. He sent gifts to his sixteen godchildren every Christmas and on their respective birthdays. He exchanged some two hundred letters with my Aunt Kick when she was away at school, and after she sailed to Europe he was perhaps the most persistent of her many suitors. She refused his marriage proposal graciously.

He was forever devising favors he could provide for the people he loved. When Lem visited Poland with Jean and Eunice in 1962, he attended a state dinner at the former Radziwill palace, which the Communists had nationalized. He noticed that the silverware and glasses were engraved with the Radziwill coat of arms and immediately thought to retrieve a family heirloom for Jackie's brother-in-law Stas Radziwill. Stas had fled the Communist takeover with nothing but the clothes on his back. Now Lem, eating with Stas's family silver, could think only of reclaiming some sentimental heirloom for the former prince.

Since stealth and dishonesty were practically beyond Lem's capacity—intrigue gave him the sweats—his painful anxiety as he purloined his silverware setting became fodder for many future recountings of the story. Sweating profusely and flanked on either side and across the table by Communist officials, he was terrified of precipitating an international incident, or landing in a Communist prison. Moments after he completed the theft, dinner was served, and Lem had to sneak the forks and knives back out of his pockets in order to eat. After the meal, Lem surreptitiously repeated the crime and soon discreetly displayed his loot to Jean and Eunice. Their competitive spirits stoked, Jean rushed back to the table to fill her purse with pieces of the rose porcelain dishes and chinaware, while Eunice instructed her puzzled KGB escort to wait outside while she ducked into a bathroom to stuff a Royal Doulton crystal vase, embossed with the Radziwill crest, into her hatbox.

Back in Palm Beach, the three smugglers hosted a luncheon for Stas, arranging the pilfered flatware and china around his place at the head of the table. When Stas noticed the coat of arms he angrily exploded, "They're selling my mother's silver!" When he learned the truth, he wept. Uncle Jack, however, who was president at the time, was furious. He scolded Lem and his sisters for risking embarrassment to our country, but Lem was unrepentant. He had done something that brought happiness to a friend.

With no pretensions of his own, Lem was quick to spot them in others, and to dispatch them with humor. When he caught my Aunt Jean posing for a newspaper photographer at a public event, he shouted to her, "They'll never use it, Jeannie; did you ever hear of the [Harry] Truman sisters?" His laughter was a great leveler. When we first entered the White House soon after President Kennedy's inaugural, Lem's smile outshone the klieg lights. Standing between Pat Lawford and Eunice Shriver, he stole a line from Prissy (Butterfly McQueen) in his favorite movie, *Gone with the Wind*. Throwing his arms open wide, Lem shouted, "Lordy, we sure is rich now!" Lem set the joyful tone that would prevail in the place where he would spend his weekends for the next three years.

While generous in spirit, Lem was notoriously tight with a buck. It killed him to tip more than 15 percent. He was also the guru of the re-gift. When Jack Paar gave him an expensive five-dollar Havana cigar, he took one puff, then clipped off the burnt end with a scissors and presented it as a gift to his doorman. He gave Chris Lawford a Chagall lithograph that he'd received as a birthday gift from philanthropist Mary Lasker the previous year, and was busted when Mary Lasker saw the print hanging on the wall during a dinner visit at the Lawford home.

While I was out of the country on a wilderness trip, a ten-foot boa constrictor that I had stowed in his house escaped into Lem's bedroom wardrobe. When Lem rang his longtime live-in dog-walker, Ms. Gassner, to ask her to keep his small basenji out of the room lest it be eaten, she panicked and called the ASPCA, swearing to Lem in her curt German accent that he must choose between her and "zee snake." Lem chose Ms. Gassner, but canceled the ASPCA and called the Bronx Zoo to pick up the boa instead, meaning to declare the snake as a tax-deductible gift.

Everything Lem Billings did became an adventure. I was always anticipating the unusual when in his company, whether we were touring a Manhattan museum; bidding in a heated auction for antique duck decoys in Cape Cod—or for live pigeons in Pennsylvania; hacking our way through a Colombian jungle; riding horses across the llanos of Latin America; chasing poachers in the Northern Frontier District of Kenya; floating down an unexplored Peruvian river; coon hunting in Alabama; or bodysurfing in California, Haiti, or Malindi. But excitement pursued Lem in even the most mundane circumstances.

In a simple act of kindness, Lem once spent our last two dollars to buy a sandwich from a street vendor for a gaunt derelict outside the Plaza Hotel. He offered it to the man, who was sorting through a garbage can in search of food. But instead of being grateful, the hobo was enraged by Lem's minor philanthropy, and pursued us, roaring obscenities and making threatening gestures. Since Lem had spent our cab fare on this mitzvah, we had to flee on foot. While the derelict appeared down and out, he was surprisingly spry and, unlike Lem, seemed to suffer from neither asthma nor emphysema. No matter how fast we walked, he kept pace, cursing Lem to bloody hell. When we finally arrived at Eighty-Eighth Street, Lem insisted we turn west (instead of east) to avoid leading the man to his door. It was a Saturday, and the museum crowd was gathered in front of the Guggenheim, so we had a large audience as we clambered on top of a Fifth Avenue bench and I boosted Lem over a high stone retaining wall, finally making our escape into the bushes of Central Park.

Lem gave me a deep appreciation for the nexus between American folk art and our democracy. He loved the plain beauty of a Vermont wooden duck decoy or an Amish pie chest. He was eloquent in describing how the American inclinations for innovation and ingenuity found expression in works fashioned by unschooled artists using primitive tools and local materials. I spent many hours with him, scraping paint from antique chairs; attending auctions for decoys, weathervanes, and scrimshaw; and searching through deteriorating Amish barns and houses to evaluate their restoration potential and get a feeling for the history and lives of the simple rural families who helped build our democracy. He taught me the language of design and gave me an eye for texture, color, and craftsmanship. From him I learned the difference between Colonial, Georgian, Victorian, and Tudor architecture, and how to read a building's age by the width of its clapboard. He taught me to look for elegant lines and graceful forms in structures and furniture.

Lem was never boring because he was never bored. He was riveted by details, large and small, and by every comical or ironic human idiosyncrasy. He burned with curiosity, and was never silent when he had a question. In turn, he had a gift for making all his endeavors sound interesting. His life seemed to me a series of wonderful stories filled with light, laughter, and grandeur. Lem had all the old-fashioned virtues—honor, duty, courage, loyalty, honesty, and generosity. At his funeral, Eunice Shriver spoke to our generation about the treasure of Lem's friendship: "No matter how long you live, if you can give to one person the gift of friendship Lem gave to us, your life will be worthwhile."

Eunice said that Lem was a living example of the teachings of Jesus Christ. Christ's message, she said, "was really quite simple: love one another even unto death, and form friendships that are eternal—full of laughter, commitment, adventure, challenge, unselfishness, wisdom, sacrifice, and vitality. That is really what Lem's life epitomized. His devotion never failed; his unselfish interest in others never slackened. At fifteen years of age, at fifty-five, at sixty, he was always giving himself, his time, his thought, his attention to others, to their ambitions and their desires and their troubles and their fears. He bore us all on his shoulders. He comforted us all in our ups and downs. He was never away when needed. He was always present when asked.

"Lem did not talk about Christianity," Eunice said, "arguing about good and evil, moralizing about people and events. He was just practicing Christianity: sacrificing himself to help and serve others. And he enjoyed himself and made everyone happy."

I feel so lucky to have known Lem, perhaps the most important influence in my life. I was fortunate to have had the experience, both as a child and as a young man, of feeling loved unconditionally. Lem taught me that love is not just an emotion; it is effort, sometimes painful, and always forgiving. That gift has helped me love in my own relationships, and has made me a better spouse, father, and friend. I think of Lem every day and ask his help in making myself a vessel for the love that he gave me, so that it might flow to my children, and to all the children whose lives I touch.

I have always felt that God must love the Kennedys in some special way, the proof being that He gave us such a friend.

• • • • •

CHAPTER 12

My Mother

And though she be but little, she is fierce.
—WILLIAM SHAKESPEARE, *A MIDSUMMER NIGHT'S DREAM*

IT WOULD BE MANY YEARS BEFORE I CAME TO APPRECIATE THE GIFTS my mother gave me, or the energies she devoted to keeping our family of eleven children together and on course after my father's death. From my angle, her love didn't always feel unconditional. Her approach was what today people would call "tough love," for which I proved a tough audience. Her exceptional qualities were mainly invisible to me as a child. And just as my own children delight in exhaustively inventorying my every perceived departure from my expressed values, I was hypervigilant to evidence of maternal hypocrisy. I honed a sharp eye for contrasting her occasionally demanding treatment of her staff, vendors, and contractors, with her advocacy for justice and compassion on a global scale.

My mother, Ethel Skakel Kennedy, divided the world into friend and foe. Generally she judged the latter by harsher standards, and yet she sometimes discarded time-honored friendships for minor infractions. I faulted her for being mercurial and arbitrary. I rebelled against the rigidity of her rules and what I saw as their biased and inconsistent enforcement. I noted her umbrage at every injustice but her own—and no wonder, I reasoned, since she regarded self-examination as sissified narcissism. And on those rare occasions when my mother noticed the proverbial log in her own eye, she was indisposed to apologize or acknowledge her mistakes. I censured her also for the way she spent money lavishly and impulsively at one moment while economizing on trifles the next.

I seem to have been at odds with my mother since birth. Her flurries of temper appeared to me haphazard and desultory, and, of all of us siblings, most often directed toward me. My rebellious nature, and my inclination for pointing out her caprices may have sharpened her disfavor. My involvement with drugs after my father's death certainly inflamed it. But her perpetual annoyance at me seemed less a rational response to my mischief than the outcome of some volatile chemical reaction; my mere presence seemed to agitate her. My younger brother Michael provoked the opposite reflex: his proximity triggered in her a calming, soporific effect. When my appearance pitched my mom into a fit, my siblings knew to summon Michael to pacify her.

Anyhow, I was pretty handy in supplying my mother with provocation. From the moment I climbed from my crib, chaos followed me like a loyal hound: I was perpetually late and disorganized; my nails were filthy from digging in the dirt; my pants were grass-stained, wet, and muddy from wading in streams, and torn from barbed wire, brambles, and climbing trees; my shirt would not stay tucked, and my hair was in ceaseless rebellion.

To make matters worse, my school record was pitiful. I entered first grade at Our Lady of Victory at age five with no kindergarten, which put me way behind the other thirty-five kids in my classroom. My children have diagnosed me with ADHD, and I certainly had a hard time sitting still and concentrating on work during those agonizingly endless school days. I broke thermometers and rolled balls of mercury down my desk. I doodled hawks and iguanas, and daydreamed about lizards and Hungarian homing pigeons. I fidgeted, studied the clock, and tried to catch a glimpse through my classroom window of the vultures roosting near MacArthur Reservoir. I couldn't wait to get outdoors: my mind was in the woods and in the creeks, wondering if I'd caught anything in my traps and snares. Only occasionally did I glance at the incomprehensibly droning nun at the head of the class. I couldn't sit still: my little desk quaked from my knees flapping and banging like a mechanized bellows. After school I ducked my homework, employing whatever perjuries the ruse required, and raced outdoors to play with David and Michael, or to hike up Pimmit Run to chase snakes and salamanders.

Dismayed by the Catholic Church's growing conservatism after the death of Pope John XXIII, my father took us out of Our Lady of Victory in 1964, when I was in the fifth grade and my brother Joe in the sixth. We transferred to Sidwell Friends, where my fellow students were already adept at Algebra II, a subject as unfathomable to me as ancient hieroglyphs and seemingly less useful. Math was not the only subject in which I was *non compos mentis*. I almost flunked biology, at which I ought to have excelled. The curriculum, with its emphasis on daily quizzes on ribosomes and chlorophyll and whatnot, seemed designed to suck the fun out of science.

My problems went beyond poor academic performance. I triggered violent anxieties in other adults besides my mother. I provoked such frustrations in my seventh-grade tutor, for example, that he bloodied my nose. Father Richard McSorley was a pacifist priest and antiwar leader whose legendary composure survived the Bataan death march and brutal internment by the Japanese. In the 1960s he helped organize the early anti–Vietnam War movement. His disciplined pacifism helped inspire a young Bill Clinton, whom he met on a train in Europe, to join the antiwar movement. Somehow, I managed to disrupt McSorley's Zen-like commitment to nonviolence soon after my mother hired him to help me catch up with my classes. I drove the unflappable Jesuit to such frustration that he slammed me with a roundhouse. I made little effort to stanch the gory cascade, partly because I was vaguely enjoying Father McSorley's discomfort as he made hash of explaining the bloodbath to my mother; while she didn't mind throttling us herself, she seemed less certain about delegating the license. Mainly I hoped to prolong the homework break.

My mother and I were so at odds that I asked to be sent to boarding school at age thirteen, when my father—who had the same pacifying effect on my mother as did Michael—was leaving for his presidential campaign. I did not want to live at home unshielded by him from what I considered my mother's petty tyrannies. As it turned out, I bridled equally at the monastic tyrannies of the series of boys' schools to which she sent me—and from which I was routinely expelled. I had a photographic memory that allowed me to memorize poetry by the ream (and later to deliver speeches without notes), but it didn't seem to help at school, because academics were incomprehensible to me from the kickoff.

After two years of testing the patience of the Quakers at Sidwell, I ended up back with the Catholics, this time at Georgetown Prep, under oversight of the Jesuits who, during the Inquisition, had practically invented discipline. I learned a little Latin and a lot about how to take a punch—it turned out Father McSorley was not the only Jesuit who could throw one. Responding to a spitball that hit his blackboard when his back was turned to our class, Father Dugan, a former Golden Gloves champ, walked the rows of desks, punching every boy in my class. "I don't know who did it," he said on returning to the blackboard, "but I know I got him!"

The trouncings did not improve my performance. My ninth-grade Georgetown Prep report card said I was the most "disorganized student in the history of the school." In 1979, as a gag to celebrate my mom's fiftieth birthday, my sister Kerry and her best friend (and my future wife), Mary Richardson, replaced my father's framed presidential letter collection with disciplinary letters from my principals and teachers. These generally closed with a variation on one refrain: "Bobby doesn't seem happy here, and we'd be happy if he were somewhere else." I didn't get straight with academics until someone mercifully held me back to repeat tenth grade. That year I began a decade of self-medicating (more about that in a moment), and the drugs, it turned out, helped me sit still and focus. I leapt to the front of my class, where I remained thereafter.

My academic volte-face did not solve my problems with my mother, and I seldom lasted longer than a few days at home when I returned from distant schools for vacation. My homecomings were like the arrival of a squall. With me around to provoke her, my mother didn't stay angry very long—she went straight to rage. Her moods were like milk on a hot stove: one moment everything seemed fine and a second later the stove had disappeared. To her credit, when the tornado cleared, she went just as rapidly back to kindness, as if nothing had happened. By then, however, I was usually long gone.

Following my father's death in June 1968, I spent the summer with Lem in Africa, and that fall I entered ninth grade at Millbrook School in New York's Hudson Valley. I recall Millbrook as a happy, and occasionally idyllic, place. I chose the school because it operated a certified zoo and several of the boys practiced falconry. We spent every free moment climbing trees, trapping and training wild hawks, fashioning tack out of leather, and hunting with our hawks in the forests and farm fields surrounding the school. It wasn't all fun; I was regularly in fistfights at Millbrook. The school was something of a right-wing redoubt, alma mater of the Buckley boys, conservative luminaries William F. and his older brother, James—who in my sophomore year was elected to the U.S. Senate from New York. James would become the plaintiff in *Buckley v. Valeo*, the 1976 landmark Supreme Court case that reshaped modern campaign finance law and opened the door to the corporate purchase of U.S. democracy.

On Cape Cod the following summer, on July 4, 1969, I was headed home from a goodbye bash for my friend Philip Kirby's older brother, Charley, who was on his way to Vietnam as a recent draftee. Charley Kirby had no love for South Vietnam's military dictator, Nguyen Van Thieu, and was not eager to risk his life in Nixon's scheme to keep Thieu's corrupt regime in power. A local rock band was celebrating his reticence with loud antidraft songs, and when a half dozen police cruisers arrived to quiet the music, a mini antiwar protest erupted and Charley ended up in jail with broken nose and knuckles. Charley's friend Jeff O'Neil picked me up as I was hitchhiking home from the melee, and offered me a tab of Orange Sunshine.

Up until that point in my life, in conformance with King Frederick II's proscription against inebriation among falconers, I had resolved never to use drugs or alcohol. In fact, I hadn't even tasted coffee. However, I had recently gathered from my favorite comic book, *Turok, Son of Stone*, that hallucinogens might allow me to see dinosaurs, which I greatly desired. Jeff O'Neil assured me that this was a near certainty, so I swallowed the LSD, which more than delivered on his promise. Buildings melted like wax candles; trees bowed and swayed on a windless night; bright lights with long comet tails lent Hyannisport the cheery aura of Christmas in July.

Acid had come to the Cape that night, and many Irish kids were taking it for the first time in protest of Charley Kirby's induction. A group of them were painting a giant North Vietnamese flag across the post office intersection. I was too much of a patriot to join that enterprise, but I shared their appetite for rebellion. Every day the news offered ample evidence that the older generation was wrong about all the big issues—civil rights, women's rights, Vietnam, and poverty. Nixon and his cronies in the military-industrial complex were busy transforming America into an imperialist warfare state, and I was happy to reason that adults were also mistaken about drugs.

Still tripping, I rode into Hyannis with two older kids and struggled in a Main Street diner with a plate of lively white noodles that squirmed and squeaked as I stabbed at them with my fork. I became suddenly appreciative of the impossibly complex choreography of minute movements required by my mouth and its various parts in order to chew and swallow food. Abandoning that endeavor, I looked up to see a picture hanging behind the counter of my father, Uncle Jack, and Jesus. All of them had their hands folded in prayer. Until that moment everything had been a delight; my soul was happy with this strange adventure, and I was laughing along with my friends. Now things turned sour. I greatly admired all three of those fellows, and I doubted that any of them would have approved of hallucinogens.

A pall came over me. What was I doing? My father had been practically a teetotaler, a straight arrow. His personal life was beyond reproach. He had sacrificed his life to a higher purpose, and here I was, high on drugs. I left my friends and walked the three miles back to Hyannisport, swearing that I would never do drugs again. By then it was morning and I was in a funk, wondering how I was going to explain my all-night absence and cope with my exhaustion. A few blocks from my home, I ran into a group of boys who prescribed a line of methedrine, and that snort miraculously solved all my problems for the day.

Two days later I flew to South America with renewed resolve to trudge the straight and narrow. I worked during July and August as a ranch hand in the Colombian llanos, returning that fall to Millbrook, where my resolution wavered before a tidal wave of rebellion and intoxicants. My generation was developing its own counterculture. That summer's Woodstock concert—just across the Hudson from Millbrook—was our constitutional convention; rock and roll was our generational anthem. I marched in Earth Day protests in New York, read underground newspapers and Mr. Natural comics, and listened to FM radio. I thought of drugs as the fuel of the insurrection. With this attitude, it's a wonder it took me as long as nine months to completely wear out my welcome with the rock-ribbed conservatives at Millbrook.

In early July 1970, Bobby Shriver and I were busted for marijuana possession in Massachusetts. It was then the height of Nixon's War on Drugs. Andy Moes, an undercover federal DEA agent posing as a cabdriver, had worked all summer to befriend me and my cousins, providing us with free rides at a time when none of us had cars. On the day in question, Moes offered Bobby and me a ride to Cohasset to retrieve a trained merlin falcon I had lost a week before. My arm was broken and in a cast from an earlier attempt to recapture the bird from the surprisingly brittle branches of a tall elm. Moes smoked a joint with us in the car. Our arrests made World War III–sized headlines. To me it seemed like a slow-motion nightmare, since I would have rather cut my arm off than tarnish my family—particularly at a time when everyone was reeling in the wake of Chappaquiddick. The Barnstable County Court agreed to dismiss the charges if Bobby and I stayed out of trouble for one year; but my mother's indictment was pretty awful: "You dragged the family name through mud."

The dragging wasn't done. The next summer I provided the sequel to that horror show. I was arrested yet again, this time during a roundup of teens loitering in downtown Hyannis—a waterfront tourist strip—and accused of spitting at a police officer. It's not the most interesting story, but where else am I going to get the chance to set the record straight? In fact, I never committed the crime, and the accusation seemed even more godawful than my drug bust the previous summer, given that my father had been the country's top law enforcer, and the Kennedys, for a variety of reasons including our Irish heritage, had strong bonds with the police. Here's what happened: I came out of a Main Street ice cream parlor carrying two ice cream cones, to find a beefy summer cop named Fredrick Ahern swearing at my friend, Kim Kelly, for sitting on the hood of a station wagon. When I explained that it was my car and she had my permission to sit on it, he ordered me to drop the cones, cuffed me, and put me under arrest for "sauntering and loitering."

Bobby Shriver bailed me out after a night in jail, and I went directly to court. It would only stir things up, I calculated, if I called home and got lawyers involved. I didn't mind the night in jail, but I was terrified of the press, and had spent the time hatching a plan that might not disgrace my family again. I was going to keep the whole affair below the radar by quietly paying my fine and ducking out before anyone noticed. The clerk at the Barnstable courthouse was an avuncular Irishman, the kind of fellow whose judgment I figured I could trust. Like most people in Massachusetts government, he was a rabid Kennedy fan.

I knew that loitering meant standing around, and he explained that sauntering meant walking slowly. These did not seem to me the sort of ghastly crimes that would earn me public humiliation; nevertheless, I was afraid of publicity if the case went to trial. The well-meaning clerk recommended I plead "nolo contendere," which he said meant, "I don't admit fault, but I'd rather pay a fine than go to trial." That sounded like just what I was looking for, so I took the plea in open court, whereupon something unexpected happened. The judge turned to the arresting officer and asked him to read the charges. That's when I heard for the first time Ahern's craven stretcher that I'd spat ice cream at him. I longed to give him the rough side of my tongue, but I was now more desperate than ever that the press not get word of this. I settled for telling the judge, "That's a lie!" paid my fine, and made a beeline for the door. On exiting the courthouse, however, I found a feeding frenzy of reporters shouting questions at me.

That second arrest sealed my fate at home. As soon as I could pack a knapsack I lit out for the territories, and spent the remainder of the summer of 1971 with my friends Johnny Kelly and Conrad Lauer, hopping freight trains across the American West. In late August I hitchhiked from the Kansas City railyard to Martha's Vineyard, where I called my mother to ask for money to come home. She advised me to get a job. For the next twelve years, each time I went home to see my little brothers and sisters, I feuded with my mother. I went to two more schools, Pomfort, a Connecticut boarding school, and Palfrey, a Boston day school that allowed me to spend most of my senior year living with Maryknoll priests in the Peruvian altiplano. In the summers I began experimenting with drugs in earnest.

At first, the drugs were all fun, but addiction is like being locked in a cage with a dancing gorilla. As the years passed, the fun wore off, and I learned that the gorilla liked to dance whether I wanted to or not. I was a functional addict—which is not nearly as good as it sounds—and I never had blackouts, meaning I remember every dumb thing I ever did. Despite the exhilaration and merriment, I was conscience-stricken from the start. I wanted to stop. Generally, I followed the script from my first night on LSD—drugs followed by remorse and earnest vows to reform, followed, incomprehensibly, by more drugs. I found myself drugging without my own permission. I continuously tried to quit, often stopping for weeks or months at a time. I switched from one drug to another and occasionally to alcohol over the fourteen years of my addiction, but all that hopping around was just rearranging deck chairs on the *Titanic*.

Addiction bewildered me. I was hardheaded, stoic, proud of my ability to bear pain, and iron-willed in other spheres of my life. I gave up candy for Lent at age twelve and didn't consume sweets again until I was in college. The following year I gave up dessert, and didn't eat it again until my freshman year of college, when I was playing rugby and rowing crew, and eating everything I could see to gain muscle mass. But my willpower was bafflingly ineffective against the battling impulse of addiction. The promises I made sincerely would soon stop binding me, and the intervals between promise and drugs got shorter over time, dropping from months to weeks to days, and finally to hours. This inability to keep contracts with myself was the most demoralizing feature of my addiction. Living against conscience in one part of my life spilled over to others, and I had to continuously lower my standards to keep pace with my declining conduct. Addiction is a disease of isolation, and although I was often surrounded by people, over time my world, along with my ambitions, grew smaller and smaller.

Susan Sontag observed that addicts have a unique opportunity for redemption. Unlike other infirmities, addicts have the benefit of a clearly delineated path out of the hell where their addiction has led them. While I tried a variety of strategies to get sober, I avoided the Twelve Step programs because of my concerns about privacy and anonymity. In September 1983, at age twenty-nine, I drew national headlines for being busted for possessing a small amount of heroin in South Dakota. And so, with anonymity no longer an issue, I began to attend Twelve Step meetings. It's difficult to characterize a Twelve Step program. It certainly wasn't a religion or a cult, as I initially suspected; in fact, it was the opposite of orthodoxy. There was no leader, no priest, no hierarchy, hardly any rules, and no cosmology—everyone has their own conception of God, or none at all. I found that it also wasn't group therapy, as I had expected. Insight doesn't cure addiction any more than it does diabetes; an addict needs to attend meetings and be of service. Those actions somehow act as a functional treatment for the disease. It's difficult to structure a scientific study to prove the efficacy of Twelve Step programs, but the proof is self-evident; I can't think of another place where you can find large cohorts of addicts and alcoholics with ten, twenty, thirty, and forty years of sobriety. There is some miracle that occurs in those rooms and church basements that has to do with service to other alcoholics. Ironically, service to others, it turns out, gives us the power to emancipate ourselves from our destructive compulsions.

Bill Wilson, founder of Alcoholics Anonymous—the world's first Twelve Step program—had this revelation during a business trip to Cleveland. With a spiritual awakening already behind him and three months of sobriety under his belt, Wilson had total confidence in his ability to stay permanently sober. The spiritual experience had miraculously lifted his uncontrollable lifelong impulse to booze. But following the collapse of the critical business deal that brought him to Cleveland, an overwhelming compulsion to drink suddenly overcame him. Wilson somehow intuited that the only way he could stay sober that day was to seek out and help another alcoholic. Using a pay phone and yellow pages, he found Dr. Bob Smith, a hopeless alcoholic, in a local hospital, and both men managed to stay sober that night by sharing their experiences of strength and hope. That germ of a meeting hatched a movement that has kept millions of alcoholics and addicts sober by giving them an organized framework for service to each other.

I believe that I am genetically hardwired to drink and drug myself to death. All my earthly efforts to dodge that destiny are bootless. Only a spiritual fire can overcome that relentless biological drive, and, as Bill Wilson learned, one can't live off the laurels of a spiritual awakening. It provides us only a one-day reprieve. The flame must be renewed through daily service. Practically the moment I immersed myself in the Twelve Steps my compulsion to drink and drug miraculously lifted, and the sobriety that had eluded all my best efforts suddenly became effortless. The Twelve Steps brought order out of the chaos of my life, and turned all the mayhem of my addiction into capital that I could use to help other alcoholics and addicts. All of that wasted time, the squandered potential, the humiliation and failure that accompanied my addiction and hounded me with regret and pitiful shame, suddenly became assets that, through honest sharing, I could use to help fellow sufferers. In light of my previous exertions to stay sober, my recovery seemed to me as much a miracle as if I could suddenly walk on water. The experience of effortlessly overcoming an irresistible impulse—against which I had earnestly and energetically struggled for over a decade—fortified my spiritual awakening with iron-clad faith.

After all my unceasing efforts to escape the slavery of addiction, it was astounding to realize that in order to maintain my freedom I had only to practice this simple program. It irked me that the daily meetings were time consuming, but I had to measure that one daily hour against the demands of taking drugs—an enterprise that required a daily commitment of many hours and endless energy and ingenuity, accompanied by personal, professional, and mortal hazards. I also learned that in order to stay sober I needed to find daily ways to be of service to others and constantly endeavor to lead a moral, rigorously honest, and upright life. Shame and secrets fuel addiction, and we only win back self-esteem by doing estimable things. This new sober life did not require me to be perfect, only that I progress along spiritual lines.

My successful struggle to recover prompted an intensive process of self-examination that caused me to acknowledge my preeminent contribution to the difficulties of my life. "If there is a nagging problem in your life," a friend told me, "it's probably you." By then

I could hear that kind of criticism without getting defensive. I was learning to listen. Hopelessness and desperation had made me teachable, and I became determined to clear up the wreckage of battered relationships that lay in the wake of my addiction.

One summer day, about a year into my sobriety, I asked to accompany my mother on her morning swim. We stroked past the sandbar far out into Nantucket Sound, and, bobbing there in the Atlantic, I made my amends. I apologized to her for not being a better son. I told her I was sorry for my part in our many conflicts: for the anguish, anxiety, and embarrassment I had caused her; for falling short of my father's ideals, of hers, and of my own; for failing to be the person whom both she and I wanted me to be. I wasn't looking for a reciprocal apology from her, and she didn't offer one. She accepted my amends graciously, and over the following years our relationship slowly healed. I was able to speak intimately and laugh with her about once-painful episodes of our past. She confided in me and began increasingly to trust my judgment. I was happy when she asked for and relied upon my advice in personal, political, legal, and financial arenas. She frequently attended my speeches on environmental and energy issues and became my biggest booster.

I was also able to recognize that my mother's passing storms of nettlesome temper were mainly the fruit of her own personal miseries, and I began to see the extraordinary qualities in her character. Her ferocious outbursts were always followed by good humor and kindness. Though I hadn't always noticed it, my mother had never faltered as my biggest champion. She always put her children first, while understanding that "I love you" and "No" could be part of the same sentence. Jinx Hack, her secretary from 1965 to 1967, who recalls the stint as "the best job ever," says that my mom only got angry at her once—when Jinx, overwhelmed by a flood of visiting celebrities, forgot to pick me up at school. My mother, returning from carpooling my sisters, scolded her, "I don't care if it is the Pope or the president, my children come first." I knew that in the event of a serious calamity my mother would always be there for me.

There was nobody kinder when you were in a jam—particularly if you were injured. My mother was more concerned about our wounds and broken bones than we were. When someone tumbled from a roof, horse, cliff, or tree, or smashed up a car, she would never ask, "Who was at fault?" but, "Are you okay?" I can't count the times she made the familiar drive to the Georgetown Hospital emergency room and waited patiently—often sleeping at the hospital when necessary—until we were discharged.

Over the course of her life, my mother's genuine concern for anyone who is sick or injured has become legendary. Given her partisanship, this concern has been remarkably ecumenical: both racist governor George Wallace and civil rights leader Vernon Jordan told me how much it meant to them that my mother was among the first at their bedsides after they were shot. Her close relationship with CIA director John McCone was rooted in the quiet care and devotion she gave to his dying wife. After Sunday mass in Virginia, she regularly drove us to visit Jack's PT boat crewman Barney Ross, an invalid from a motorcycle accident, or to feed shut-ins at senior homes and soup kitchens. Every summer day she brought a tuna sandwich to Putt, a disabled World War I vet made mute by poison gas. She attended every funeral for every friend who died, and was disciplined about writing thank-you notes and long heartfelt consolation messages to people who suffered loss, many of whom she had never met.

In sobriety, I began to appreciate the extraordinary things my mother gave to our family. As I got older and had six kids of my own (most of whom regard me with a kind of genial disdain, or, as my twelve-year-old son, Aidan, explained, as a "comical artifact"), I became mystified about how she had managed. How did she possibly get eleven mutinous kids to show up for dinner on time? To ask to be excused? To pray together before and after each meal? To assemble nightly for Bible readings and rosaries? How did she keep our dinner table conversation elevated and inspiring? How did she, in all the hurly-burly, get us to play games of history and memorize poems and read all those books to broaden our intellect? Or to take

summer jobs in environments that exposed us to the hardships endured by less fortunate communities? And how did she keep her
strength amid the havoc and tragedy? When people ask how she
coped with the violent and untimely deaths that claimed both her
parents, her three brothers, her husband, and two sons, she says
with a smile, "Everyone takes their licks." As she told me, "We feel
like we ought to be able to write our own scripts to our lives, and
sometimes we feel disappointed in God when life rewrites the plot.
The key is acceptance and gratitude. We need to practice wanting
what we've got, not what we wish we had."

Following our talk in the ocean, I made a practice of engaging
in a mental exercise whenever I became annoyed at my mother's
occasional prickly caprices: I tried to look at my mother's life and
to understand the challenges she had faced, and to see her ornery
aspects as coping mechanisms for dealing with the anarchy and tu-
mult of her own chaotic childhood and the parade of tragedies that
followed. When I did that instead of judging her, I could admire
her extraordinary qualities: her perseverance, courage, and faith in
something larger than herself; her love for competitive sports, athleticism, and humor; her fearlessness, both moral and physical; her
irreverence toward authority, orthodoxy, or convention; her passionate ardor for a great story and her strange combination of piety
accompanied by a skeptical mind; I could admire how she moder-
ated her affinities for chaos and spontaneity with self-discipline and
resolve. As time passed, our worldviews became more aligned. She
had been through her own interior struggles, as well as the pre-
dictable reassessments of childhood religious faith that come from
enduring the exigencies of a long and sometimes difficult life. All
the richness and diversity of human experience no longer fit into
the neat, tight box of Sacred Heart doctrine.

In her years as a zealous young Catholic, my mother's religion
had been ritualistic and superstitious, magical and otherworldly—
more tribalism than ethics. She did not overburden her faith with
reason or self-examination. Like all orthodoxies, hers was rigid
and authoritarian. But after my father's death she reoriented her

religion around his ethical concerns, and particularly around his notions of justice and compassion for society's outcasts, the sick and the poor. She became less accepting of the Church hierarchy's claims to power and infallibility, and more fearless in pursuing the ethical precepts of Christ's Gospels, which she never equated with the noblest vision of America. Her own life became a cavalcade of causes along my father's trajectories: justice for African-Americans, Native Americans, farmworkers, people with intellectual and physical disabilities, victims of torturers, death squads, and despots, and, ultimately, LGBT people everywhere.

My mother believes that God wants us to enjoy His gifts and to fight to make sure everyone else can enjoy them, too; she came to conclude that the Gospels sometimes require her to make herself a nuisance to the powerful. Traveling across the United States and to distant countries, she stuck her nose into any controversy where she might serve as a champion for the disenfranchised. She confronted dictators and walked picket lines. She called congressmen and governors and senators and corporate CEOs and cabinet members and heads of state on behalf of the poor. She became an authentic force for good in her own right.

When the Nixon administration in 1970 tried to forcibly expel several hundred Native American activists who had occupied Alcatraz Island, my mother made her way to the old prison to show her support. She helped the protesters prepare communal meals, held one mother's hand as she delivered a baby, and called the White House to ask President Nixon to relax rules that blockaded food and water.

My mother was with César Chávez when he broke a thirty-day fast in 1970, and again following his twenty-four-day hunger strike in 1972. "I went to visit him in his cell," she said, "and I remember thinking that it was like jailing Saint Francis." In 1983, my mother rallied Rory, Kathleen, Chris, Kerry, and me to join Chávez on a three-day fast, and she was with him again following a thirty-six-day hunger strike in 1988. At the Chávez family's request, she led forty thousand supporters in his funeral march in 1993, walking hand in hand with César's widow, Helen, and UFW cofounder Dolores Huerta.

In 1983, my mother drove my younger siblings, Rory, thirteen, and Douglas, fourteen, to picket the South African embassy in D.C. to protest apartheid, and then watched her youngest children get arrested. "I was proud of the two of them. And I was so proud the Catholic Church was in the forefront of protest."

My mother joined Rory and Douglas on another picket line on a cold rainy night in October 1989 when Jiang Zemin, general secretary of the Chinese Communist Party, visited the United States. They were at the Chinese embassy, protesting the Tiananmen Square massacre and the human rights crackdown that followed. My mother worked closely with the CoMadres in Guatemala throughout the 1980s, when Reagan's CIA was conducting its wars against the poor in Central America. In Managua, Nicaragua, she told a group of mothers of the disappeared, "The American people are with you and on the side of the poor, even though our government is still allied with the oligarchs."

She was tough on left-wing dictators as well. In 1987, she visited Solidarity leaders in Poland with my Uncle Teddy, my sister Kerry, and twenty-six cousins. "The Communist leadership didn't want us to come and didn't want to hear about Lech Walesa or Solidarity, which they refused to recognize. We went anyhow," she recalls. "We stayed a week. We embarrassed the dickens out of the government, every day and every which way we could. It was very effective. It was also fun."

After Poland my mother accompanied Joe, Courtney, and Michael to support democratic movements aginst Communist regimes in Hungary, Albania, and Czechoslovakia (now the Czech Republic). That trip included a humanitarian mission. My brothers had launched a project to greenhouse environmentally and socially responsible companies in the former Eastern Bloc countries. Those were the tension-filled days when Communist regimes across Eastern Europe were imploding. The revolutionary playwright Václav Havel was leading a popular revolt against Czechoslovakia's

Communist regime. According to Joe, "She became the loudest voice, even though it was Michael's and my trip. People flocked to her. Her curiosity and genuine interest in people's lives electrified the crowds. She had that effect on everyone she touched; kings, presidents, housekeepers, and street cleaners."

In 2001, my mother visited me when I was jailed for thirty days in a maximum-security prison in Puerto Rico for committing a civil-disobedience in protest of the naval bombing of the impoverished island of Vieques. She told me how proud she was.

Five years later, I traveled with her to South Africa to see Nelson Mandela and spend eight days celebrating the anniversary of my father's 1966 "Ripples of Hope" speech at the University of Cape Town. "How could anybody spend twenty-seven years in prison," my mother marveled after a long meeting with Mandela, "and then emerge devoid of anger, with that extraordinary commitment to peace and strong principles?" She compared him to Christ. "He took all that personal pain and suffering and used it to heal people," she said. "Isn't that what Jesus did?"

The year before, at some personal risk, she accompanied Kerry and a delegation from RFK Human Rights to appeal to Kenya's mercurial tyrant, Daniel arap Moi, whose henchmen had imprisoned and tortured several associates of the RFK Human Rights program. Moi's partisans had murdered my friend Tom Mboya in 1969, and when Moi became president, in 1978, he began transforming his nation into a full-blown dictatorship. At the time of my mother's trip, Moi was blocking Kenya's leading environmental and human rights advocate, Wangari Maathai, from traveling to the United States to accept the RFK Human Rights Award. Moi had also jailed and tortured Maathai's attorney. When my mother complained about Maathai's travel ban during her private audience with the Kenyan tyrant, Moi, imposing at six foot six, wagged his scepter at my mother in a menacing gesture. "You are meddling with the internal affairs of our country," he roared, "and taking the side of common criminals."

"We both knew Moi was a dangerous man who cared little for

international opinion," Kerry remembers. "I was a bit nervous," she recalls. "But Moi's threatening brandishments made Mummy more brash." "Well, you know, Mr. President," she said matter-of-factly, "your people did torture our human rights award winner's lawyer." Moi replied, "Oh, no, no, no, you don't understand. This guy was a criminal." Ignoring the oscillating scepter, my mother replied, "I hear what you're saying, but actually he is a respected member of the Kenyan bar." She reminded the fuming Moi that Kenya had at one time been the flagship of new democracies, and gently scolded him for tarring his country's hard-earned reputation with harsh treatment of his own people. By the time Kerry and my mother left, Moi was apoplectic, and he lambasted my mother for days in the Kenyan press. But he nevertheless relented. Having rebuffed all previous entreaties, he released Wangari Maathai's lawyer and several other human rights activists. "She mustered the courage to stand up and speak out," recalls Kerry. "It taught me a lesson."

My mother made similar visits to Fidel Castro in Cuba to free political prisoners. Each time she visited the old tyrant, he released detainees into her custody. "I know, I know," she recalls him saying, "you have a list. Give it to me."

My mother dismisses any talk of her activism requiring courage. "I think anyone who had been in my position would have done the same thing. I was lucky because I had help bringing up the children, and I had the wherewithal to go to the countries and see what was going on." She likes to say that "celebrity is currency," and that people who have fame or abundance ought to spend them down trying to make the world a kinder place. "Because of Daddy, I had access to people in power," she explained, "I felt that gave me an obligation to inform and sometimes to confront people when they seemed to be abusing their authority."

In recent years, my mother has made more than a dozen human rights pilgrimages to deliver food and medical supplies, repatriate refugees, afflict the tyrannical, and comfort the afflicted. By standing up to bigotry, corporate misbehavior, and confronting cowardly or venal public officials, whether on the left or the right, she has won freedom for prisoners of conscience all over the world. Farm-workers continued to be a special cause. In 2012, at my mother's re-quest, I marched with her in South Florida with Lucas Benitez, the charismatic leader of the Coaltion of Immokalee Workers, a labor union he helped organize to redress the shockingly commonplace abuses, including the use of slavery in America's tomato fields. We visited homes and labor camps, where we found dozens of workers shoehorned into filthy trailers. In a typical house five men slept in shifts on a single soiled mattress on the floor. There were no show-ers, and turbid water trickled out of washbasin faucets. "It's not just inhumane," said my mother angrily, "it's inhuman." She marveled that "despite the indignity of these conditions, these gentle, humble people have never lost their dignity." Like César Chávez, Benitez and his followers hewed to disciplined restraint and nonviolence in their protests.

At Benitez's request, my mom called up McDonald's CEO Jim Skinner and persuaded him to arrange for McDonald's suppliers to pay their tomato workers an additional penny per pound, an increase that would allow them to live in minimal decency. "It's hard to say no to her," Skinner said later. My mother was picketing Chipotle and Taco Bell and the supermarket chain Target, to pro-test the abuse of farmworkers by their suppliers. A particular target of my mothere's ire was billionaire Ken Langone, who sat on the board of the hedge fund that owns those restaurants. Unlike Mc-Donald's, Taco Bell had refused to part with the penny-per-pound raise that its tomato pickers were requesting. My mother refused to eat at those restaurants for a year. "It really upset me," she says, "because I love Taco Bell." Kerry and I subsequently joined our mother, Benitez, and about a thousand tomato pickers for a two-mile protest march against Target. When the march ended and I was leaving to catch my plane, I watched my eighty-three-year-old mother, in her wheelchair, with forty ragged farmworkers heading for Target's firecracker-hot Immokalee, Florida, parking lot, where she would spend the rest of the day on a picket line.

In 2012, my mother was the first member of our family to

endorse Barack Obama. It was a particularly difficult choice for her, since she was so close to the Clintons. She continued to love them, but she felt that in an era of racial, tribal, and religious tension following President George W. Bush's Mideast wars, Obama's election would send a message to the world that America was back on track and living up to its promise to be an exemplary nation.

Despite her seriousness about human rights and justice issues, my mother's dominant qualities continue to be her humor and her irreverence and sense of fun. And few things are more entertaining than a close look at her daily life. The show begins with her pack of noisome dogs. My mother worships them, and perhaps for this reason subjects them to no real training or discipline. She includes them in every activity, from morning mass to sleeping in her bed at night. As a child, it never occurred to me to question why all our oversized dogs needed to join us every day on the small sailboat for our picnic lunch. Their presence must have helped re-create that potent mix of fun, adventure, and bedlam that shaped her youth.

In the summer of 1992, a decidedly tipsy dinner guest gushed to my mother, over dinner, "I'm so happy to be invited here." Her husband endeavored to bridge the awkward silence that followed by saying, "If you behave, honey, you'll be invited back." Without missing a beat, my mother assured her dryly, "If you misbehave, you'll be invited back sooner." That pretty much summarizes my mom's gestalt. A needlepoint pillow in her home recites her guiding principle: "If You Obey All the Rules, You Miss All the Fun."

In December of that same year, a crazy man had attacked her on the frozen C & O Canal, near Fletcher's Boathouse. While she was ice-skating, he emerged from the woods and chased her up the canal. Her skates gave her the advantage as he hounded her across the ice, but slowed her dramatically once she got to Fletcher's gravel parking lot. As my mother and her pursuer arrived at her car almost simultaneously, she leapt in and threw on the door locks. Furious, he dived onto her car and pounded the windshield, but now the assailant was playing in her sandbox. Always a bold driver, my mother, driving in her skates, executed a series of doughnuts in the gravel parking lot, interspersed with violent accelerations and sudden stops, and finally ejected him forcefully onto the gravel. "That part was really fun!" she confesses. "Running for my life in my skates? Not so much!"

In the autumn of 1984 my mother hosted my eldest son Bobby's christening at Hickory Hill. Ena and Aunt Jackie were his godmothers. Father Jerry Creedon, the family priest (and former Irish football champion), said mass while dogs and children swirled around in typical chaos. Following the baptism, Jackie sat on a cast-iron easy chair on the patio with a large slobbering dog, watching carefully as she ate. Humorist Art Buchwald was there in his tennis shorts. Studying the scene, I heard him remark to his wife, "Wasn't it nice of Ethel to put a dog on Jackie's right?"

Ann replied, "Yes. Girl, dog, girl, dog."

Every year I love and appreciate my mother more. Her occasional mood swings and sharp rebukes no longer trouble me. They are part of the endearing personality of an extraordinary character, whose many colorful qualities, and her place at the center of history, make her one of the funniest, most exciting people I know. During the summer, my mother sits dutifully at a long children's table on the outside porch as the sun sets and the Atlantic shimmers behind her, presiding over dinnertime with dozens of little grandchildren, competing with them at word, history, and math games. Afterward, they alternate playing backgammon with her. She takes no prisoners in any of these competitions. "I think it's more fun that way, because then, when they do win they know that they have earned it. It helps them learn to concentrate, and of course be lucky."

"Also," she adds, "I don't like to lose."

To the extent that my mother gave us her humor, her faith, her athleticism, irreverence, and competitive spirit, I'm grateful. She tried to give us the sense that we mustn't be satisfied with "making a big pile for ourselves and whoever dies with the most stuff wins." Our lives, she taught us, should serve a higher purpose. From my dad she came to understand that religious piety required something more than daily mass. It meant caring for the alienated and disenfranchised, and helping our country live up to its ideals of peace and justice for all. My mother will leave behind an impressive surplus of genetic material in the form of numerous children and grandchildren, but her greatest legacy is the conviction, shared by virtually all her living descendants, that we must go forward as a community, that we can't advance ourselves as a nation by leaving our poorer brothers and sisters behind. So long as any American is suffering from poverty or injustice, the American experience is diminished for all of us. Both our faith and our national heritage require us to fight for the poor, even if we are going to lose, and even if it means breaking the law. Like the martyrs and saints, whose stories she relentlessly read to us as children, we had to fight for a kind, moral world—for justice. Like them, we should be proud to endure ridicule, prison, and the loss of status and power. And in these moral adventures we should never relent. "Kennedys never give up," she chided us. "We have to die with our boots on!"

[ACKNOWLEDGMENTS]

• • • • •

MY WIFE, CHERYL HINES, IS MY MOST VALUED EDITOR. VIRTUALLY ALL her myriad notes are integrated into the final book. For her wisdom, judgment, patience, support, and a million other reasons, I thank her.

Peter Guzzardi helped me cut and craft a 500,000-word manuscript down to 120,000 words.

My assistant Fredrick Spendlove worked long days and late, late nights to keep this project on track, and still found time to orchestrate my life during four intense months of writing.

Lori Morash and Val Chamberlain deciphered my chicken-scratch scrawl, typed my handwritten manuscript and helped me fact-check, spell-check, and edit it with reliable insight, humor, and patience. Their advice was always invaluable.

For photos, I thank Harry Benson, Jacques Lowe, Lawrence Schiller, Bill Eppridge, Ron Galella, Nat Fein, Paul Fusco, and Stanley Tretick. Lawrence Schiller bent over backward to find photos and obtain authorization.

Thanks to Tim Duggan and HarperCollins for encouraging me to write this book and to Jonathan Jao for his patience and faith in this project.

Kris Dahl is my brilliant agent, whose friendship and sage advice I treasure.

Thanks to the many RFK and JFK researchers who helped me. Lisa Pease, Jim Douglass, Jim DiEugenio, David Talbot, Karen Croft, Ted Sorensen, Arthur Schlesinger, Burke Marshall, and Dick Russell, Antony Summers, Malcolm Blunt, Larry Hancock, Peter Dale Scott, and John Newman. Thanks also to the attorneys and investigators who served on the House Select Committee on Assassinations, particularly Dan Hardaway, Professor G. Robert Blakey, and Robert Tarenbaum.

Thanks to Dick Russell for fact-checking my manuscript, and Ginger Miles, who transcribed dozens of interviews for me, quickly and efficiently.

• • • • •

Chapter 1: Grandpa

2 hanged for the offense: "Hedge schools" were established in secret to bypass Ireland's laws forbidding Catholics from setting up schools of their own. Different age groups attended the same master, and most hedge schools had classes in reading, writing, and arithmetic; some taught Greek and Latin. The penal laws ended in 1782.

7 Grandpa . . . made headlines: "I am happier being the father of nine children and making a hole-in-one than I would be as their father making a hole-in-nine." One English paper dubbed him "the father of his country." Peter Collier and David Horowitz, *The Kennedys: An American Drama* (New York: Encounter Books, 1984), pp. 126–27.

9 "Kennedy neither imported nor sold any liquor": David Nasaw, *The Patriarch: The Remarkable Life and Turbulent Times of Joseph P. Kennedy* (New York: Penguin Press, 2012), p. 79.

9 "All sorts of biographers and journalists": Nora Ephron, "Four or Five Things You Don't Know About Me." Opinion article in the *New York Times*, February 24, 2008.

9 "With exceedingly few exceptions": Edward M. Kennedy, *True Compass: A Memoir* (New York: Twelve, 2009), pp. 41–42.

10 "Feel it duty": Nasaw, *The Patriarch*, p. 391.

11 "We cannot dominate": Arthur Schlesinger, Jr., *Robert Kennedy and His Times* (2002 repr.; Boston: Houghton Mifflin, 1978), p. 71.

12 "to the best . . . guise of benign interest": Schlesinger, *Robert Kennedy and His Times*, p. 69.

12 "Wherever the standard": John Quincy Adams's July 4, 1821, speech quoted in *American Conservative*, July 4, 2013.

12 "It is not too difficult": Nasaw, *The Patriarch*, p. 660.

12 Chalmers Johnson quote: Interview with Harry Kreisler, March 7, 2007. http://globetrotter.berkeley.edu/people7/CJohnson/cjohnson07-con1 .html.

13 Joseph Kennedy's letter to Ted Kennedy: Peter S. Canellos, *Last Lion: The Fall and Rise of Teddy Kennedy* (New York: Simon & Schuster, 2009), p. 18.

14 Joseph Kennedy's last speech: "Our Foreign Policy—Its Casualties and Prospects," quoted in Nasaw, *The Patriarch*, p. 659.

18 "Do not comply": Ronen Shamir, *Managing Legal Uncertainty: Elite Lawyers in the New Deal* (Durham, NC: Duke University Press, 1995), p. 67.

18 "Business Plot": For details, see Jules Archer, *The Plot to Seize the White House*, and Butler's *War Is a Racket*.

18 Henry Wallace quotes: Henry A. Wallace, "The Danger of American Fascism," *New York Times*, April 9, 1944.

20 "We are reaping": Schlesinger, *Robert Kennedy and His Times*, p. 23.

21 SEC "the most ably administered": Nasaw, *The Patriarch*, p. 236.

21 *Wall Street Journal* on Kennedy: Ibid.

22 "There has scarcely been": Doris Kearns Goodwin, *The FitzGeralds and the Kennedys: An American Saga* (New York: Simon & Schuster, 1987), p. 495.

22 "Stalin Delano Roosevelt": Robert Dallek, *Franklin D. Roosevelt: A Political Life* (New York: Viking, 2017), p. 225.

22 "abandoned themselves in orgies": Manchester quoted in "Hillary Haters and the Roosevelts," blog by Robert F. Kennedy, Jr., May 25, 2011.

22 "The majority of those who rail": Marquis William Childs, "They Hate Roosevelt," *Harper's Magazine*, May 1936.

23 "Anyone who has moved in the circles": Joseph P. Kennedy, *I'm For Roosevelt* (New York: Reynal & Hitchcock, 1936), p. 7.

24 "My father has believed": Edward M. Kennedy, ed., *The Fruitful Bough, A Tribute to Joseph P. Kennedy* (for private distribution, 1965), p. 214.

27 Thomas Campbell on Joseph Kennedy: Ibid., p. 18.

30 Carroll Rosenbloom on Joseph Kennedy: Ibid., p. 110.

30 "There were wrongs which needed attention": Ibid., pp. 213–14.

Chapter 2: Grandma

54 "If I was president": "Excerpts from the Eulogy by Sen. Kennedy," *New York Times*, January 25, 1995.

Chapter 3: The Skakels

57 "The unity of freedom," John F. Kennedy, State of the Union speech, January 14, 1963.

57 Some of the background on the Skakel family is drawn from Jerry Oppenheimer, *The Other Mrs. Kennedy* (New York: St. Martin's Press, 1994).

81 The story of the *Virginia* is also recounted in ibid., p. 211.

Chapter 4: The White House

91 "Nobody asked my brother": Mary Ann Watson, *The Expanding Vista: American Television in the Kennedy Years* (Durham, NC: Duke University Press, 1994), p. 6.

94 Transcript, "JFK's Speech on His Religion": https://www.npr.org/templates/story/story.php?storyId=16920600.

99 "For man holds in his mortal hands . . . mankind's final war": John F. Kennedy, Inaugural Address, John F. Kennedy Presidential Library and Museum, www.jfklibrary.org.

101 "We have the power": John F. Kennedy, Address before the 18th General Assembly of the United Nations, September 20, 1963.

101 "He kept the peace": Robert F. Kennedy, Jr., "John F. Kennedy's Vision of Peace," *Rolling Stone*, November 20, 2013.

102 Hugh Sidey: Sheldon M. Stern, *Averting the "Final Failure": John F. Kennedy and the Secret Cuban Missile Crisis Meetings* (Redwood City, CA: Stanford University Press, 2003), p. 35.

102 "Our objective must be the destruction": Richard Hofstadter, *The Paranoid Style in American Politics* (New York: Vintage Books, 2008), p. 128.

102 "Men acquainted with the battlefield": Maxwell Taylor Kennedy, *Danger's Hour* (New York: Simon & Schuster, 2008), p. 226.

103 "We arm to parlay": Aleksandr Fursenko and Timothy Naftali, *Khrushchev's Cold War: The Inside Story of an American Adversary* (New York: W. W. Norton, 2006), p. 347.

103 "at the summit than at the brink": Arthur M. Schlesinger, Jr., *A Thousand Days: John F. Kennedy in the White House* (2002, repr.; Boston: Houghton Mifflin, 1965), p. 305.

103 "I had no cause for regret": From Khrushchev's memoirs, quoted in Schlesinger, *Robert Kennedy and His Times*, p. 442.

103 "The Nation behaves well": www.trcp.org/2012/01/14/treat-natural-resources-as-assets/.

107 "You're lucky": https://www.npr.org/2013/ . . . /walking-enthusiasts-to
-retrace-steps-of-1963-kennedy-mar.

107 "Bob would have completed": Edwin Guthman, *That Shining Hour.*

108 Eisenhower farewell address: http://www.presidency.ucsb.edu/ws
/index.php?pid=12086.

109 Patrice Lumumba: Lumumba's assassination is described in a number
of books. The most definitive recent account may be found in David
Talbot, *The Devil's Chessboard* (New York: HarperCollins, 2015).

109 "The only way to keep the Cold War out": Jean H. Baker, *The
Stevensons: A Biography of an American Family* (New York: W. W. Norton,
1996), p. 424.

109 President Truman: In a column that appeared in the *Washington Post* on
December 22, 1963, ten days after my uncle was assassinated, former
president Harry S. Truman wrote: "I have never had any thought that
when I set up the CIA, that it would be injected into peacetime cloak
and dagger operations. . . . For some time, I have been disturbed by the
way CIA has been diverted from its original assignment. It has become
an operational and at times a policy-making arm of government."

110 top-secret National Security Council directive: https://www.cia.gov
/library/readingroom/ . . . /CIA-RDP86B00269R000500040001-1.p.

111 For more on the Iran coup, see Talbot, *The Devil's Chessboard.*

112 Hoover Commission urging Eisenhower to limit CIA: Schlesinger,
Robert Kennedy and His Times, p. 457.

112 JFK as senator on Dulles brothers: Ibid., p. 120.

112 Secret budgets: Series of articles by Dana Priest and William M. Arkin,
Secret budgets: "Top Secret America," http://projects.washingtonpost.com
/top-secret-america/articles/a-hidden-world-growing-beyond-control/.

114 Nixon's secret memo on Castro: Jim Rasenberger, *The Brilliant Disaster:
JFK, Castro, and America's Doomed Invasion of Cuba's Bay of Pigs* (New
York: Scribner, 2011), p. 21.

116 CIA secret 1960 estimate on Castro: Schlesinger, *Robert Kennedy and His
Times*, p. 453.

117 Kennedy to Dave Powers on the Bay of Pigs: David Talbot, *Brothers:
The Hidden History of the Kennedy Years* (New York: Free Press/Simon &
Schuster, 2007), p. 47.

117 "splinter the CIA": Kennedy, "John F. Kennedy's Vision of Peace."

117 "Bobby should be in the CIA": Quoted in Philip A. Goduti, Jr., *Kennedy's
Kitchen Cabinet and the Pursuit of Peace* (Jefferson, NC: McFarland, 2009),
p. 113.

119 Che Guevara to Dick Goodwin: Jack Calhoun, *Gangsterismo: The United
States, Cuba and the Mafia, 1933 to 1966.*

119 Dulles on "that little Kennedy": Talbot, *The Devil's Chessboard*, p. 1.

120 JFK to Cheddi Jagan: Schlesinger, *Robert Kennedy and His Times*, p. 626.

120 "Let us never negotiate": Thurston Clarke, *Ask Not: The Inauguration of John F. Kennedy and the Speech That Changed America* (New York: Penguin Books, 2005).

124 "the people themselves": Ibid., 470.

126 "Robert Kennedy's most conspicuous folly": Schlesinger, *A Thousand Days*, p. 534.

126 Frank Sturgis: Chief Phil Doherty, *3 Killers at Dallas?* (XLibris, 2014), p. 195; see also Richard D. Mahoney, *Sons and Brothers: The Days of Jack and Bobby Kennedy* (New York: Arcade, 1999), p. 175.

127 "Kennedy's Cubans": James DiEugenio, *Destiny Betrayed: JFK, Cuba, and the Garrison Case*, 2nd edition (New York: Skyhorse Publishing, 2012), p. 51.

127 Blakey on assassins: The same men were also named as part of a JFK assassination team by Tony Cuesta, an operative for the CIA and the Alpha 66 Cuban exile group, in a voluntary confession to Cuban authorities shortly before being released from a Cuban prison in 1978. Diaz was a political assassin who had worked as head of security at a Havana casino run by Mafia boss Santo Trafficante prior to Castro's takeover.

129 Frank Church on CIA as "a rogue elephant": *Congressional Record*, U.S. Government Printing Office, 1975.

130 "We had been operating a damned Murder Inc." Leo Janos, "The Last Days of the President: LBJ in Retirement," *Atlantic*, July 1973, p. 39.

131 "he acknowledged to a Vincent Bugliosi": Vincent Bugliosi, *Reclaiming History: The Assassination of President John F. Kennedy* (New York: W. W. Norton & Co., 2007), p. 786.

132 "Halpern should have already been suspect to Hersh": James DiEugenio, review of David Talbot's book, *Brothers*, August 3, 2007, https://kennedysandking.com/robert-f-kennedy-reviews/review-david -talbot-s-brothers.

134 Operation Northwoods: James W. Douglass, *JFK and the Unspeakable: Why He Did It and Why It Matters* (Maryknoll, NY: Orbis Books, 2009), pp. 96–98. For full text, see National Security Archive, https:// nsarchive2.gwu.edu/news/20010430/northwoods.pdf.

Chapter 5: Hickory Hill

134 Andrew Glass: Keith Badman, *Marilyn Monroe: The Final Years* (New York: Thomas Dunne Books/St. Martin's Press, 2010), p. 75.

144 Robert Kennedy on Army–McCarthy Hearings: Philip A. Goduti, Jr., *Robert F. Kennedy and the Shaping of Civil Rights, 1960–1964* (Jefferson, NC: McFarland, 2013), p. 16.

144 Roy Cohn on Robert Kennedy/Henry Jackson: Lester David, *Ethel: The Story of Mrs. Robert F. Kennedy* (New York: World Publishing, 1971), p. 62.

147 Joey Gallo in the office: Schlesinger, *Robert Kennedy and His Times*, p. 164.

148 Kennedy–Giancana exchange: Ibid., p. 165.

148 Seymour Hersh allegations: Seymour Hersh, *The Dark Side of Camelot* (Boston: Little, Brown, 1997); excerpt at www.theatlantic.com.

148 the slander was Sam Halpern's contrivance: Talbot, *Brothers*, pp. 122–23.

158 "the brother-in-law of an admitted horse thief": Oppenheimer, *The Other Mrs. Kennedy*, p. 248.

175 "Lumumba and Sukarno": Evan Thomas, *The Very Best Men: Four Who Dared—The Early Years of the CIA* (New York: Simon & Schuster, 2006), pp. 232–33.

175 Kennedy on CIA and Sukarno: Oliver Stone and Peter Kuznick, citing Roger Hilsman, in *The Untold History of the United States* (New York: Gallery Books/Simon & Schuster, 2012), p. 349.

177 On Sukarno coup: Ralph McGehee, *Deadly Deceits: My 25 Years in the CIA* (New York: Sheridan Square Publications, 1983), pp. 57–58.

177 Robert Martens on Indonesia massacre: Kathy Kadane, "Ex-Agents Say CIA Compiled Death Lists for Indonesians"; at namebase.org/kadane .html. Originally published in the *San Francisco Examiner*, May 20, 1990.

179 Indonesian summary on RFK trip: Schlesinger, *Robert Kennedy and His Times*, p. 568.

181 Embassy's deputy chief: Ibid., p. 566.

181 central revelation from his journey: Ibid., p. 573.

Chapter 6: Attorney General

188 Robert Kennedy on U.S. Communist Party: Evan Thomas, *Robert Kennedy: His Life* (New York: Simon & Schuster, 2009), p. 169.

191 Hoffa to Partin on Robert Kennedy: Walter Sheridan, *The Fall and Rise of Jimmy Hoffa* (New York: Saturday Review Press, 1972), p. 216. See also Talbot, *Brothers*, p. 120.

192 Ragano message from Hoffa: Frank Ragano and Selwyn Rabb, *Mob Lawyer* (New York: Charles Scribner's Sons, 1994), pp. 3, 359.

192 "No, José, he is going to be hit": Anthony Summers, *Conspiracy* (New York: McGraw-Hill, 1980), p. 254.

193 "Don't worry about that little Bobby": Ibid., p. 256.

196 "send the FBI in every time": Anthony Summers, "The Secret Life of J. Edgar Hoover," *Guardian*, December 31, 2011, https://www .theguardian.com/film/2012/jan/01/j-edgar-hoover-secret-fbi.

198 RFK–Barnett exchange: Schlesinger, *Robert Kennedy and His Times*, pp. 318–19.

199 "If things get rough": Mahoney, *Sons and Brothers*.

201 Arthur Hanes: Michael Beschloss, *Presidential Courage: Brave Leaders and How They Changed America, 1789–1989* (New York: Simon & Schuster, 2007), p. 264.

201 George Wallace 1963 inaugural address: www.blackpast.org/1963 -george-wallace-segregation-now-segregation-forever.

202 "The point," my father later said: Schlesinger, *Robert Kennedy and His Times*, p. 338.

202 Wallace on King: Diane McWhorter, *Carry Me Home: Birmingham, Alabama: The Climactic Battle of the Civil Rights Revolution* (New York: Simon & Schuster, 2001), p. 448.

202 "I can't fight bayonets": Schlesinger, *Robert Kennedy and His Times*, p. 342.

203 Historian Andrew Cohen: Andrew Cohen, *Two Days in June: John F. Kennedy and the 48 Hours That Made History* (New York: Random House, 2014).

203 "I shall ask the Congress . . . to make a commitment": JFK Civil Rights Address, June 11, 1963: dubois.fas.harvard.edu/ . . . /JFK%20Civil%20 Rights%20Speech%20June%2011,%2019.

203 "There comes a time": Steven Levingston, *Kennedy and King: The President, the Pastor, and the Battle over Civil Rights* (New York: Hachette Books, 2017).

203 Joseph L. Rauh, Jr.: Schlesinger, *Robert Kennedy and His Times*, p. 349.

204 "We have lost the South": http://billmoyers.com/2014/07/02/when-the -republicans-really-were-the-party-of-lincoln/.

206 "My father always told me": "Those SOBs," *Atlantic*, JFK Commemorative Issue (Fall 2013), https://www.theatlantic.com /magazine/archive/2013/08/those-sobs/309478/.

207 President Kennedy's news conference: https://www.jfklibrary.org.

208 "I understand better every day": Schlesinger, *A Thousand Days*, p. 641.

208 JFK to United Auto Workers, May 8, 1962: https://www.youtube.com /watch?v=tNvqjn7z5h8.

Chapter 7: JFK: In Pursuit of Peace

213 Jack "had obviously studied all the questions": Nikita Khrushchev, *Memoirs of Nikita Khrushchev*, ed. Sergei Khrushchev, vol. 3, *Statesman, 1953–1964* (University Park: Penn State University Press, 2013), p. 305, kiatipis.org/Writers/N/Nikita.Khrushchev/Memoirs-of-Nikita -Khrushchev[Vol3].pdf.

215 U.S. version of the "Final Solution": http://prospect.org/article/did-us-military-plan-nuclear-first-strike-1963.

215 "And we call ourselves": Douglass, *JFK and the Unspeakable*, p. 237.

215 "These people are crazy": https://en.wikipedia.org/wiki/Curtis_LeMay, citing United States Strategic Bombing Survey. Summary Report (Pacific War). Washington, DC, July 1, 1946.

216 "There is little value": https://www.youtube.com/watch?v=8Jb8KqaerhQ.

217 "They call for a 'man on horseback'": Talbot, *Brothers*, p. 77.

217 *Seven Days in May*: Ibid., pp. 148–50.

218 The account of General Clay and the Berlin Wall crisis is drawn from James W. Douglass's *JFK and the Unspeakable*, pp. 109–13.

219 "particularly stupid risk": John F. Kennedy, *The Letters of John F. Kennedy*, ed. Martin W. Sandler (New York: Bloomsbury Press, 2013), p. 230.

221 JFK speech to the UN: www.presidency.ucsb.edu/ws/?pid=8352.

222 Khrushchev's first letter to JFK: *The Kennedy–Khrushchev Letters*, ed. Thomas Fensch (The Woodlands, TX: New Century Books, 2013).

222 JFK's letter to Khrushchev: Goduti, *Kennedy's Kitchen Cabinet and the Pursuit of Peace*, pp. 97–98.

223 Khruschev's second letter to JFK: in Fensch, *The Kennedy–Khrushchev Letters*.

224 Kennedy on Eisenhower and Laos: Kenneth P. O'Donnell and David F. Powers, *"Johnny, We Hardly Knew Ye"* (Boston: Little, Brown, 1970), p. 244.

224 Kennedy trip to Vietnam in 1951: Robert Dallek, *An Unfinished Life: John F. Kennedy, 1917–1963* (Boston: Little, Brown), 2003.

224 "The chickens are coming home": Schlesinger, *Robert Kennedy and His Times*, p. 759.

224 MacArthur warning: Schlesinger, *A Thousand Days*, p. 339.

224 "made a hell of an impression": Schlesinger, *Robert Kennedy and His Times*, p. 760.

225 Kennedy on Laos and Cuba: Schlesinger, *A Thousand Days*, p. 339.

225 "essentially the creation of the United States": *Pentagon Papers*, vol. 2 (Boston: Beacon Press), p. 22.

226 Eisenhower gave his "domino theory" speech on April 7, 1954.

227 "I don't recall anyone": recorded interview by L. J. Hackman, November 13, 1959, p. 47; cited in Schlesinger, *Robert Kennedy and His Times*, p. 761.

227 Taylor on need in Vietnam: *Pentagon Papers*, vol. 2, p. 108.

227 As CIA historian: John M. Newman, *JFK and Vietnam: Deception, Intrigue, and the Struggle for Power* (New York: Grand Central Publishing, 1992), p. 401.

227 "There are limits": Willliam E. Leuchtenburg, *The American President: From Teddy Roosevelt to Bill Clinton* (New York: Oxford University Press, 2015), p. 411.

227 JFK on Vietnam to Schlesinger: Schlesinger, *A Thousand Days*, p. 547.

228 JFK to O'Donnell: A. J. Langguth, *Our Vietnam: The War 1954–1975* (New York: Simon & Schuster, 2000), p. 208.

228 "malignancy": Richard Starnes, "'Arrogant' CIA Disobeys Orders in Viet Nam," *Washington Daily News*, October 2, 1963, p. 3, cited in Douglass, *JFK and the Unspeakable*.

229 McNamara on JFK's intentions: Schlesinger, *Robert Kennedy and His Times*, p. 710.

229 JFK on Vietnam to Walter Cronkite: David Kaiser, *American Tragedy: Kennedy, Johnson, and the Origins of the Vietnam War* (Cambridge, MA: Belknap Press, 2000), p. 246.

229 NSAM 263: cited in Douglass, *JFK and the Unspeakable*, p. 188.

229 Kennedy on "how we can bring Americans out": George Johnson, *The Kennedy Presidential Press Conferences* (Earl M. Coleman Enterprises, 1978), p. 585.

229 Malcolm Kilduff: Douglass, *JFK and the Unspeakable*, p. 304 (based on personal interview).

230 LBJ to Lodge: Tom Wicker, *JFK and LBJ: The Influence of Personality Upon Politics* (Chicago: Ivan R. Dee, 1968), p. 185.

230 Kennedy on Indochina, 1954: Schlesinger, *A Thousand Days*, p. 322.

232 Khrushchev on missiles: E. J. Carter, *The Cuban Missile Crisis* (Chicago: Heinemann Library, 2004).

233 "Now I know how Tojo felt": Robert Kennedy, *Thirteen Days* (New York: Signet, 1969), p. 31.

234 Kennedy on Russian retaliation: Ernest R. May and Philip D. Zelikow, eds., *The Kennedy Tapes: Inside the White House During the Cuban Missile Crisis* (New York: W. W. Norton, 2002).

234 Curtis LeMay: Sheldon M. Stern, *Averting "The Final Failure": John F. Kennedy and the Secret Cuban Missile Crisis Meetings* (Redwood City, CA: Stanford University Press, 2003), p. 85.

235 JFK to O'Donnell: Thomas, *Robert Kennedy: His Life*, p. 217.

236 RFK diary entry: Ibid., p. 225.

237 "The thought that disturbed him": Kennedy, *Thirteen Days*, p. 106.

238 Dobrynin's follow-up memo: Dobrynin's cable, cited in Douglass, *JFK and the Unspeakable* (p. 27), was upon its declassification by the Soviet Foreign Ministry reprinted in translation in Richard Ned Lebow and Janice Gross Stein, *We All Lost the Cold War* (Princeton, NJ: Princeton University Press, 1994), pp. 523–26.

239 William Harvey insubordination and transfer: Talbot, *The Devil's Chessboard*, pp. 472–73.

239 LeMay comment: Schlesinger, *Robert Kennedy and His Times*, p. 524.

239 "This is the night I should go to the theater": Kennedy, *Thirteen Days*, p. 110.

240 LeMay quote: Talbot, *Brothers*, p. 172.

240 Lemay on "greatest defeat": Ibid., p. 172.

240 Ellsberg on coup atmosphere: Ibid.

240 "You would have been impeached": May and Zelikow, *The Kennedy Tapes*, p. 219.

240 "JFK had a great capacity": Talbot, *Brothers*, p. 171.

241 Khrushchev on hard-liners: Schlesinger, *Robert Kennedy and His Times*, p. 529.

241 Khrushchev on Kennedy: Ibid., p. 531, citing Khrushchev's memoirs, *Khrushchev Remembers*, and *Last Testament*.

242 Kennedy secret letter to Khrushchev: *Foreign Relations of the United States, 1961–1963 Volume XI, Cuban Missile Crisis and Aftermath* (Washington, DC: United States Printing Office, 1997), p. 760.

243 David Ormsby-Gore: Schlesinger, *Robert Kennedy and His Times*, p. 530.

243 "What, if any": cited as part of a quote from Ted Sorensen in C. A. J. Coady and Igor Primoratz, eds., *Military Ethics* (New York: Routledge, 2016).

243 Wiesner memo: Marcus Raskin, "JFK and the Culture of Violence," *American Historical Review*, April 1992, p. 497.

243 Wiesner and fallout: O'Donnell and Powers, *"Johnny, We Hardly Knew Ye,"* p. 285.

244 Kennedy on fallout dangers: John F. Kennedy, televised address, July 26, 1963, cited in Douglass, *JFK and the Unspeakable*, p. 51.

245 "One of the ironic things": Norman Cousins, *The Improbable Triumvirate: John F. Kennedy, Pope John, Nikita Khrushchev* (New York: W. W. Norton, 1972), pp. 113–14.

246 Kennedy's American University speech, June 10, 1963: www.presidency .ucsb.edu/ws/?pid=9266.

248 "The full text of the speech": Theodore C. Sorensen, *Kennedy* (New York: Harper & Row, 1965), p. 733.

248 Khrushchev on speech: Schlesinger, *A Thousand Days*, p. 904.

249 "By moving so swiftly": Richard Reeves, *President Kennedy: Profile of Power* (New York: Simon & Schuster, 1994), p. 554.

250 "That's the only thing": Ibid., p. 550.

250 Kennedy on treaty: Kennedy, "John F. Kennedy's Vision of Peace."

251 "A full-scale nuclear war": https://hautevitrine.com/2011/12/10/the -survivors-would-envy-the-dead/.

251 Nuclear first strike proposal in 1963: See Douglass, *JFK and the Unspeakable*, pp. xxvii and 238–42, citing numerous sources.

251 Kennedy to Mormon Assembly: Schlesinger, *A Thousand Days*, p. 980.

252 JFK's last letter to Khrushchev: Raymond L. Garthoff, *A Journey Through the Cold War: A Memoir of Containment and Coexistence* (Washington, DC: Brookings Institution Press, 2001), p. 167.

253 Sergei Khrushchev, "Commentary on 'Thirteen Days,'" *New York Times*, February 4, 2001, p. 17. Cited in Douglass, *JFK and the Unspeakable*, p. 53.

253 Khrushchev's reaction to JFK assassination: Pierre Salinger, *With Kennedy* (Garden City, NY: Doubleday, 1966), p. 335.

254 Castro on Khrushchev reading messages: Ibid., p. 68, citing Castro's January 11, 1992, address.

254 Helms memo: Ibid., p. 69, citing Helms Memorandum of June 5, 1963, "Reported Desire of the Cuban Government for Rapprochement with the United States."

258 Kennedy's speech on U.S.–Cuba policy: *Public Papers of the Presidents; John F. Kennedy, 1963*, pp. 875–76.

258 "There are few subjects": Jean Daniel, "Unofficial Envoy: Report from Two Capitals," *New Republic*, December 14, 1963, p. 16.

259 "I believe Kennedy is sincere": Jean Daniel, "When Castro Heard the News," *New Republic*, December 7, 1963.

261 Castro to Daniel: Dick Russell, *The Man Who Knew Too Much* (New York: Carroll & Graf, 1992), p. 567.

261 Castro on being blamed: Talbot, *Brothers*, p. 253.

Chapter 8: A Farewell to Camelot

264 Dallas's then mayor, Earle Cabell, was a CIA agent: https://whowhat why.org/2017/0 8/02/dallas-mayor-jfk-assassination-cia-asset/.

264 Manchester on Dallas: William Manchester, *The Death of a President November 20–25, 1963* (New York: Arbor House, 1985, paperback edition), p. 44. Other portions of this chapter are derived from Manchester's epic story of that week.

264 "any white man who did": Alice L. George, *The Assassination of John F. Kennedy: Political Trauma and American Memory* (New York: Routledge, 2013), p. 81.

265 "The Kennedy administration struck": Russ Baker, *Family of Secrets* (New York: Bloomsbury Press, 2009), p. 94.

266 JFK to Jackie: Talbot, *Brothers*, p. 242.

267 Hoover phone call/Robert Kennedy to Manchester: Schlesinger, *Robert Kennedy and His Times*, p. 608.

268 Robert Kennedy and McCone: Talbot, *Brothers*, pp. 6–7.

268 Robert Kennedy's call to Williams and Johnson: Ibid., p. 6; confirmed

by Lamar Waldron and Thom Hartmann, *Legacy of Secrecy: The Long Shadow of the JFK Assassination* (Berkeley, CA: Counterpoint Press, 2009), p. 157.

274 Jackie to Mikoyan: Manchester, *The Death of a President*, p. 610.

279 "a hero of peace": William Manchester, *The Death of a President* (New York: Harper & Row, 1988), p. 594.

280 LBJ asking Congress to pass Civil Rights Act, November 27, 1963: www.pbs.org/ladybird/epicenter/epicenter_doc_speech.html.

Chapter 9: Senator Robert F. Kennedy

286 "The innocent suffer": Schlesinger, *Robert Kennedy and His Times*, p. 617.

286 Edith Hamilton: Ibid., p. 618.

295 "blind, inexplicable, meaningless courage": "Our Climb Up Mt. Kennedy," Robert Kennedy, *Life*, April 9, 1965, https://books.google.com/books?id=RVMEAAAAMBAJ.

306 "a declaration of aggression against the Kennedys": Schlesinger, *Robert Kennedy and His Times*, p. 631.

306 "contrast with such things as the Berlin Wall": Kathleen Kennedy Townsend, "Robert F. Kennedy Advocated Ending the Ban on Travel to Cuba," *Washington Post*, April 23, 2009.

308 "If Johnson had to choose": Richard N. Goodwin, *Remembering America: A Voice from the Sixties* (Boston: Little, Brown, 1988).

316 Campaign appearances at reservations: Thurston Clarke, *The Last Campaign: Robert F. Kennedy and 82 Days That Inspired America* (New York: Henry Holt, 2008), p. 155.

318 Kennedy to students: Schlesinger, *Robert Kennedy and His Times*, p. 746.

319 *Rand Daily Mail*: Quoted in ibid., p. 748.

Chapter 10: The Final Campaign

323 Nixon on victory in Vietnam: Conrad Black, *Richard M. Nixon: A Life in Full* (New York: PublicAffairs, 2007), p. 480.

323 Americans and torture in Vietnam: Schlesinger, *Robert Kennedy and His Times*, p. 740.

325 Johnson's prediction and Kennedy's reaction: Ibid., p. 768.

326 Kennedy speech to Senate: Ibid., p. 773.

327 Tom Wicker, and Robert Kennedy response: Ibid., p. 824.

327 Kennedy to Schlesinger: Ibid., p. 823.

327 Kennedy on Tet: Jack Newfield, *RFK: A Memoir* (New York: Thunder's Mouth Press/Nation Books, 2003), p. 205.

328 Comment to Tom Watson, Jr.: Schlesinger, *Robert Kennedy and His Times*, p. 844.

330 Newfield column: Quoted in Jack Newfield, *Somebody's Gotta Tell It: A Journalist's Life on the Lines* (New York: St. Martin's Press, 2002), p. 189.

330 "deeply fearful": Schlesinger, *Robert Kennedy and His Times*, p. 895.

331 Announcement of candidacy: Robert F. Kennedy, "Announcement of Candidacy for President," March 16, 1968, www.4president.org /speeches/rfk1968announcement.htm.

331 Kennedy to Newfield and Johnson: Schlesinger, *Robert Kennedy and His Times*, p. 854.

331 "I know my brother thinks": Jack Newfield, *Robert Kennedy: A Memoir*, p. 252.

334 "Tragedy is a tool": Robert F. Kennedy, paraphrasing Sophocles.

334 Robert Kennedy speech at Kansas State University: Edwin O. Guthman and C. Richard Allen, eds., *RFK: Collected Speeches* (New York: Viking, 1993), pp. 325–26.

335 RFK to Jimmy Breslin: Schlesinger, *Robert Kennedy and His Times*, p. 863.

335 Tuscaloosa speech: Joseph A. Palermo, "RFK, Barack Obama, and Blackwater," *HuffPost*, January 4, 2008, https://www.huffingtonpost .com/joseph-a . . . /barack-obama-rfk-and-blac_b_79751.html.

335 Kennedy in Watts: Clarke, *The Last Campaign*, p. 58.

337 *New York Times* and *Meet the Press*: Ibid., p. 12.

337 Creighton University speech: Schlesinger, *Robert Kennedy and His Times*, p. 888.

338 Speech in Cleveland: Newfield, *RFK: A Memoir*, p. 249.

338 Black Panthers: Schlesinger, *Robert Kennedy and His Times*, p. 909.

340 Speech after Dr. King's death: Robert Torricelli and Andrew Carroll, eds., *In Our Own Words: Extraordinary Speeches of the American Century* (New York: Washington Square Press, 1999), p. 273.

341 Hosea Williams: Schlesinger, *Robert Kennedy and His Times*, p. 879.

343 Westbrook Pegler: Ibid., p. 808, citing Finis Farr, *Fair Enough: The Life of Westbrook Pegler*.

343 Clyde Tolson: Ibid., p. 808, recounting an interview Manchester conducted with William C. Sullivan of the FBI, who worked with Tolson.

343 John J. Lindsay: Ibid., p. 900.

344 Kennedy on campaign security: David, *Ethel*, p. 193.

345 For the first time: Jack Newfield, *RFK: A Memoir* (New York: Berkley Publishing Group, 1978).

345 Keats excerpt: Helen Vendler, *The Odes of John Keats* (Cambridge, MA: Belknap Press of Harvard University Press, 2003), p. 177.

346 Charles Evers: Pierre Salinger, *An Honorable Profession: A Tribute to Robert F. Kennedy* (New York: Doubleday, 1993), p. 66.

Chapter 11: Lem

364 Eunice Shriver: David Pitts, *Jack and Lem: The Untold Story of an Extraordinary Friendship* (New York: Carroll & Graf, 2007), introduction.

365 Kennedy on Billings's breath: Ibid.

366 "Stayed at a Youth Hostel": Ibid.